Integral Inequalities and Applications

Mathematics and Its Applications (*East European Series*)

Volume 57

Integral Inequalities and Applications

by

Drumi Bainov
Academy of Medicine,
Sofia, Bulgaria

and

Pavel Simeonov
Plovdiv University,
Plovdiv, Bulgaria

Translated by R. A. M. Hoksbergen

SPRINGER-SCIENCE+BUSINESS MEDIA, B.V.

Library of Congress Cataloging-in-Publication Data

Baĭnov, D. (Dimitŭr)
Integral inequalities and applications / Drumi Bainov and Pavel Simeonov.
p. cm. -- (Mathematics and its applications. East European series ; 57)
ISBN 978-90-481-4154-8
1. Integral inequalities. I. Simeonov, P. S. (Pavel S.) II. Title. III. Series: Mathematics and its applications (Kluwer Acdemic Publishers). East European series ; v. 57.
QA295.B32 1992
515'.46--dc20 92-9022

ISBN 978-90-481-4154-8 ISBN 978-94-015-8034-2 (eBook)
DOI 10.1007/978-94-015-8034-2

Printed on acid-free paper

SERIES EDITOR'S PREFACE

'Et moi, ..., si j'avait su comment en revenir, je n'y serais point allé.'

Jules Verne

The series is divergent; therefore we may be able to do something with it.

O. Heaviside

One service mathematics has rendered the human race. It has put common sense back where it belongs, on the topmost shelf next to the dusty canister labelled 'discarded non-sense'.

Eric T. Bell

Mathematics is a tool for thought. A highly necessary tool in a world where both feedback and non-linearities abound. Similarly, all kinds of parts of mathematics serve as tools for other parts and for other sciences.

Applying a simple rewriting rule to the quote on the right above one finds such statements as: 'One service topology has rendered mathematical physics ...'; 'One service logic has rendered computer science ...'; 'One service category theory has rendered mathematics ...'. All arguably true. And all statements obtainable this way form part of the raison d'être of this series.

This series, *Mathematics and Its Applications*, started in 1977. Now that over one hundred volumes have appeared it seems opportune to reexamine its scope. At the time I wrote

> "Growing specialization and diversification have brought a host of monographs and textbooks on increasingly specialized topics. However, the 'tree' of knowledge of mathematics and related fields does not grow only by putting forth new branches. It also happens, quite often in fact, that branches which were thought to be completely disparate are suddenly seen to be related. Further, the kind and level of sophistication of mathematics applied in various sciences has changed drastically in recent years: measure theory is used (non-trivially) in regional and theoretical economics; algebraic geometry interacts with physics; the Minkowsky lemma, coding theory and the structure of water meet one another in packing and covering theory; quantum fields, crystal defects and mathematical programming profit from homotopy theory; Lie algebras are relevant to filtering; and prediction and electrical engineering can use Stein spaces. And in addition to this there are such new emerging subdisciplines as 'experimental mathematics', 'CFD', 'completely integrable systems', 'chaos, synergetics and large-scale order', which are almost impossible to fit into the existing classification schemes. They draw upon widely different sections of mathematics."

By and large, all this still applies today. It is still true that at first sight mathematics seems rather fragmented and that to find, see, and exploit the deeper underlying interrelations more effort is needed and so are books that can help mathematicians and scientists do so. Accordingly MIA will continue to try to make such books available.

If anything, the description I gave in 1977 is now an understatement. To the examples of interaction areas one should add string theory where Riemann surfaces, algebraic geometry, modular functions, knots, quantum field theory, Kac-Moody algebras, monstrous moonshine (and more) all come together. And to the examples of things which can be usefully applied let me add the topic 'finite geometry'; a combination of words which sounds like it might not even exist, let alone be applicable. And yet it is being applied: to statistics via designs, to radar/sonar detection arrays (via finite projective planes), and to bus connections of VLSI chips (via difference sets). There seems to be no part of (so-called pure) mathematics that is not in immediate danger of being applied. And, accordingly, the applied mathematician needs to be aware of much more. Besides analysis and numerics, the traditional workhorses, he may need all kinds of combinatorics, algebra, probability, and so on.

In addition, the applied scientist needs to cope increasingly with the nonlinear world and the extra mathematical sophistication that this requires. For that is where the rewards are. Linear models are honest and a bit sad and depressing: proportional efforts and results. It is in the nonlinear world that infinitesimal inputs may result in macroscopic outputs (or vice versa). To appreciate what I am hinting at: if electronics were linear we would have no fun with transistors and computers; we would have no TV; in fact you would not be reading these lines.

There is also no safety in ignoring such outlandish things as nonstandard analysis, superspace and anticommuting integration, p-adic and ultrametric space. All three have applications in both electrical engineering and physics. Once, complex numbers were equally outlandish, but they frequently proved the shortest path between 'real' results. Similarly, the first two topics named have already provided a number of 'wormhole' paths. There is no telling where all this is leading - fortunately.

Thus the original scope of the series, which for various (sound) reasons now comprises five subseries: white (Japan), yellow (China), red (USSR), blue (Eastern Europe), and green (everything else), still applies. It has been enlarged a bit to include books treating of the tools from one subdiscipline which are used in others. Thus the series still aims at books dealing with:

- a central concept which plays an important role in several different mathematical and/or scientific specialization areas;
- new applications of the results and ideas from one area of scientific endeavour into another;
- influences which the results, problems and concepts of one field of enquiry have, and have had, on the development of another.

A significant part of hard analysis, sometimes distressingly large to a more geometric and algebraic type of thinker like myself, is about inequalities and obtaining (growth) estimates on the functions appearing in differential, integral, and other equations. There are many types of inequalities of importance. One type is known as Gronwall type inequalities, named after its ancestral example (Gronwall 1919) which says that if $0 \leqslant u(t) \leqslant \int_a^t (c + du(s))ds$ for $t \in [a, a+h]$, then for the function $u(t)$ one has the estimate $u(t) \leqslant che^{dh}$ for t in the same interval. There are many generalizations and analogues.

Gronwall type inequalities are important, and have applications to questions of stability, uniqueness of solutions, dichotomy, asymptotic behaviour etc.

Roughly from 1980 on, a great deal of attention has been paid to more general equations which go by such names as equations with delays, functional differential equations, equations with deviating arguments, impulse differential equations. Among many others, the authors themselves have contributed nontrivially to these subjects. There are analogues of Gronwall type inequalities in these settings with similar uses.

This book is a comprehensive treatise on both the 'classical' and these new Gronwall type inequalities. Besides the inequalities themselves it also describes how they are used and applied.

The shortest path between two truths in the real domain passes through the complex domain.

J. Hadamard

La physique ne nous donne pas seulement l'occasion de résoudre des problèmes ... elle nous fait pressentir la solution.

H. Poincaré

Never lend books, for no one ever returns them; the only books I have in my library are books that other folk have lent me.

Anatole France

The function of an expert is not to be more right than other people, but to be wrong for more sophisticated reasons.

David Butler

Bussum, March 1992

Michiel Hazewinkel

Contents

Preface

'No one will comprehend the incomprehensible.'
Koz'ma Prutkov[1]

Let $u\colon [\alpha, \alpha + h] \to \mathbf{R}$ be a continuous function satisfying the inequality

$$0 \le u(t) \le \int_\alpha^t [a + bu(s)]\, ds, \qquad \text{for } t \in [\alpha, \alpha + h],$$

where a, b are nonnegative constants. Then $u(t) \le ahe^{bh}$ for $t \in [\alpha, \alpha + h]$. This result was proved by T.H. Gronwall in 1919, and is the prototype for the study of many integral inequalities of Volterra type, and also for obtaining explicit bounds of the unknown function. Among the publications on this subject, the papers of Bellman [20] (1943) and Bihari [33] (1956) is well known and have found wide application. Therefore integral inequalities of this type are usually associated with the names of Gronwall, Bellman, and Bihari. Other names to be mentioned in connection with the further development of the theory of integral inequalities are: R.P. Agarval, N.B. Azbelev, D.D. Bainov, P.R. Beesack, S.G. Deo, U.D. Dhongade, V. Lakshmikantham, S. Leela, B.G. Pachpatte, J. Popenda, P.S. Simeonov, Z.B. Tsalyuk, D. Willett, J.S.W. Wong, E.C. Young.

Integral inequalities are a necessary tool in the study of various classes of equations. This justifies the intensive investigations on integral inequalities, and the appearance of hundreds of publications on it. A part of the results can be found in the monographs of E.F. Beckenbach and R. Bellman (1965), V. Lakshmikantham and S. Leela (1969), W. Walter (1970), P.R. Beesack (1975), A.N. Filatov and L.V. Sharova (1976), A.A. Martynyuk and R. Gutowski (1979), R. Rabczuk (1979), D.Ya. Mamedov, S. Ashirov, and S. Atdaev (1980), and J. Schröder (1980).

In the majority of these monographs, separate aspect of the application of the inequalities considered in them are brought forward.

In relation with the study of various classes of equations (integro-differential equations, equations with deviating argument, equations with maxima, difference equations, impulse differential equations, etc.) from 1980 onwards new types of inequali-

[1] Works of Koz'ma Prutkov, Publ. House 'Soviet Russia', 1981, p. 79.

ties, connected with these equations, have been investigated. The results were often published in marginal journals, inaccessible to the possibly interested scholars. This is one reason why we contemplated the idea of writing this book.

We have aimed at the following goals:

- to give a systematic exposition of the classical results about integral inequalities and of the methods which are most often used in investigating these;
- to consider some types of integral inequalities and their analogues which have appeared in the mathematical literature in recent years, in relation with the investigation of new classes of equations;
- to illustrate the application of integral inequalities in the study of solutions of various classes of equations.

The book consists of three chapters. In chapters 1 and 2 we consider the classical linear and nonlinear integral inequalities. Chapter 3 is devoted to various classes of integral inequalities of Gronwall type, and their analogues, which find application in the theory of integro-differential equations, partial differential equations, differential equations with deviating argument, impulse differential equations, etc. Each chapter ends with a section in which the principal manners of application of the inequalities considered in the chapter are illustrated.

The authors express their gratitude to Kluwer Academic Publishers, especially Prof. M. Hazewinkel and Dr. D. Larner for the opportunity to publish this book.

They also sincerely thank Dr. V. Covachev for translating the book into English and for his help in preparing the manuscript, the Ministry of Science and Higher Education of the Republic of Bulgaria for support during the writing of the book, and Drs. R.A.M. Hoksbergen for polishing the translation as well as wordprocessing.

This publication has been made possible by the support of the Bulgarian Ministry of Science and Higher Education under Grant MM-7.

Pleven,
12 August 1991

D.D. Bainov
P.S. Simeonov

Notation

$\mathbf{R}$	$=$	the set of all real numbers
$\mathbf{R}^n$	$=$	real n-dimensional vector space
$[a,b]$	$=$	the closed interval $a \le x \le b$
(a,b)	$=$	the open interval $a < x < b$
$\langle a,b\rangle$	$=$	an arbitrary interval
$\mathbf{R}_+ = [0,\infty)$	$=$	the set of all nonnegative real numbers
$[a;b]$	$=$	the closed n-dimensional interval
$\operatorname{Int} I$	$=$	the interior of a set I
$\lvert x\rvert$	$=$	the norm of the element $x \in \mathbf{R}^n$
$A = (a_{ij})$	$=$	the matrix with entries a_{ij}
$\lvert A\rvert = \sup_{\lvert x\rvert=1} \lvert Ax\rvert$	$=$	the norm of the matrix A
$\mathbf{R}^{n\times m}$	$=$	the set of all matrices with real entries, and having n rows and m columns
$f_x = \frac{\partial f}{\partial x} = \left(\frac{\partial f_i}{\partial x_j}\right)_{\substack{i=1,\dots,m\\ j=1,\dots,n}}$	$=$	Jacobi matrix of the function $f\colon \mathbf{R}^n \to \mathbf{R}^m$, $x \to f(x)$
$\binom{n}{s}$	$=$	$\frac{n!}{s!(n-s)!}$
$x(a^+)$	$=$	$\lim_{t\to a+} x(t)$
$x(a^-)$	$=$	$\lim_{t\to a-} x(t)$
$\dot{x}$	$=$	$x' = \frac{dx}{dt}$
$\int_a^b \int_c^d \dots dx\,dy$	$=$	$\iint_{\substack{a\le x\le b\\ c\le y\le d}} \dots dx\,dy$

A function $f\colon \mathbf{R}_+ \to \mathbf{R}_+$ is said to be

i) subadditive if $f(x+y) \le f(x) + f(y)$, $x,y \in \mathbf{R}_+$;
ii) superadditive if $f(x+y) \ge f(x) + f(y)$, $x,y \in \mathbf{R}_+$;
iii) submultiplicative if $f(xy) \le f(x)f(y)$, $x,y \in \mathbf{R}_+$;
iv) supermultiplicative if $f(xy) \ge f(x)f(y)$, $x,y \in \mathbf{R}_+$.

A function $f\colon [a,b] \to \mathbf{R}$ is said to be convex (concave) in $[a,b]$ if $f(tx+(1-t)y) \le tf(x)+(1-t)f(y)$ (respectively, $f(tx+(1-t)y) \ge tf(x)+(1-t)f(y)$) for all $x,y \in [a,b]$ and $t \in [0,1]$.

For the symbols $\mathbf{Z}$, $\mathbf{N}_\alpha$, $\mathbf{N}_{\alpha,\beta}$, $\sum$, $\prod$, Δu, $\Delta^k u$ see page 157; $PC(\mathbf{R}_+,\mathbf{R}^n)$, $PC^1(\mathbf{R}_+,\mathbf{R}^n)$ see page 182; $\mathcal{F}$ see page 67; $H_{r,w}$ see page 89; $\propto$ see page 94.

CHAPTER I

Scalar linear integral inequalities

In this chapter we consider linear integral inequalities of Gronwall type for scalar functions of one variable. These inequalities are important because of their many applications and because many methods for investigating them can be successfully used in the study of inequalities of other types.

1. Gronwall-type inequalities

We begin by giving one of the simplest and most frequently used integral inequalities.

THEOREM 1.1. (Bellman, 1943) *Let $u(t)$ and $b(t)$ be nonnegative continuous functions for $t \geq \alpha$, and let*

$$u(t) \leq a + \int_\alpha^t b(s)u(s)\,ds, \qquad t \geq \alpha, \tag{1.1}$$

where $a \geq 0$ is a constant. Then

$$u(t) \leq a \exp\left(\int_\alpha^t b(s)\,ds\right), \qquad t \geq \alpha. \tag{1.2}$$

PROOF. Let $a > 0$. Then (1.1) implies the inequality

$$\frac{b(\tau)u(\tau)}{a + \int_\alpha^\tau b(s)u(s)\,ds} \leq b(\tau), \qquad \tau \geq \alpha.$$

Integrating this from α to t yields

$$\log\left[a + \int_\alpha^t b(s)u(s)\,ds\right] - \log a \leq \int_\alpha^t b(s)\,ds.$$

Together with (1.1) this implies (1.2).

Let $a = 0$. Then $u(t) < \epsilon + \int_\alpha^t b(s)u(s)\,ds$ for any $\epsilon > 0$. Hence

$$u(t) \leq \epsilon \exp\left(\int_\alpha^t b(s)\,ds\right),$$

and letting $\epsilon \to 0$ we find $u(t) \leq 0$. □

COROLLARY 1.1. *Let $u(t)$ and $b(t)$ be nonnegative continuous functions for $t \geq \alpha$, and let*

$$u(t) \leq ae^{-\gamma(t-\alpha)} + \int_\alpha^t e^{-\gamma(t-s)} b(s)u(s)\,ds, \qquad t \geq \alpha, \tag{1.3}$$

where $a \geq 0$ and γ are constants. Then

$$u(t) \leq a \exp\left(-\gamma(t-\alpha) + \int_\alpha^t b(s)\,ds\right), \qquad t \geq \alpha. \tag{1.4}$$

PROOF. Setting $w(t) = e^{\gamma t}u(t)$ we obtain from (1.3)

$$w(t) \leq ae^{\gamma\alpha} + \int_\alpha^t b(s)w(s)\,ds, \qquad t \geq \alpha.$$

By theorem 1.1 this implies $w(t) \leq ae^{\gamma\alpha} \exp\left(\int_\alpha^t b(s)\,ds\right)$, which gives (1.4). □

We have given the classical proof of theorem 1.1, because of its importance. We can also give a proof along the following scheme: from (1.1) we derive a linear differential inequality, multiply it by the integrating factor of the corresponding linear differential equation, and integrate the result. We then also obtain the required estimate. Since this approach is often used, we will first prove a *comparison result* for linear differential inequalities.

LEMMA 1.1. *Let $b(t)$ and $f(t)$ be continuous functions for $t \geq \alpha$, let $v(t)$ be a differentiable function for $t \geq \alpha$, and suppose*

$$\begin{aligned} v'(t) &\leq b(t)v(t) + f(t), \qquad t \geq \alpha, \\ v(\alpha) &\leq v_0. \end{aligned} \tag{1.5}$$

Then, for $t \geq \alpha$,

$$v(t) \leq v_0 \exp\left(\int_\alpha^t b(s)\,ds\right) + \int_\alpha^t f(s) \exp\left(\int_s^t b(\tau)\,d\tau\right) ds. \tag{1.6}$$

PROOF. Conditions (1.5) imply that

$$[v'(s) - b(s)v(s)] \exp\left(\int_s^t b(\tau)\,d\tau\right) \leq f(s) \exp\left(\int_s^t b(\tau)\,d\tau\right), \qquad s \geq \alpha,$$

or

$$\frac{d}{ds}\left[v(s) \exp\left(\int_s^t b(\tau)\,d\tau\right)\right] \leq f(s) \exp\left(\int_s^t b(\tau)\,d\tau\right).$$

Integration over s from α to t gives

$$v(t) - v(\alpha) \exp\left(\int_\alpha^t b(\tau)\,d\tau\right) \leq \int_\alpha^t f(s) \exp\left(\int_s^t b(\tau)\,d\tau\right) ds,$$

which implies (1.6) since $v(\alpha) \leq v_0$. □

REMARK 1.1. Note that the righthand side of (1.6) coincides with the unique solution of the equation

$$v'(t) = b(t)v(t) + f(t), \qquad t \geq \alpha, \tag{1.7}$$

for which

$$v(\alpha) = v_0. \tag{1.8}$$

Equation (1.7) is called the *comparison differential equation* of the inequality (1.5). The *comparison initial value problem* (1.7)–(1.8) is obtained by replacing in (1.5) $\leq$ by $=$. □

REMARK 1.2. The result of lemma 1.1 remains valid if $\leq$ is replaced by $\geq$ in both (1.5) and (1.6). □

REMARK 1.3. If the functions $b(t)$ and $f(t)$ are continuous for $t \leq \alpha$ and

$$v'(t) \leq b(t)v(t) + f(t), \qquad t \leq \alpha, \tag{1.9}$$

then

$$v(t) \geq v(\alpha) \exp\left(\int_\alpha^t b(s)\,ds\right) + \int_\alpha^t f(s) \exp\left(\int_s^t b(\tau)\,d\tau\right) ds, \qquad t \leq \alpha. \tag{1.10}$$

Moreover, this result remains valid if in (1.9) $\leq$ is replaced by $\geq$, and in (1.10) $\geq$ is replaced by $\leq$. □

Using lemma 1.1 it is easy to prove

THEOREM 1.2. (Reid, 1930) *Let $u(t)$ and $b(t)$ be nonnegative continuous functions in $J = [\alpha, \beta]$, and suppose*

$$u(t) \leq a + \int_{t_0}^t b(s)u(s)\,|ds|, \quad t \in J,$$

where $t_0 \in J$ and a is a constant. Then

$$u(t) \leq a \exp\left(\int_{t_0}^t b(s)\,|ds|\right), \qquad t \in J. \qquad \square$$

We will use lemma 1.1 in the proof of the following assertion.

THEOREM 1.3. *Let $a(t)$, $b(t)$, and $u(t)$ be continuous functions in $J = [\alpha, \beta]$, and let $b(t)$ be nonnegative in J. Suppose*

$$u(t) \leq a(t) + \int_\alpha^t b(s)u(s)\,ds, \quad t \in J.$$

Then

$$u(t) \leq a(t) + \int_\alpha^t a(s)b(s) \exp\left(\int_s^t b(\tau)\,d\tau\right) ds, \qquad t \in J. \tag{1.11}$$

PROOF. Set $v(t) = \int_\alpha^t b(s)u(s)\,ds$. Then

$$u(t) \leq a(t) + v(t), \tag{1.12}$$

$$v'(t) = b(t)u(t) \leq b(t)v(t) + a(t)b(t), \qquad t \in J,$$
$$v(\alpha) = 0.$$

From lemma 1.1 we find that

$$v(t) \leq \int_\alpha^t a(s)b(s)\exp\left(\int_s^t b(\tau)\,d\tau\right) ds.$$

This and (1.12) imply (1.11). □

COROLLARY 1.2. *Let, under the conditions of theorem 1.3, the function $a(t)$ also be nondecreasing in J. Then*

$$u(t) \leq a(t)\exp\left(\int_\alpha^t b(s)\,ds\right), \qquad t \in J.$$

PROOF. In fact, (1.11) implies that

$$u(t) \leq a(t) + a(t)\int_\alpha^t b(s)\exp\left(\int_s^t b(\tau)\,d\tau\right) ds =$$
$$= a(t)\left\{1 - \int_\alpha^t \frac{d}{ds}\left[\exp\left(\int_s^t b(\tau)\,d\tau\right)\right] ds\right\} = a(t)\exp\left(\int_\alpha^t b(\tau\,d\tau\right). \qquad \square$$

COROLLARY 1.3. *Let $b(t)$ and $u(t)$ be continuous functions in $J = [\alpha, \beta]$, let $b(t)$ be nonnegative in J, and suppose*

$$u(t) \leq a + \int_\alpha^t b(s)u(s)\,ds, \qquad t \in J,$$

where a is a constant. Then

$$u(t) \leq a\exp\left(\int_\alpha^t b(s)\,ds\right) ds, \qquad t \in J. \qquad \square$$

REMARK 1.4. The conclusion of corollary 1.3 shows that in theorem 1.1 and corollary 1.1 we may omit the requirement that $u(t)$ and a be nonnegative. □

COROLLARY 1.4. *Let $u(t)$ be a continuous function in $J = [\alpha, \beta]$, and suppose*

$$u(t) \leq a + \int_\alpha^t bu(s)\,ds, \qquad t \in J,$$

where $b \geq 0$ and a are constants. Then

$$u(t) \leq ae^{b(t-\alpha)}, \qquad t \in J. \qquad \square$$

COROLLARY 1.5. (Chandirov, 1958) *Let $a(t)$, $b(t)$, $c(t)$, and $u(t)$ be continuous functions in $J = [\alpha, \beta]$, let $b(t)$ and $c(t)$ be nonnegative in J, and suppose*

$$u(t) \leq a(t) + \int_\alpha^t [b(s)u(s) + c(s)]\, ds, \qquad t \in J. \tag{1.13}$$

Then

$$u(t) \leq \left[\sup_{s\in[\alpha,t]} a(s) + \int_\alpha^t c(s)\, ds \right] \exp\left(\int_\alpha^t b(s)\, ds \right), \qquad t \in J. \tag{1.14}$$

PROOF. The estimate (1.14) follows from corollary 1.2, since (1.13) implies

$$u(t) \leq A(t) + \int_\alpha^t b(s)u(s)\, ds,$$

where $A(t) = \sup_{s\in[\alpha,t]} a(s) + \int_\alpha^t c(s)\, ds$ is a nondecreasing function in J. □

The following theorem gives the best possible estimate for a function $u(t)$ satisfying (1.13).

THEOREM 1.4. (Chandirov, 1970) *Let $a(t)$, $b(t)$, $c(t)$, and $u(t)$ be continuous functions in $J = [\alpha, \beta]$, let $b(t)$ be nonnegative in J, and suppose*

$$u(t) \leq a(t) + \int_\alpha^t [b(s)u(s) + c(s)]\, ds, \qquad t \in J.$$

Then

$$u(t) \leq a(t) + \int_\alpha^t [a(s)b(s) + c(s)] \exp\left(\int_s^t b(\tau)\, d\tau \right) ds, \qquad t \in J. \tag{1.15}$$

PROOF. For $v(t) = \int_\alpha^t [b(s)u(s) + c(s)]\, ds$ we have

$$v'(t) \leq b(t)v(t) + a(t)b(t) + c(t), \qquad t \in J,$$
$$v(\alpha) = 0.$$

Hence, from lemma 1.1 and the fact that $u(t) \leq a(t) + v(t)$ we can derive (1.15). □

COROLLARY 1.6. *If in theorem 1.4 the function $a(t)$ is nondecreasing in J, then*

$$u(t) \leq a(t) \exp\left(\int_\alpha^t b(\tau)\, d\tau \right) + \int_\alpha^t c(s) \exp\left(\int_s^t b(\tau)\, d\tau \right) ds, \qquad t \in J. \qquad \square$$

COROLLARY 1.7. *If in theorem 1.4 $a(t) \equiv a$, then*

$$u(t) \leq a \exp\left(\int_\alpha^t b(\tau)\, d\tau \right) + \int_\alpha^t c(s) \exp\left(\int_s^t b(\tau)\, d\tau \right) ds, \qquad t \in J. \qquad \square$$

COROLLARY 1.8. (Quade, 1942; Zadiraka, 1968) *Let $u(t)$ be a continuous function for $t \geq \alpha$, and suppose*

$$u(t) \leq a e^{-\gamma(t-\alpha)} + \int_\alpha^t e^{-\gamma(t-s)} [bu(s) + c]\, ds, \qquad t \geq \alpha, \tag{1.16}$$

where $b \geq 0$, a, c, and $\gamma \neq b$ are constants. Then

$$u(t) \leq a e^{(b-\gamma)(t-\alpha)} + \frac{c}{\gamma - b}[1 - e^{(b-\gamma)(t-\alpha)}], \qquad t \geq \alpha.$$

PROOF. Set $w(t) = e^{\gamma t}u(t)$ in (1.16) and apply corollary 1.7 to the resulting inequality. □

COROLLARY 1.9. (Filatov, 1971) *Let $u(t)$ be a continuous function in $J = [\alpha, \beta]$, and suppose*

$$u(t) \leq a + \int_\alpha^t [bu(s) + c]\, ds, \qquad t \in J,$$

where $b > 0$, a, and c are constants. Then

$$u(t) \leq \frac{c}{b}(e^{b(t-\alpha)} - 1) + ae^{b(t-\alpha)}, \qquad t \in J. \qquad \square$$

COROLLARY 1.10. (Two-sided estimate; Rudakov, 1970)) *Let $a(t)$, $b(t)$, $c(t)$, and $u(t)$ be continuous functions in $J = [\alpha, \beta]$, let $b(t)$ be nonnegative in J, and suppose*

$$u(t) \leq u(r) + \int_r^t [b(s)u(s) + c(s)]\, |ds| \tag{1.17}$$

for all $t, r \in J$. Then

$$\left[u(\alpha) - \int_\alpha^t c(s) \exp\left(\int_\alpha^s b(\tau)\, d\tau\right) ds\right] \exp\left(-\int_\alpha^t b(s)\, ds\right) \leq u(t), \qquad t \in J, \tag{1.18}$$

$$u(t) \leq \left[u(\alpha) + \int_\alpha^t c(s) \exp\left(-\int_\alpha^s b(\tau)\, d\tau\right) ds\right] \exp\left(\int_\alpha^t b(s)\, ds\right), \qquad t \in J. \tag{1.19}$$

PROOF. From (1.17) for $\alpha = r \leq t \leq \beta$ we can derive (1.15) with $a(t) \equiv u(\alpha)$. From corollary 1.7 we then obtain (1.19). From (1.17) for $\alpha \leq t \leq r = \beta$ we can derive

$$u(t) \leq u(r) + \int_t^r [b(s)u(s) + c(s)]\, ds \equiv v(t).$$

Then $v(r) = u(r)$ and

$$v'(t) \geq -b(t)v(t) - c(t), \qquad t \leq r.$$

By remark 1.3 this implies the estimate

$$u(t) \leq v(t) \leq u(r) \exp\left(-\int_r^t b(s)\, ds\right) - \int_r^t c(s) \exp\left(-\int_s^t b(\tau)\, d\tau\right) ds.$$

Replacing t by α and subsequently r by t we arrive at

$$u(\alpha) \leq u(t) \exp\left(-\int_t^\alpha b(s)\, ds\right) - \int_t^\alpha c(s) \exp\left(-\int_s^\alpha b(\tau)\, d\tau\right) ds, \qquad t \geq \alpha,$$

which is equivalent to (1.18). □

COROLLARY 1.11. (Generalized Bellman inequality; Bellman, 1953) *Let $u(t)$ and $b(t) \geq 0$ be continuous functions in $J = [\alpha, \beta]$, and suppose*

$$u(t) \leq u(r) + \int_r^t b(s)u(s)\, |ds|$$

for all $t, r \in J$. *Then for* $\alpha \le t \le \beta$,

$$u(\alpha)\exp\left(-\int_\alpha^t b(s)\,ds\right) \le u(t) \le u(\alpha)\exp\left(\int_\alpha^t b(s)\,ds\right). \qquad \square$$

The following assertion is related to work of Giuliano (1946), Kharlamov (1955), Willett (1965), and Beesack (1975).

THEOREM 1.5. *Let* $u(t)$ *and* $b(t)$ *be continuous functions in* $J = [\alpha, \beta]$, *and let* $a(t)$ *and* $q(t)$ *be Riemann integrable functions in* J. *Suppose that* $b(t)$ *and* $q(t)$ *are nonnegative in* J.

i) *If*

$$u(t) \le a(t) + q(t)\int_\alpha^t b(s)u(s)\,ds, \qquad t \in J, \tag{1.20}$$

then

$$u(t) \le a(t) + q(t)\int_\alpha^t a(s)b(s)\left(\int_s^t b(\tau)q(\tau)\,d\tau\right)ds, \qquad t \in J. \tag{1.21}$$

Moreover, equality holds in (1.21) *for a subinterval* $J_1 = [\alpha_1, \beta_1]$ *of* J *if equality holds in* (1.20) *for* $t \in J_1$.

ii) *If* $\le$ *is replaced by* $\ge$ *in both* (1.20) *and* (1.21), *the result remains valid.*

iii) *Both* i) *and* ii) *remain valid if* $\int_\alpha^t$ *is replaced by* $\int_t^\beta$ *and* $\int_s^t$ *by* $\int_t^s$ *throughout.*

PROOF. For $v(t) = \int_\alpha^t b(s)u(s)\,ds$ we have

$$\begin{aligned} u(t) &\le a(t) + q(t)v(t), && t \in J,\\ v'(t) &\le b(t)q(t)v(t) + a(t)b(t), && t \in J,\\ v(\alpha) &= 0. \end{aligned}$$

By lemma 1.1, these relations imply (1.21).

The conditions for equality are obvious, and the proof of ii) is similar.

The proof of iii) is based on remark 1.3. E.g., in the case $\le$ we have

$$\begin{aligned} u(t) &\le a(t) + q(t)\int_t^\beta b(s)u(s)\,ds \equiv a(t) + q(t)v(t),\\ v'(t) &\ge -b(t)q(t)v(t) - a(t)b(t), \qquad v(\beta) = 0, \end{aligned}$$

so that remark 1.3 implies

$$v(t) \le -\int_\beta^t a(s)b(s)\exp\left(-\int_s^t b(\tau)q(\tau)\,d\tau\right)ds, \qquad \alpha \le t \le \beta.$$

Hence

$$u(t) \le a(t) + q(t)\int_t^\beta a(s)b(s)\exp\left(\int_t^s b(\tau)q(\tau)\,d\tau\right)ds, \qquad t \in J.$$

The proof for the case $\ge$ is similar. $\square$

REMARK 1.5. Note that the hypotheses guarantee the existence of all integrals appearing in the proof, as well as the equality $v'(t) = b(t)u(t)$ for all $t \in J$. Of course, this is more than we strictly need. If the integrals are understood in the sense of Lebesgue, the hypothesis can be relaxed to: a, b, q, u are measurable functions such that $bu, ba, bq \in L(J)$. (In fact, this is the version given by Willett.) The inequalities (and equalities) are then to be understood to hold almost everywhere (a.e.), and the stated condition for equality is sufficient and necessary. Similar remarks apply to theorems 1.3 and 1.4, as well as to all our subsequent theorems below. However, for the sake of simplicity we will state the continuous versions only. □

COROLLARY 1.12. (Gollwitzer, 1969) *Let $y(t)$ and $b(t)$ be continuous functions in $J = [\alpha, \beta]$, let $q(t)$ be a nonnegative function in J, and suppose $b(t)$ is also nonnegative in J.*

i) *If a function $u(t)$ in J is such that*

$$u(t) \geq y(r) - q(t) \int_r^t b(s)y(s)\,ds, \qquad \alpha \leq r \leq t \leq \beta, \tag{1.22}$$

then

$$u(t) \geq y(r) \exp\left(-q(t) \int_r^t b(s)\,ds\right), \qquad \alpha \leq r \leq t \leq \beta. \tag{1.23}$$

ii) *The result in* i) *remains valid if in both* (1.22) *and* (1.23) $\geq$ *is replaced by* $\leq$.

PROOF. To prove i) we write (1.22) as

$$y(r) \leq u(t) + q(t) \int_r^t b(s)y(s)\,ds, \qquad \alpha \leq r \leq t \leq \beta,$$

and use on the interval $\alpha \leq r \leq t$ part iii) of theorem 1.5. In it, the functions u, b, a, and q should be replaced by, respectively, the functions y, b, and constants $u(t)$ and $q(t)$. □

COROLLARY 1.13. (Sardarly, 1965) *Let $u(t)$, $a(t)$, $b(t)$, and $q(t)$ be continuous functions in $J = [\alpha, \beta]$, let $c(t, s)$ be a continuous function for $\alpha \leq s \leq t \leq \beta$, let $b(t)$ and $q(t)$ be nonnegative in J, and suppose*

$$u(t) \leq a(t) + \int_\alpha^t [q(t)b(s)u(s) + c(t,s)]\,ds, \qquad t \in J.$$

Then for $t \in J$,

$$u(t) \leq a(t) + \int_\alpha^t c(t,s)\,ds + \\ + q(t) \int_\alpha^t b(s) \left[a(s) + \int_\alpha^s c(s,\tau)\,d\tau\right] \exp\left(\int_s^t b(\tau)q(\tau)\,d\tau\right) ds. \qquad \square$$

COROLLARY 1.14. (Gamidov, 1969) *Let $u(t)$, $a(t)$, $b_i(t)$, and $q_i(t)$ $(i = 1, \dots, n)$ be continuous functions in $J = [\alpha, \beta]$, let $b_i(t)$ and $q_i(t)$ be nonnegative in J, and suppose*

$$u(t) \leq a(t) + \sum_{i=1}^{n} q_i(t) \int_\alpha^t b_i(s)u(s)\,ds, \qquad t \in J. \tag{1.24}$$

Then for $t \in J$,

$$u(t) \leq a(t) + Q(t) \int_\alpha^t a(s) \sum_{i=1}^{n} b_i(s) \exp\left(\int_s^t Q(\tau) \sum_{i=1}^{n} b_i(\tau)\,d\tau\right) ds, \tag{1.25}$$

where $Q(t) = \sup_{i=1,\dots,n} q_i(t)$.

PROOF. The inequality (1.24) implies that

$$u(t) \leq a(t) + Q(t) \int_\alpha^t \sum_{i=1}^{n} b_i(s)u(s)\,ds, \qquad t \in J,$$

and theorem 1.5 now implies (1.25). □

The following theorem is concerned with inequality (1.24) in the frequently occurring case when all functions figuring in it are nonnegative. By induction we can obtain an estimate for $u(t)$ which is better than (1.25), and even better than corresponding estimates of Willett (1965) and Dhongade and Deo (1973).

THEOREM 1.6. (Kong and Zhang, 1989) *Let $a(t)$, $b_i(t)$, and $q_i(t)$ $(i = 1, \dots, n)$ be nonnegative continuous functions in $J = [\alpha, \beta]$, let $u(t)$ be a continuous function in J, and suppose*

$$u(t) \leq a(t) + \sum_{i=1}^{n} q_i(t) \int_\alpha^t b_i(s)u(s)\,ds, \qquad t \in J.$$

Then

$$u(t) \leq A_n[a](t), \qquad t \in J, \tag{1.26}$$

where the function $A_n[a]$ is defined by

$$\begin{aligned} A_0[x] &= x, \\ A_{k+1}[x] &= A_k[x] + A_k[q_{k+1}] \int_\alpha^t b_{k+1} A_k[x] \exp\left(\int_s^t b_{k+1} A_k[q_{k+1}]\,d\tau\right) ds. \end{aligned} \tag{1.27}$$

To prove theorem 1.6 we need the following lemma.

LEMMA 1.2. *The functions $A_k[x]$ $(k = 0, \dots, n)$ defined by (1.27) satisfy the following conditions for $x(t) \geq 0$, $y(t) \geq 0$:*

i) $A_k[x + y] = A_k[x] + A_k[y]$;
ii) $A_k[xy](t) \leq (A_k[x]y)(t)$ *if $y(t)$ is nondecreasing.*

PROOF. Clearly, i) and ii) are true for $k=0$. Suppose i) and ii) are true for $k=i$ ($i=0,\dots,n-1$), i.e.

$$A_i[x+y]=A_i[x]+A_i[y], \qquad A_i[xy]\le A_i[x]y.$$

Then

$$\begin{aligned}A_{i+1}[x+y] &= A_i[x+y]+A_i[q_{i+1}]\int_\alpha^t b_{i+1}A_i[x+y]\exp\left(\int_s^t b_{i+1}A_i[q_{i+1}]\,d\tau\right)ds=\\ &= A_i[x]+A_i[q_{i+1}]\int_\alpha^t b_{i+1}A_i[x]\exp\left(\int_s^t b_{i+1}A_i[q_{i+1}]\,d\tau\right)ds+\\ &\quad + A_i[y]+A_i[q_{i+1}]\int_\alpha^t b_{i+1}A_i[y]\exp\left(\int_s^t b_{i+1}A_i[q_{i+1}]\,d\tau\right)ds=\\ &= A_{i+1}[x]+A_{i+1}[y],\end{aligned}$$

$$\begin{aligned}A_{i+1}[xy] &= A_i[xy]+A_i[q_{i+1}]\int_\alpha^t b_{i+1}A_i[xy]\exp\left(\int_s^t b_{i+1}A_i[q_{i+1}]\,d\tau\right)ds\le\\ &\le\left\{A_i[x]+A_i[q_{i+1}]\int_\alpha^t b_{i+1}A_i[x]\exp\left(\int_s^t b_{i+1}A_i[q_{i+1}]\,d\tau\right)ds\right\}y=\\ &= A_{i+1}[x]y.\end{aligned}$$

This proves that i) and ii) are true for $k=i+1$. □

PROOF OF THEOREM 1.6. This proof is also by induction. For $n=1$ (1.26) becomes

$$u(t)\le a(t)+q_1(t)\int_\alpha^t b_1(s)a(s)\exp\left(\int_s^t b_1(\tau)q_1(\tau)\,d\tau\right)ds.$$

By theorem 1.5 this is true.

Suppose (1.26) holds for $n=k$ ($1\le k\le n-1$). Then, in view of the induction assumption,

$$u\le\left[a+q_{k+1}\int_\alpha^t b_{k+1}u\,ds\right]+\sum_{i=1}^k q_i\int_\alpha^t b_iu\,ds$$

implies

$$u\le A_k\left[a+q_{k+1}\int_\alpha^t b_{k+1}u\,ds\right].$$

Since

$$\int_\alpha^t b_{k+1}u\,ds$$

is nondecreasing in t, i) and ii) imply

$$u\le A_k[a]+A_k[q_{k+1}]\int_\alpha^t b_{k+1}u\,ds.$$

Now theorem 1.5 implies

$$u\le A_k[a]+A_k[q_{k+1}]\int_\alpha^t b_{k+1}A_k[a]\exp\left(\int_s^t b_{k+1}A_k[q_{k+1}]\,d\tau\right)ds=A_{k+1}[a]. \qquad \square$$

The following three theorems are based on lemma 1.1.

THEOREM 1.7. *Let $u(t)$, $b(t)$, and $c(t)$ be continuous functions in $J = [\alpha, \beta]$, let $b(t)$ and $c(t)$ be nonnegative in J, and suppose*

$$u(t) \le a + \int_\alpha^t b(s)u(s)\,ds + \int_\alpha^\beta c(s)u(s)\,ds, \qquad t \in J, \tag{1.28}$$

where a is a constant. If

$$q = \int_\alpha^\beta c(s)\exp\left(\int_\alpha^s b(\tau)\,d\tau\right)\,ds < 1,$$

then

$$u(t) \le \frac{a}{1-q}\exp\left(\int_\alpha^t b(s)\,ds\right), \qquad t \in J. \tag{1.29}$$

PROOF. Set $v(t)$ equal to the righthand side of (1.28). Then

$$v'(t) \le b(t)v(t), \qquad v(\alpha) = a + \int_\alpha^\beta c(s)u(s)\,ds.$$

By lemma 1.1 this implies

$$u(t) \le v(t) \le v(\alpha)\exp\left(\int_\alpha^t b(s)\,ds\right), \qquad t \in J.$$

But

$$v(\alpha) = a + \int_\alpha^\beta c(s)u(s)\,ds \le a + v(\alpha)\int_\alpha^\beta c(s)\exp\left(\int_\alpha^s b(\tau)\,d\tau\right)\,ds.$$

Consequently $v(\alpha) \le \frac{a}{1-q}$, which implies the result. □

THEOREM 1.8. *Let $u(t)$ and $a(t)$ be nonnegative continuous functions for $t \ge \alpha$, let $k(t,s)$ and its partial derivative $k_t(t,s)$ be nonnegative continuous functions for $\alpha \le s \le t$, and suppose*

$$u(t) \le a(t) + \int_\alpha^t k(t,s)u(s)\,ds, \qquad t \ge \alpha. \tag{1.30}$$

Then

$$u(t) \le a(t) + \int_\alpha^t A(s)\exp\left(\int_s^t B(\tau)\,d\tau\right)\,ds, \qquad t \ge \alpha, \tag{1.31}$$

where

$$A(t) = k(t,t)a(t) + \int_\alpha^t k_t(t,s)a(s)\,ds,$$
$$B(t) = k(t,t) + \int_\alpha^t k_t(t,s)\,ds.$$

PROOF. Putting $v(t) = \int_\alpha^t k(t,s)u(s)\,ds$, we find $v(\alpha) = 0$ and

$$v'(t) = k(t,t)u(t) + \int_\alpha^t k_t(t,s)u(s)\,ds.$$

Since $v(t)$ is nondecreasing and $u(t) \le a(t) + v(t)$, we have

$$v'(t) \le \left[k(t,t) + \int_\alpha^t k_t(t,s)\,ds\right] v(t) + \left[k(t,t)a(t) + \int_\alpha^t k_t(t,s)a(s)\,ds\right] =$$
$$= B(t)v(t) + A(t), \qquad t \ge \alpha.$$

Applying lemma 1.1 we arrive at (1.31). □

COROLLARY 1.15. (Norbury and Stuart, 1987) *Let $u(t)$, $k(t,s)$ and its partial derivative $k_t(t,s)$ be nonnegative continuous functions for $\alpha \le s \le t$, and suppose*

$$u(t) \le a + \int_\alpha^t k(t,s)u(s)\,ds, \qquad t \ge \alpha,$$

where $a \ge 0$ is a constant. Then

$$u(t) \le a \exp\left(\int_\alpha^t k(t,s)\,ds\right), \qquad t \ge \alpha. \tag{1.32}$$

PROOF. Since $A(t) = aB(t)$, by integrating (1.31) we arrive at

$$u(t) \le a \exp\left(\int_\alpha^t B(s)\,ds\right), \qquad t \ge \alpha,$$

which implies (1.32) since $\int_\alpha^t B(s)\,ds \equiv \int_\alpha^t k(t,s)\,ds$, $t \ge \alpha$. □

The estimate (1.32) was obtained in [138] (Theorem 2.1) under more general assumptions than those of corollary 1.15, but still including the existence and nonnegativity of $k_t(t,s)$. This last requirement can be replaced by the condition that $k(t,s)$ be nondecreasing in t, as in the following theorem.

THEOREM 1.9. (Movlyankulov and Filatov, 1972) *Let $u(t)$ be a continuous function in $J = [\alpha,\beta]$, let $b(t)$ be a nonnegative continuous function in J, let $k(t,s)$ be a nonnegative continuous function for $\alpha \le s \le t \le \beta$, and suppose*

$$u(t) \le a(t) + b(t)\int_\alpha^t k(t,s)u(s)\,ds, \qquad t \in J. \tag{1.33}$$

Then

$$u(t) \le A(t) \exp\left(B(t)\int_\alpha^t K(t,s)\,ds\right), \qquad t \in J, \tag{1.34}$$

where

$$A(t) = \sup_{s\in[\alpha,t]} a(s), \qquad B(t) = \sup_{s\in[\alpha,t]} b(s), \qquad K(t,s) = \sup_{\tau\in[s,t]} k(\tau,s).$$

PROOF. The functions $A(t)$, $B(t)$, and $K(t,s)$ are nondecreasing in $t \in J$. Thus, (1.33) implies that for all $\alpha \le t \le \tau \le \beta$,

$$u(t) \le A(\tau) + B(\tau)\int_\alpha^t K(\tau,s)u(s)\,ds.$$

Consequently, corollary 1.3 implies

$$u(t) \le A(\tau) \exp\left(B(\tau)\int_\alpha^t K(\tau,s)\,ds\right),$$

and for $\tau = t$ we obtain (1.34). □

REMARK 1.6. Inequality (1.34) becomes (1.32) when $b(t) \equiv 1$ and $a(t) \equiv a$ is an arbitrary constant. □

2. Inequalities of the type of Chu and Metcalf

The estimates for $u(t)$ contained in theorems 1.1–1.5 can be obtained by applying the comparison lemma 1.1. These estimates are exact in the sense that they cannot be improved upon, and their righthand sides are solutions of corresponding integral equations (see theorem 1.5i). Thus the idea arises of the possibility of estimating a function $u(t)$ satisfying an inequality of the type (1.30) by the solution of the corresponding integral equation. This possibility is based on the following theorem.

THEOREM 2.1. (Chu and Metcalf, 1967) *Let $u(t)$ and $a(t)$ be continuous functions in $J = [\alpha, \beta]$, let $k(t,s)$ be a nonnegative continuous function in the triangle $\Delta = \{(t,s) \in \mathbf{R}^2\colon \alpha \le s \le t \le \beta\}$, and suppose*

$$u(t) \le a(t) + \int_\alpha^t k(t,s)u(s)\,ds, \qquad t \in J. \tag{2.1}$$

Then

i)

$$u(t) \le v(t), \qquad t \in J, \tag{2.2}$$

where $v(t)$ is a solution of the equation

$$v(t) = a(t) + \int_\alpha^t k(t,s)v(s)\,ds, \qquad t \in J. \tag{2.3}$$

ii) *The solution $v(t)$ of* (2.3) *is unique and can be obtained as the sum of a Neumann series:*

$$v(t) = v_0(t) + v_1(t) + \cdots + v_n(t) + \cdots,$$

where $v_0(t) = a(t)$, $v_n(t) = \int_\alpha^t k(t,s)v_{n-1}(t)\,ds$, $n = 1, 2, \ldots$. □

We will prove this comparison theorem 2.1 in chapter III under more general assumptions. Equation (2.3) is called the *comparison integral equation* corresponding to inequality (2.1). This result has been generalized by Beesack (1969) to the case when $u, a \in L^2(J)$ and $k \in L^2(\Delta)$, and it has been proved that the theorem is still true if in (2.1) and (2.2) the sign $\le$ is replaced by $\ge$.

Note that theorems 1.1–1.5 can be proved using theorem 2.1, since the comparison integral equations corresponding to the inequalities under consideration are of the form (2.3). On the other hand, the comparison integral equation corresponding to inequality (1.28) is not of this form, and this is the reason why the exact estimate (1.29) cannot be obtained via theorem 2.1.

In the following comparison theorem we consider a comparison integral equation which is more general than (2.3).

THEOREM 2.2. (Daletskii and Krein, 1970) *Let J be a (finite or infinite) interval in $\mathbf{R}$. Let $k(t,s)$ be a nonnegative function in J^2 such that the integral operator*

$$K[u](t) = \int_J k(t,s)u(s)\,ds$$

leaves invariant the space $C(J)$ of bounded continuous functions in J and has, in this space, spectral radius less than one. Suppose the function $u \in C(J)$ satisfies the inequality

$$u(t) \leq a(t) + \int_J k(t,s)u(s)\,ds,$$

where $a \in C(J)$. Then

$$u(t) \leq v(t), \qquad t \in J,$$

where $v \in C(J)$ is the unique solution of the integral equation

$$v(t) = a(t) + \int_J k(t,s)v(s)\,ds, \qquad t \in J.$$

This theorem follows from the theorem on integral inequalities in a space with a cone (§ 19). □

REMARK 2.1. Note that the condition on the spectral radius in theorem 2.2 is sometimes difficult to verify. In such cases, however, it may be expedient to know that it suffices to show that the operator K (or some power K^n of it) is a contraction. (Recall that a linear operator T acting on a Banach space B with norm $\|\cdot\|$ is a *contraction* if there is a constant $q \in [0,1)$ such that

$$\|Tv\| \leq q\|v\| \qquad \text{for any } v \in B.) \qquad \square$$

The following theorems are based on theorem 2.2.

THEOREM 2.3. (Coppel, 1965) *Let $u(t)$ be a bounded continuous function in $J = [\alpha, \infty)$, and suppose*

$$u(t) \leq ae^{-\gamma(t-\alpha)} + \int_\alpha^\infty be^{-\gamma|t-s|}u(s)\,ds, \qquad t \in J,$$

where $a \geq 0$, $b \geq 0$, and $\gamma > 0$ are constants and $b < \gamma/2$. Then

$$u(t) \leq \frac{a}{b}(\gamma - \delta)e^{-\delta(t-\alpha)}, \qquad t \in J, \tag{2.4}$$

where $\delta = \sqrt{\gamma^2 - 2b\gamma}$.

PROOF. Let $C(J)$ be the Banach space of functions v which are bounded and continuous in $J = [\alpha, \infty)$ with norm $\|v\| = \sup_{t\in J}|v(t)|$.

Consider the linear operator $K\colon C(J) \to C(J)$ defined by

$$K[v](t) = \int_\alpha^\infty be^{-\gamma|t-s|}v(s)\,ds, \qquad t \in J.$$

If $v \in C(J)$ and $\|v\| = L$, it is easy to see that

$$|K[v](t)| \leq \int_\alpha^t bLe^{-\gamma(t-s)}\,ds + \int_t^\infty bLe^{\gamma(t-s)}\,ds \leq \frac{2b}{\gamma}L = \frac{2b}{\gamma}\|v\|,$$

whence we conclude that $K[v] \in C(J)$ and that K is a contraction with $q = 2b/\gamma < 1$. Hence $u(t) \leq v(t)$, where $v(t) \in C(J)$ is the unique solution of the integral equation

$$v(t) = ae^{-\gamma(t-\alpha)} + K[v](t), \qquad t \in J.$$

By a straightforward calculation we can verify that $v(t)$ is equal to the righthand side of (2.4). □

The following theorem can be proved in a similar manner.

THEOREM 2.4. (Coppel, 1965) *Let $u(t)$ be a continuous function for $\alpha \leq t \leq \beta$, and suppose*

$$u(t) \leq ae^{-\gamma(\beta-t)} + \int_\alpha^\beta be^{-\gamma|t-s|}u(s)\,ds, \qquad \alpha \leq t \leq \beta, \tag{2.5}$$

where $a \geq 0$, $b \geq 0$, and $\gamma > 0$ are constants and $b < \gamma/2$. Then

$$u(t) \leq \frac{a}{b}(\gamma - \delta)e^{-\delta(\beta-t)}, \qquad \alpha \leq t \leq \beta, \tag{2.6}$$

where $\delta = \sqrt{\gamma^2 - 2b\gamma}$. □

REMARK 2.2. The estimate (2.6) is not sharp, but it is simpler than the sharp estimate obtained from the solution of the *comparison integral equation* corresponding to (2.5). □

THEOREM 2.5. *Let $u(t)$ be a bounded continuous function in $J = [\alpha, \infty)$, let $g(t)$ be a nonnegative continuous function in J, and suppose*

$$u(t) \leq ae^{-\gamma(t-\alpha)} + \int_\alpha^\infty be^{-\gamma|t-s|}g(s)u(s)\,ds, \qquad t \in J,$$

where $a \geq 0$, $b \geq 0$, and $\gamma > 0$ are constants. If there are numbers $\tau > 0$ and $m = \sup_{t\in J}\int_t^{t+\tau} g(s)\,ds < \infty$ such that

$$q_0 = \frac{2bm}{1 - e^{-\gamma\tau}} < 1,$$

then there is a $\mu \in (0, \gamma)$ such that

$$q(\mu) \equiv bm\tau \left[\frac{1}{1 - e^{-(\gamma-\mu)\tau}} + \frac{1}{1 - e^{-(\gamma+\mu)\tau}}\right] < 1 \tag{2.7}$$

and

$$u(t) \leq \frac{a}{1 - q(\mu)}e^{-\mu(t-\alpha)}, \qquad t \in J. \qquad \square$$

To prove theorem 2.5 we need the following lemma.

LEMMA 2.1. (Coppel, 1978) *Let $g(t)$ be a nonnegative locally integrable function for $t \geq \alpha$ such that for some fixed $\tau > 0$,*

$$\frac{1}{\tau}\int_t^{t+\tau} g(s)\,ds \leq m < \infty \qquad \text{for all } t \geq \alpha.$$

Then, for any $\gamma > 0$ and $t \geq \alpha$,

$$\int_\alpha^t e^{-\gamma(t-s)} g(s)\,ds \leq m\tau(1-e^{-\gamma\tau})^{-1},$$

$$\int_t^\infty e^{-\gamma(s-t)} g(s)\,ds \leq m\tau(1-e^{-\gamma\tau})^{-1}. \qquad \square$$

PROOF OF THEOREM 2.5. Since $q(\mu)$ is a continuous function for $\mu \in (0,\gamma)$ and $\lim_{\mu\downarrow 0} q(\mu) = q_0 < 1$, there is a $\mu \in (0,\gamma)$ satisfying (2.7).

Now consider the operator K defined (on the space $C(J)$ of bounded continuous functions in J) by

$$K[u](t) = \int_\alpha^\infty be^{-\gamma|t-s|} g(s)u(s)\,ds, \qquad t \in J.$$

For $u \in C(J)$, $\|u\| = \sup_{t\in J}|u(t)| = L$, we find, by lemma 2.1,

$$|K[u](t)| \leq \int_\alpha^t bLe^{-\gamma(t-s)} g(s)\,ds + \int_t^\infty bLe^{-\gamma(s-t)} g(s)\,ds \leq \frac{2bm\tau}{1-e^{-\gamma\tau}}\|u\|,$$

i.e. $\|K[u]\| \leq q_0\|u\|$. Consequently, K is a contraction and the *comparison integral equation*

$$v(t) = ae^{-\gamma(t-\alpha)} + K[v](t)$$

has a unique solution. Moreover,

$$u(t) \leq v(t) = r(t) + K[r](t) + \cdots + K^n[r](t) + \cdots,$$

where $r(t) = ae^{-\gamma(t-\alpha)}$. But since $r(t) \leq ae^{-\mu(t-\alpha)} = R(t)$, we have

$$u(t) \leq R(t) + K[R](t) + \cdots + K^n[R](t) + \cdots.$$

Using lemma 2.1 we find the estimate

$$K[R](t) = \int_\alpha^t abe^{-\gamma(t-s)}\cdot e^{-\mu(s-\alpha)} g(s)\,ds + \int_t^\infty abe^{-\gamma(s-t)}\cdot e^{-\mu(s-\alpha)} g(s)\,ds =$$

$$= abe^{-\mu(t-\alpha)}\left[\int_\alpha^t e^{-(\gamma-\mu)(t-s)} g(s)\,ds + \int_t^\infty e^{-(\gamma+\mu)(s-t)} g(s)\,ds\right] \leq$$

$$\leq bm\tau\left[\frac{1}{1-e^{-(\gamma-\mu)\tau}} + \frac{1}{1-e^{-(\gamma+\mu)\tau}}\right] ae^{-\mu(t-\alpha)} = q(\mu)R(t).$$

Hence

$$u(t) \leq (1 + q(\mu) + \cdots + q^n(\mu) + \cdots)R(t) = \frac{a}{1-q(\mu)}e^{-\mu(t-\alpha)}. \qquad \square$$

3. Applications

Here we illustrate the application of integral inequalities of Gronwall and Bellman type to the solution of certain problems involving ordinary differential equations.

3.1. Uniqueness of the solution. Consider the initial value problem

$$x' = f(t, x), \tag{3.1}$$
$$x(t_0) = x_0. \tag{3.2}$$

It is equivalent to the integral equation

$$x(t) = x_0 + \int_{t_0}^{t} f(s, x(s))\, ds. \tag{3.3}$$

Suppose that $t_0 \in \mathbf{R}_+$, $x_0 \in \mathbf{R}^n$, and $f\colon \mathbf{R}_+ \times \mathbf{R}^n \to \mathbf{R}^n$ is a continuous function which is locally Lipschitz continuous with respect to x in $\mathbf{R}^n$, i.e.

$$|f(t,x) - f(t,y)| \le L(t)|x - y|, \qquad t \in \mathbf{R}_+, \quad x, y \in \mathbf{R}^n, \tag{3.4}$$

where $L\colon \mathbf{R}_+ \to \mathbf{R}_+$ is a continuous function.

Let $\phi(t)$ and $\psi(t)$ be two solutions of (3.1) defined on an interval $J = \langle \alpha, \beta \rangle$ such that $t_0 \in J$. From (3.3) and (3.4) we obtain

$$|\phi(t) - \psi(t)| \le |\phi(t_0) - \psi(t_0)| + \int_{t_0}^{t} L(s)|\phi(s) - \psi(s)|\, |ds|, \qquad t \in J.$$

Theorem 1.2 now implies that

$$0 \le |\phi(t) - \psi(t)| \le |\phi(t_0) - \psi(t_0)| \exp\left(\int_{t_0}^{t} L(s)\, |ds| \right), \qquad t \in J.$$

Hence $\phi(t_0) = \psi(t_0)$, which implies $\phi(t) \equiv \psi(t)$, $t \in J$, i.e. the solution of the initial value problem (3.1), (3.2) is unique.

In the sequel we will denote this unique, maximally continued, solution of the initial value problem (3.1), (3.2) by $x(t; t_0, x_0)$.

3.2. Differentiability of the solution. Consider the initial value problem

$$x' = f(t, x, \lambda), \qquad x(t_0) = x_0, \tag{3.5}$$

where $t, t_0 \in \mathbf{R}_+$, $x, x_0 \in \mathbf{R}^n$, the parameter $\lambda \in \mathbf{R}^m$, and $f\colon \mathbf{R}_+ \times \mathbf{R}^n \times \mathbf{R}^m \to \mathbf{R}^n$ is a continuous function which is locally Lipschitz continuous with respect to $(x, \lambda) \in \mathbf{R}^n \times \mathbf{R}^m$, i.e. there is a nonnegative continuous function $L(t)$ such that

$$\begin{gathered} |f(t,x,\lambda) - f(t,y,\mu)| \le L(t)(|x - y| + |\lambda - \mu|), \\ t \in \mathbf{R}_+, \qquad x, y \in \mathbf{R}^n, \qquad \lambda, \mu \in \mathbf{R}^m. \end{gathered} \tag{3.6}$$

We denote by $x(t; t_0, x_0, \lambda)$ the unique, maximally continued solution of (3.5). Suppose the solutions $x(t) = x(t; t_0, x_0, \lambda_0)$ and $y(t) = y(t; t_0, y_0, \mu_0)$ of (3.5) are defined in intervals J_x and J_y, respectively. Then for $t, t_0 \in [a, b] \subset J_x \cap J_y$ we obtain from

(3.6),

$$\begin{aligned}|x(t)-y(t)| &\le |x_0-y_0| + \int_{t_0}^{t} |f(s,x(s),\lambda_0)-f(s,y(s),\mu_0)|\,|ds| \le \\ &\le |x_0-y_0| + |\lambda_0-\mu_0| \int_{t_0}^{t} L(s)\,|ds| + \int_{t_0}^{t} L(s)|x(s)-y(s)|\,|ds| \le \\ &\le |x_0-y_0| + |\lambda_0-\mu_0| \int_{a}^{b} L(s)\,ds + \int_{t_0}^{t} L(s)|x(s)-y(s)|\,|ds|,\end{aligned}$$

and theorem 1.2 implies that

$$\begin{aligned}|x(t)-y(t)| &\le \left[|x_0-y_0| + |\lambda_0-\mu_0| \int_a^b L(s)\,ds\right] \exp\left(\int_{t_0}^{t} L(s)\,|ds|\right) \le \\ &\le \left[|x_0-y_0| + |\lambda_0-\mu_0| \int_a^b L(s)\,ds\right] \exp\left(\int_a^b L(s)\,ds\right).\end{aligned}$$

The last estimate shows that the solution $x(t;t_0,x_0,\lambda_0)$ is a continuous function of the initial value $x_0 \in \mathbf{R}^n$ and the parameter $\lambda \in \mathbf{R}^m$.

3.3. Estimate for the Cauchy matrix of a perturbed linear equation. Consider the linear equations

$$x' = A(t)x, \tag{3.7}$$

$$y' = [A(t)+B(t)]y, \tag{3.8}$$

where $t \in J = (\alpha,\beta)$, $x,y \in \mathbf{R}^n$, and $A, B \in C(\mathbf{R}, \mathbf{R}^{n\times n})$.

Suppose the *Cauchy matrix* $X(t,s)$ of equation (3.7) satisfies an estimate

$$|X(t,s)| \le \phi(t)\psi(s), \qquad t,s \in J, \quad s \le t, \tag{3.9}$$

where $\phi,\psi\colon J \to \mathbf{R}_+$ are positive continuous functions. We will derive several estimates for the Cauchy matrix $Y(t,s)$ of the perturbed equation (3.8), using (3.9).

Using the formula for variation of parameters, for an arbitrary solution $y(t)$, $t \in J$, of (3.8) we find

$$y(t) = X(t,s)y(s) + \int_s^t X(t,\tau)B(\tau)y(\tau)\,d\tau, \qquad t,s \in J, \quad s \le t.$$

Thus,

$$|y(t)| \le \phi(t)\psi(s)|y(s)| + \int_s^t \phi(t)\psi(\tau)|B(\tau)||y(\tau)|\,d\tau.$$

Hence the function $u(t) = |y(t)|/\phi(t)$ satisfies the inequality

$$u(t) \le \psi(s)|y(s)| + \int_s^t \phi(\tau)\psi(\tau)|B(\tau)|u(\tau)\,d\tau.$$

From theorem 1.1 we now obtain the estimate

$$|y(t)| \le M(t,s)|y(s)|, \tag{3.10}$$

where

$$M(t,s) = \phi(t)\psi(s)\exp\left(\int_s^t \phi(\tau)\psi(\tau)|B(\tau)|\,d\tau\right). \tag{3.11}$$

From inequality (3.10) and the equality $y(t) = Y(t,s)y(s)$ we can now immediately derive the following assertions:

THEOREM 3.1. *The Cauchy matrix* $Y(t,s)$ *of equation* (3.8) *satisfies the estimate*

$$|Y(t,s)| \le M(t,s),$$

where $M(t,s)$ *is given by* (3.11). □

COROLLARY 3.1. *If*

$$|X(t,s)| \le Ke^{\alpha(t-s)}, \qquad t,s \in J, \quad s \le t, \tag{3.12}$$

where $K \ge 1$ *and* α *are constants, then*

$$|Y(t,s)| \le Ke^{\alpha(t-s)+\int_s^t K|B(\tau)|\,d\tau}, \qquad t,s \in J, \quad s \le t. \qquad \square \tag{3.13}$$

COROLLARY 3.2. *If the estimate* (3.12) *is valid in the interval* $J = \mathbf{R}_+$, *and there is a constant* $\delta > 0$ *such that*

$$\sup_{\tau\in\mathbf{R}_+} |B(\tau)| \le \delta, \tag{3.14}$$

then

$$|Y(t,s)| \le Ke^{(\alpha+K\delta)(t-s)}, \qquad 0 \le s \le t. \qquad \square \tag{3.15}$$

COROLLARY 3.3. *If the estimate* (3.12) *is valid in the interval* $J = \mathbf{R}_+$, *and there is a constant* $M > 0$ *such that*

$$\int_0^\infty |B(\tau)|\,d\tau \le M, \tag{3.16}$$

then

$$|Y(t,s)| \le Ke^{KM} \cdot e^{\alpha(t-s)}, \qquad 0 \le s \le t. \qquad \square \tag{3.17}$$

3.4. Stability of a perturbed linear equation. Before inferring some consequences of the estimates obtained above we will give a few definitions of stability.

Suppose that $x = \phi(t)$, $t \in \mathbf{R}_+$, is a solution of (3.1), and suppose that for any $(t_0, x_0) \in \mathbf{R}_+ \times \mathbf{R}^n$ the initial value problem (3.1), (3.2) has a unique solution $x(t; t_0, x_0)$.

DEFINITION 3.1. The solution $x = \phi(t)$ of (3.1) is said to be

S1) *stable* if for any $\epsilon > 0$ and $t_0 \in \mathbf{R}_+$ there is a $\delta = \delta(\epsilon, t_0)$ such that $|x_0 - \phi(t_0)| < \delta$, $x_0 \in \mathbf{R}^n$, implies that

$$|x(t; t_0, x_0) - \phi(t)| < \epsilon \quad \text{for } t \ge t_0;$$

S2) *uniformly stable* if δ in S1) can be chosen independently of t_0;

S3) *attractive* if for each $t_0 \in \mathbf{R}_+$ there is a $\lambda = \lambda(t_0) > 0$, and for each $\epsilon > 0$ and $x_0 \in \mathbf{R}^n$, $|x_0| < \lambda$, there is a $\sigma = \sigma(t_0, \epsilon, x_0) > 0$ such that $|x(t; t_0, x_0) - \phi(t)| < \epsilon$ for $t \geq t_0 + \sigma$;

S4) *uniformly attractive* if for some $\lambda > 0$ and each $\epsilon > 0$ there is a $\sigma = \sigma(\epsilon) > 0$ such that $|x(t; t_0, x_0) - \phi(t)| < \epsilon$ for all $x_0 \in \mathbf{R}^n$, $|x_0| < \lambda$, $t_0 \in \mathbf{R}_+$, and all $t \geq t_0 + \sigma$;

S5) *asymptotically stable* if it is stable and attractive;

S6) *uniformly asymptotically stable* if it is uniformly stable and uniformly attractive;

S7) *exponentially stable* if there are constants $K \geq 1$, $\alpha > 0$, and $\rho > 0$ such that if $|x_0 - \phi(t_0)| < \rho$ and $t_0 \in \mathbf{R}_+$, then

$$|x(t; t_0, x_0) - \phi(t)| \leq K|x_0 - \phi(t_0)|e^{-\alpha(t-t_0)} \quad \text{for } t \geq t_0;$$

S8) *unstable* if for some $\epsilon > 0$ and $t_0 \in \mathbf{R}_+$, and each $\delta > 0$ there are $x_0 \in \mathbf{R}^n$, $|x_0| < \delta$, and $t > t_0$ such that

$$|x(t; t_0, x_0) - \phi(t)| \geq \epsilon.$$

Note [30], [64] that if the zero solution $x \equiv 0$ of the linear equation (3.7) is stable (uniformly stable, etc.), then all solutions of (3.7) are such. In this case the equation (3.7) itself is said to be *stable* (*uniformly stable*, etc.).

It can be seen from the estimates of $Y(t, s)$ obtained above that certain types of stability of equation (3.7) are preserved under 'small' perturbations $B(t)$. More precisely, the following theorem holds.

THEOREM 3.2. *If equation* (3.7) *is exponentially stable, i.e. the estimate* (3.12) *holds with* $\alpha < 0$, *then equation* (3.8) *is exponentially stable for perturbations* $B(t)$ *satisfying* (3.14) *with a sufficiently small* $\delta > 0$.

If equation (3.7) *is uniformly stable, i.e. the estimate* (3.12) *holds with* $\alpha = 0$ *and the condition* (3.16) *is met, then equation* (3.8) *is also uniformly stable.* □

3.5. Exponential and ordinary dichotomy. Asymptotic equivalence of linear equations. Let $X(t)$, $t \in J$, be the fundamental matrix of (3.7), and let I be the $(n \times n)$-identity matrix.

DEFINITION 3.2. Equation (3.7) is said to have *exponential dichotomy* on J if there is a projection operator P, i.e. a matrix P for which $P^2 = P$, and positive constants K, L, α, and β such that

$$\begin{aligned} &|X(t)PX^{-1}(s)| \leq Ke^{-\alpha(t-s)}, \qquad t, s \in J, \quad s \leq t, \\ &|X(t)(I - P)X^{-1}(s)| \leq Le^{-\beta(s-t)}, \qquad t, s \in J, \quad s \leq t. \end{aligned} \tag{3.18}$$

It is said to have *ordinary dichotomy* on J if the inequalities (3.18) hold with α and β equal to zero.

Note [57] that in the case of exponential dichotomy on $J = \mathbf{R}_+$ the range of P is uniquely determined in the sense that if $X(0) = I$, then the range of P is the subspace $V_1 \subset \mathbf{R}^n$ of initial values of all bounded solutions.

On the other hand, in the case of ordinary dichotomy P is not unique. If $V_0 \subset V_1$ is the subspace of initial values of all solutions of (3.7) tending to zero as $t \to \infty$, then (3.7) has ordinary dichotomy on $\mathbf{R}_+$ with projection operator Q if and only if

$$V_0 \subset Q(\mathbf{R}^n) \subset V_1. \tag{3.19}$$

If (3.7) is an *autonomous equation*, i.e. $A(t) \equiv A$, $t \in J$, then (3.7) has exponential dichotomy on $\mathbf{R}_+$ if and only if the real parts of all eigenvalues of the matrix A are distinct from zero. It has ordinary dichotomy on $\mathbf{R}_+$ if and only if all eigenvalues of A with zero real part are semisimple.

Let M be the Banach space of locally integrable vector functions $f\colon \mathbf{R}_+ \to \mathbf{R}^n$ for which $\int_t^{t+1} |f(s)|\, ds$ is uniformly bounded in $t \geq 0$, with norm $\|f\|_M = \sup_{t\geq 0} \int_t^{t+1} |f(s)|\, ds$.

Let L be the Banach space of locally integrable vector functions $f\colon \mathbf{R}_+ \to \mathbf{R}^n$ which are Lebesgue integrable in $\mathbf{R}_+$, with norm $\|f\|_L = \int_0^\infty |f(t)|\, dt$.

Next to (3.7) we consider the inhomogeneous equation

$$x' = A(t)x + f(t). \tag{3.20}$$

The following assertions characterize the dichotomy properties of (3.7).

PROPOSITION 3.1. *The inhomogeneous equation* (3.20) *has at least one bounded solution for each $f \in M$ if and only if the homogeneous equation* (3.7) *has exponential dichotomy on* $\mathbf{R}_+$.

PROPOSITION 3.2. *The inhomogeneous equation* (3.20) *has at least one bounded solution for each $f \in L$ if and only if the homogeneous equation* (3.7) *has ordinary dichotomy on* $\mathbf{R}_+$.

One of the most important properties of exponential dichotomy is *roughness*. That is, exponential dichotomy is not lost under small perturbations of the coefficient matrix.

THEOREM 3.3. (Coppel, 1978) *Suppose that the linear equation* (3.7) *has exponential dichotomy (see* (3.18)*) on* $\mathbf{R}_+$. *If*

$$\delta = \sup_{t\in\mathbf{R}_+} |B(t)| < \frac{\alpha}{4K^2},$$

then the perturbed equation (3.8) *also has exponential dichotomy:*

$$|Y(t)QY^{-1}(s)| \leq \frac{5}{2}K^2 e^{-(\alpha-2K\delta)(t-s)}, \qquad t \geq s \geq 0,$$

$$|Y(t)(I-Q)Y^{-1}(s)| \leq \frac{5}{2}K^2 e^{-(\alpha-2K\delta)(s-t)}, \qquad s \geq t \geq 0,$$

where $Y(t)$ is the fundamental matrix of (3.8) *such that $Y(0) = I$ and the projection operator Q has the same nullspace as P.*

Moreover,

$$|Y(t)QY^{-1}(t) - X(t)PX^{-1}(t)| \le 4\alpha^{-1}K\delta, \qquad t \ge 0.$$

SKETCH OF THE PROOF. For any bounded continuous matrix function $Y(t)$ we set $\|Y\| = \sup_{t\ge 0}\|Y(t)|$, and

$$TY(t) = X(t)P + \int_0^t X(t)PX^{-1}(\tau)B(\tau)Y(\tau)\,d\tau - \int_t^\infty X(t)(I-P)X^{-1}(\tau)B(\tau)Y(\tau)\,d\tau.$$

Then

$$|TY(t)| \le Ke^{-\alpha t} + K\|Y\|\left\{\int_0^t e^{-\alpha(t-\tau)}|B(\tau)|\,d\tau + \int_t^\infty e^{-\alpha(\tau-t)}|B(\tau)|\,d\tau\right\}.$$

It follows that $TY(t)$ is again bounded and continuous, and $\|TY\| \le K + 2\alpha^{-1}K\delta\|Y\|$.

Similarly,

$$\|TY - TZ\| \le 2\alpha^{-1}K\delta\|Y - Z\|.$$

Hence T is a contraction, because $q = 2\alpha^{-1}K\delta < 1/(2K) < 1$. By the contraction mapping principle (theorem 19.1), T has a unique fixed point. Denoting this point by $Y_1(t)$, we have

$$Y_1(t) = X(t)P + \int_0^t X(t)PX^{-1}(\tau)B(\tau)Y_1(\tau)\,d\tau - \int_t^\infty X(t)(I-P)X^{-1}(\tau)B(\tau)Y_1(\tau)\,d\tau.$$

It follows that $Y_1(t)$ is differentiable, and is a matrix solution of (3.8). We set $Y_1(0) = Q$. Using the fact that $Y_1(t)P$ is also a fixed point of T, we can prove that $Y_1(t)Q$ is also a fixed point of T and that Q is a projection operator. Let $Y(t)$ be the fundamental matrix of (3.8) and $Y(0) = I$. Then for $Y_1(t) = Y(t)Q$ and $Y_2(t) = Y(t)(I-Q)$ we find the following equalities:

$$Y_1(t) = X(t)PX^{-1}(s)Y_1(s) + \int_s^t X(t)PX^{-1}(\tau)B(\tau)Y_1(\tau)\,d\tau + \\ - \int_t^\infty X(t)(I-P)X^{-1}(\tau)B(\tau)Y_1(\tau)\,d\tau, \tag{3.21}$$

$$Y_2(t) = X(t)(I-P)X^{-1}(s)Y_2(s) + \int_0^t X(t)PX^{-1}(\tau)B(\tau)Y_2(\tau)\,d\tau + \\ - \int_t^s X(t)(I-P)X^{-1}(\tau)B(\tau)Y_2(\tau)\,d\tau. \tag{3.22}$$

It follows from (3.21) and (3.22) that for any vector $\xi \in \mathbf{R}^n$,

$$|Y_1(t)\xi| \le Ke^{-\alpha(t-s)}|Y_1(s)\xi| + K\delta\int_s^\infty e^{-\alpha|t-\tau|}|Y_1(\tau)\xi|\,d\tau, \qquad t \ge s \ge 0,$$

$$|Y_2(t)\xi| \le Ke^{-\alpha(s-t)}|Y_2(s)\xi| + K\delta\int_0^s e^{-\alpha|t-\tau|}|Y_2(\tau)\xi|\,d\tau, \qquad s \ge t \ge 0.$$

Hence, by theorems 2.3 and 2.4,

$$|Y_1(t)\xi| \le \rho e^{-\beta(t-s)}|Y_1(s)\xi|, \qquad t \ge s \ge 0,$$

$$|Y_2(t)\xi| \le \rho e^{-\beta(s-t)}|Y_2(s)\xi|, \qquad s \ge t \ge 0, \tag{3.23}$$

where $\beta = \sqrt{\alpha^2 - 2K\delta\alpha}$, $\rho = (\alpha - \sqrt{\alpha^2 - 2K\delta\alpha})\delta^{-1}$.

The last step consists of proving that $Y(t)QY^{-1}(t)$ is bounded in $\mathbf{R}_+$. The estimates (3.23) then imply that (3.8) has exponential dichotomy on $\mathbf{R}_+$. □

THEOREM 3.4. (Coppel, 1978) *Suppose that the linear differential equation* (3.7) *has ordinary dichotomy on the halfline* $\mathbf{R}_+$. *If*

$$\int_0^\infty |B(t)|\, dt < \infty,$$

then the perturbed equation (3.8) *also has ordinary dichotomy on* $\mathbf{R}_+$. *Moreover, there exist an invertible matrix* T *such that* (3.7) *has dichotomy with projection operator* P *if and only if* (3.8) *has dichotomy with projection operator* TPT^{-1}. □

In fact, this follows from proposition 3.1 and theorem 3.5 below, which is also of independent interest.

THEOREM 3.5. (Coppel, 1978) *Suppose that the linear differential equation* (3.7) *has ordinary dichotomy on the halfline* $\mathbf{R}_+$. *If*

$$\int_0^t |B(t)|\, dt < \infty, \qquad \int_0^\infty |f(t)|\, dt < \infty,$$

then there is a bijective affine mapping between the bounded solutions of equation (3.7) *and the bounded solutions of the inhomogeneous perturbed equation*

$$y' = [A(t) + B(t)]y + f(t) \tag{3.24}$$

such that the difference between corresponding solutions tends to zero as $t \to \infty$.

PROOF. Let $X(t)$ be the fundamental matrix of (3.7) with $X(0) = I$. Then there are a projection operator P and a constant $K > 0$ such that

$$\begin{aligned} |X(t)PX^{-1}(s)| &\le K, \qquad t \ge s \ge 0, \\ |X(t)(I-P)X^{-1}(s)| &\le K, \qquad s \ge t \ge 0. \end{aligned} \tag{3.25}$$

Moreover, in view of the remark about (3.19), we may assume that $|X(t)P\xi| \to 0$ as $t \to \infty$ for any vector $\xi \in \mathbf{R}^n$. Choose $t_0 \ge 0$ so large that

$$\theta = K \int_{t_0}^\infty |B(t)|\, dt < 1.$$

Let C be the space of bounded continuous functions $y\colon \mathbf{R}_+ \to \mathbf{R}^n$ with finite norm $\|y\| = \sup_{t \ge 0} |y(t)|$. On C we define the operator T as

$$\begin{aligned} Ty(t) = &\int_{t_0}^t X(t)PX^{-1}(s)[B(s)y(s) + f(s)]\, ds + \\ &- \int_t^\infty X(t)(I-P)X^{-1}(s)[B(s)y(s) + f(s)]\, ds. \end{aligned}$$

Using (3.25) it is easy to prove that $TC \subset C$ and that

$$\|Ty - Tz\| \le \theta \|y - z\|, \qquad y, z \in C.$$

Let $x(t)$ be an arbitrary bounded solution of (3.7). Then the integral equation

$$y(t) = x(t) + Ty(t) \tag{3.26}$$

has, according to the contraction mapping principle, a unique bounded solution $y(t)$. It is readily verified that $y(t)$ is also a solution of (3.24). Conversely, if $y(t)$ is a bounded continuous solution of (3.24), then $x(t) = y(t) - Ty(t)$ is a bounded solution of (3.7).

Thus, (3.26) sets up a bijective affine mapping between the bounded solutions of (3.24) and (3.7). Given an $\epsilon > 0$, we can choose $t_1 \geq t_0$ so that

$$K \int_{t_1}^{\infty} [|B(s)||y(s)| + |f(s)|]\, ds < \epsilon.$$

Then

$$|Ty(t)| \leq \left| X(t)P \int_{t_0}^{t_1} [B(s)y(s) + f(s)]\, ds \right| + \epsilon < 2\epsilon$$

for all sufficiently large t, since $|X(t)P\xi| \to 0$ as $t \to \infty$ for all $\xi \in \mathbf{R}^n$. Hence $y(t) - x(t) \to 0$ as $t \to \infty$. □

Theorem 3.5 shows that between the systems (3.7) and (3.24) there is *bounded asymptotic equivalence*. Also, the following assertion is valid.

COROLLARY 3.4. (Levinson, 1946) *Suppose $A(t) \equiv A$, $t \in \mathbf{R}_+$, and suppose all solutions of* (3.7) *are bounded in $\mathbf{R}_+$. If $B \in C(\mathbf{R}_+, \mathbf{R}^{n\times n})$ and*

$$\int_0^{\infty} |B(t)|\, dt < \infty,$$

then (3.7) *and* (3.24) *are asymptotically equivalent, i.e. there is a bijective affine mapping between the solutions $x(t)$ and $y(t)$ of these equations, respectively, such that*

$$\lim_{t\to\infty} [x(t) - y(t)] = 0. \qquad \square$$

Note also that theorem 3.4 can be proved by a straightforward use of theorem 1.1 (see [64], [117]).

3.6. Stability of the centre manifold. Near an equilibrium point, a C^1 autonomous system of differential equations in $\mathbf{R}^n$ can be written in the form

$$\begin{aligned} x' &= Ax + f(x, y), \\ y' &= Bx + g(x, y), \end{aligned} \tag{3.27}$$

where $x \in \mathbf{R}^k$, $y \in \mathbf{R}^{n-k}$, $A \in \mathbf{R}^{k\times k}$, $B \in \mathbf{R}^{(n-k)\times(n-k)}$, the eigenvalues of A have nonnegative real parts and those of B have negative real parts, and f and g are C^1 (i.e. continuously differentiable) vector functions defined in a neighbourhood of $(0,0)$ that vanish, together with their first order partial derivatives, at $(0,0)$. We assume that all eigenvalues of A have zero real part. (This situation occurs most often in applications.)

A *centre manifold* for the system (3.27) is a k-dimensional C^1-submanifold of $\mathbf{R}^n$ passing through $(0,0)$ which is locally invariant and has the x-space as its tangent space at $(0,0)$. Equivalently, a centre manifold for (3.27) is the graph of a C^1 function $y = v(x)$ satisfying the partial differential equation

$$v_x(x)[Ax + f(x, v(x))] = Bv(x) + g(x, v(x)) \tag{3.28}$$

and the conditions $v(0) = 0$, $v_x(0) = 0$.

Suppose that $f(x,y)$, $g(x,y)$, and $v(x)$ are defined in $|x| < \Delta_0$, $|y| < \Delta_0$. We denote by $\omega(\Delta)$ the maximum of the least upper bounds of the norms of the derivatives f_x, f_y, g_x, g_y, v_x in the set $|x| < \Delta$, $|y| < \Delta$, so that $\omega(\Delta) \downarrow 0$ as $\Delta \downarrow 0$. The assumptions regarding the eigenvalues of A and B imply that there are constants $K \geq 1$, $\beta > 0$, $\alpha < \beta$ such that for $t \geq 0$,

$$|e^{-tA}| \leq Ke^{\alpha t}, \qquad |e^{tB}| \leq Ke^{-\beta t}.$$

LEMMA 3.1. (Palmer, 1987) *For a given γ $(0 < \gamma < \beta - \alpha)$ there is a $\Delta > 0$ such that if $x(t), y(t)$ is a solution of* (3.27) *defined in an interval* $[0,T]$ *and*

$$|x(t)| < \Delta, \qquad |y(t)| < \Delta,$$

then the solution $\tilde{x}(t)$ of the system

$$x' = Ax + f(x, v(x)) \tag{3.29}$$

with $\tilde{x}(T) = x(T)$ is defined in $[0,T]$ and the inequalities $|\tilde{x}(t)| \leq 2\Delta$ and

$$|x(t) - \tilde{x}(t)| + |y(t) - v(\tilde{x}(t))| \leq 2K|y(0) - v(x(0))|e^{-(\beta-\gamma)t} \tag{3.30}$$

hold for $0 \leq t \leq T$.

PROOF. Choose $\Delta > 0$ such that

$$2\Delta < \Delta_0, \qquad 4K^2\omega(2\Delta) < \min\{2\gamma, \beta - \gamma - \alpha, 4K^2\}. \tag{3.31}$$

Using (3.28) we find that $z(t) = y(t) - v(x(t))$ satisfies

$$\begin{aligned} z' = Bz(t) &+ g(x(t), y(t)) - g(x(t), v(x(t))) + \\ &- v_x(x(t))[f(x(t), y(t)) - f(x(t), v(x(t)))]. \end{aligned}$$

By the method of variation of parameters we find that

$$\begin{aligned} z(t) = e^{tB}z(0) + \int_0^t e^{(t-s)B}\{&g(x(s), y(s)) - g(x(s), v(x(s))) + \\ &- v_x(x(s))[f(x(s), y(s)) - f(x(s), v(x(s)))]\}\, ds. \end{aligned}$$

Using $|v(x(t))| \leq \omega(\Delta)\Delta < \Delta$ we find

$$|z(t)| \leq Ke^{-\beta t}|z(0)| + \int_0^t Ke^{-\beta(t-s)}\omega(\Delta)(1 + \omega(\Delta))|z(s)|\, ds.$$

Applying corollary 1.8 we obtain

$$|z(t)| \leq K|z(0)|e^{-[\beta - K\omega(\Delta)(1+\omega(\Delta))]t}.$$

It follows that for $0 \leq t \leq T$,

$$|y(t) - v(x(t))| \leq K|y(0) - v(x(0))|e^{-(\beta-\gamma)t}. \tag{3.32}$$

Now, let τ be the smallest nonnegative number such that $\tilde{x}(t)$ is defined in $[\tau, T]$ and satisfies $|\tilde{x}(t)| \le 2\Delta$ in it. Since $|\tilde{x}(T)| = |x(T)| < \Delta$, we have $0 \le \tau < T$. Also, for $\tau > 0$, $|\tilde{x}(\tau)| = 2\Delta$. The function $u(t) = x(t) - \tilde{x}(t)$ satisfies (by the method of variation of parameters)

$$u(t) = -\int_t^T e^{(t-s)A}\{f(x(s), y(s)) - f(x(s), v(x(s))) + \\ + f(x(s), v(x(s))) - f(x(s, v(\tilde{x}(s))) + f(x(s), v(\tilde{x}(s))) - f(\tilde{x}(s), v(\tilde{x}(s)))\}\, ds,$$

so that for $\tau \le t \le T$,

$$|u(t)| \le K \int_t^T e^{\alpha(s-t)}[\omega(\Delta)|y(s) - v(x(s))| + \omega(2\Delta)(1 + \omega(2\Delta))|u(s)|]\, ds \le \\ \le K^2(\beta - \gamma - \alpha)^{-1}\omega(\Delta)|y(0) - v(x(0))|e^{-(\beta-\gamma)t} + 2K\omega(2\Delta)\int_t^T e^{\alpha(s-t)}|u(s)|\, ds$$

(where we have used (3.32) and subsequent integration). Putting $w(t) = e^{\alpha t}|u(t)|$ and applying theorem 1.5 to the resulting inequality, we find that for $\tau \le t \le T$,

$$|x(t) - \tilde{x}(t)| \le K^2[\beta - \gamma - \alpha - 2K\omega(2\Delta)]^{-1}\omega(\Delta)|y(0) - v(x(0))|e^{-(\beta-\gamma)t} \le \\ \le \frac{1}{2}|y(0) - v(x(0))|e^{-(\beta-\gamma)t}. \tag{3.33}$$

It follows that for $\tau \le t \le T$,

$$|\tilde{x}(t)| \le |x(t)| + |x(t) - \tilde{x}(t)| \le \Delta + \frac{1}{2}(1 + \omega(\Delta))\Delta < 2\Delta. \tag{3.34}$$

In particular, $|\tilde{x}(\tau)| \le 2\Delta$, and so $\tau = 0$. Hence $\tilde{x}(t)$ is defined in $[0, T]$, and (3.33) and (3.34) hold for $0 \le t \le T$.

Finally, using (3.32) and (3.33), for $0 \le t \le T$ we obtain

$$|x(t) - \tilde{x}(t)| + |y(t) - v(\tilde{x}(t))| \le \\ \le |x(t) - \tilde{x}(t)| + |y(t) - v(x(t))| + |v(x(t)) - v(\tilde{x}(t))| \le \\ \le [1 + \omega(2\Delta)]|x(t) - \tilde{x}(t)| + |y(t) - v(x(t))| \le \\ \le \left[K + \frac{1}{2}(1 + \omega(2\Delta))\right]|y(0) - v(x(0))|e^{-(\beta-\gamma)t} \le \\ \le 2K|y(0) - v(x(0))|e^{-(\beta-\gamma)t}. \qquad \square$$

Now we fix $\gamma \in (0, \beta - \alpha)$, and let $\Delta > 0$ be such that the conditions of the lemma are satisfied.

THEOREM 3.6. (Palmer, 1987) *Small bounded solutions must lie on the centre manifold. That is, if $x(t), y(t)$ is a solution of (3.27) satisfying*

$$|x(t)| < \Delta, \qquad |y(t)| < \Delta \qquad \textit{for } t \in \mathbf{R},$$

then $y(t) = v(x(t))$ for all $t \in \mathbf{R}$.

PROOF. In (3.30) we set $t = T$, and find

$$|y(T) - v(x(T))| \leq 2K|y(0) - v(x(0))|e^{-(\beta-\gamma)T}.$$

Since this is true for all $T \geq 0$ we may apply it to the solution $x(t+s), y(t+s)$ to find that, for $t \geq s$,

$$|y(t) - v(x(t))| \leq 2K|y(s) - v(x(s))|e^{-(\beta-\gamma)(t-s)}.$$

Now we let $s \to -\infty$, and obtain the result. □

Note that we could have also proved this by using (3.32).

THEOREM 3.7. (Palmer, 1987) *Suppose the solution $x(t), y(t)$ of (3.27) is defined for all $t \geq 0$ and satisfies $|x(t)| < \Delta$, $|y(t)| < \Delta$. Then there is a solution $x_\infty(t)$ of (3.29) such that for $t \geq 0$,*

$$|x(t) - x_\infty(t)| + |y(t) - v(x_\infty(t))| \leq 2K|y(0) - v(x(0))|e^{-(\beta-\gamma)t}. \quad (3.35)$$

PROOF. Let $\tilde{x}_m(t)$ be the solution of (3.29) satisfying $\tilde{x}_m(m) = x(m)$. Lemma 3.1 implies that $\tilde{x}_m(t)$ is defined in $[0, m]$, and that the inequalities $|\tilde{x}_m(t)| < 2\Delta$ and

$$|x(t) - \tilde{x}_m(t)| + |y(t) - v(\tilde{x}_m(t))| \leq 2K|y(0) - v(x(0))|e^{-(\beta-\gamma)t} \quad (3.36)$$

hold for $0 \leq t \leq m$. It also follows that for $0 \leq t \leq m$,

$$|\dot{\tilde{x}}(t)| \leq |A\tilde{x}_m(t) + f(\tilde{x}_m(t), v(\tilde{x}_m(t)))| \leq 2(|A| + \omega(2\Delta)(1 + \omega(2\Delta)))\Delta. \quad (3.37)$$

Now the Arzelà–Ascoli theorem implies that there is a subsequence $\tilde{x}_{j_m}(t)$ which converges on compact subintervals of $\mathbf{R}_+$ to a function $x_\infty(t)$. The function $x_\infty(t)$ is a solution of (3.29), and we obtain (3.35) by limit transition in (3.36) for fixed t and m replaced by j_m. □

Chapter I: notes and comments

G. Peano [166] proved in 1885 explicitly the special case of (1.1) with $a = 0$. In his book *Differentialgleichungen reeller Funktionen*, p. 93, published in 1930, E. Kamke proved and used an inequality which is, in essence, the same as Gronwall's. Similarly, W.T. Reid, in his Ph.D. dissertation (published in 1930 in [184]), stated without proof a result which is, in essence, the same as that of Bellman, except that the continuity assumption on $b(t)$ was relaxed to Lebesgue integrability. Although Reid's result precedes that of Bellman by 13 years, it remained almost unnoticed. Only the single paper [109] mentions 'Bellman–Gronwall–Reid inequalities'.

In relation with the results in chapter 1, the following should be noted.

Theorem 1.4 is taken from [127], Thm. 1.6. *Willett's theorem* is included in the first part of theorem 1.5. Willett's paper [233] also gives a linear generalization to the case when $q(t)b(s)$ in (1.20) is replaced by $\sum_{i=1}^{n} q_i(t)b_i(s)$. Lemma 1.2 is taken from [105]. Theorem 1.7 is new, and includes as a special case Thm. 1.1.5 in [113]. Theorem 1.8 has been adapted from Sh.T. Gadimov [76]. For results similar to theorem 2.3 and lemma 2.1 see Yu.L. Daletskii and M.G. Krein [60].

For other applications see also [30], [56], [113], [125], [127], [141], [185], [215], [221], and [222].

CHAPTER II

Scalar nonlinear integral inequalities

Nonlinear equations are mainly used in a more detailed simulation of actual processes and phenomena. The nonlinear integral inequalities which are dealt with in this chapter are a natural mathematical apparatus in this simulation.

4. Integral inequalities investigated by nonstandard methods

In this section we consider various nonstandard methods for studying integral inequalities. The successful application of these methods depends only on the skill of the investigator and his/her use of suitable substitutions and transformations.

An essential element in the investigation of the integral inequalities in the following three theorems is the application of the inequalities of Hölder, Minkowski, and Jensen.

THEOREM 4.1. (Martynyuk and Kosolapov, 1978) *Let $u(t)$, $a(t)$, and $k(t)$ be nonnegative continuous functions in $J = [\alpha, \beta]$, and suppose*

$$u(t) \leq a(t) + \int_\alpha^t k(s) u^p(s)\, ds, \qquad t \in J, \tag{4.1}$$

where $0 < p < 1$. Then

$$u(t) \leq a(t) + x_0^p \left(\int_\alpha^t k^{1/q}(s)\, ds \right)^q, \qquad q = 1 - p, \tag{4.2}$$

where x_0 is the unique positive root of the equation

$$x - a - bx^p = 0, \qquad a = \int_\alpha^\beta a(s)\, ds, \qquad b = \int_\alpha^\beta \left[\int_\alpha^t k^{1/q}(s)\, ds \right]^q dt.$$

PROOF. From Hölder's inequality we obtain

$$\int_\alpha^t k(s) u^p(s)\, ds \leq \left(\int_\alpha^t k^{1/q}(s)\, ds \right)^q \left(\int_\alpha^t u(s)\, ds \right)^p.$$

Then (4.1) implies

$$u(t) \leq a(t) + \left(\int_\alpha^t u(s)\, ds \right)^p \left(\int_\alpha^t k^{1/q}(s)\, ds \right)^q, \tag{4.3}$$

and integration of (4.3) from α to β gives

$$\int_\alpha^\beta u(s)\,ds \le \int_\alpha^\beta a(s)\,ds + \left(\int_\alpha^\beta u(s)\,ds\right)^p \int_\alpha^\beta \left[\int_\alpha^t k^{1/q}(s)\,ds\right]^q dt.$$

If $x = \int_\alpha^\beta u(s)\,ds$, $a = \int_\alpha^\beta a(s)\,ds$, $b = \int_\alpha^\beta [\int_\alpha^t k^{1/q}(s)\,ds]^q\,dt$, then the last inequality shows that the nonnegative number x satisfies the inequality

$$x \le a + bx^p.$$

Analyzing this, we conclude that $x \le x_0$, where x_0 is the unique positive root of the equation $x = a + bx^p$. Hence (4.3) implies (4.2). □

THEOREM 4.2. (Willett and Wong, 1965) *Let $u(t)$, $a(t)$, $b(t)$, and $k(t)$ be nonnegative continuous functions in $J = [\alpha, \beta]$, let $p \ge 1$ be a constant, and suppose*

$$u(t) \le a(t) + b(t)\left(\int_\alpha^t k(s)u^p(s)\,ds\right)^{1/p}, \qquad t \in J. \tag{4.4}$$

Then

$$u(t) \le a(t) + b(t)\frac{\left(\int_\alpha^t k(s)\epsilon(s)a^p(s)\,ds\right)^{1/p}}{1-[1-\epsilon(t)]^{1/p}}, \qquad t \in J, \tag{4.5}$$

where $\epsilon(t) \equiv \exp(-\int_\alpha^t k(s)b^p(s)\,ds)$.

PROOF. Set $v(t) = \epsilon(t)\int_\alpha^t k(s)u^p(s)\,ds$. Then (4.4) implies that

$$u(t) \le a(t) + b(t)v^{1/p}(t)\epsilon^{-1/p}(t), \tag{4.6}$$

and since $\epsilon'(t) = -k(t)b^p(t)\epsilon(t)$,

$$v' = \epsilon k u^p - kb^p v \le k[a\epsilon^{1/p} + bv^{1/p}]^p - kb^p v.$$

Integration from α to t gives

$$v \le \int_\alpha^t k[a\epsilon^{1/p} + bv^{1/p}]^p\,ds - \int_\alpha^t kb^p v\,ds,$$

or

$$\left(v + \int_\alpha^t kb^p v\,ds\right)^{1/p} \le \left(\int_\alpha^t [ak^{1/p}\epsilon^{1/p} + bk^{1/p}v^{1/p}]^p\,ds\right)^{1/p}.$$

By Minkowski's inequality,

$$\left(\int_\alpha^t [ak^{1/p}\epsilon^{1/p} + bk^{1/p}v^{1/p}]^p\,ds\right)^{1/p} \le \left(\int_\alpha^t ka^p\epsilon\,ds\right)^{1/p} + \left(\int_\alpha^t kb^p v\,ds\right)^{1/p}.$$

Thus,

$$\left(v + \int_\alpha^t kb^p v\,ds\right)^{1/p} - \left(\int_\alpha^t kb^p v\,ds\right)^{1/p} \le \left(\int_\alpha^t ka^p\epsilon\,ds\right)^{1/p}. \tag{4.7}$$

For given $v \geq 0$ and $p \geq 1$ the function $f(x) = (v+x)^{1/p} - x^{1/p}$ is nonincreasing for $x \geq 0$. So inequality (4.7) still holds if $\int_\alpha^t kb^p v\,ds$ is replaced by a larger number. Thus,

$$x = \int_\alpha^t kb^p v\,ds = \int_\alpha^t kb^p \epsilon \left(\int_\alpha^s ku^p\,d\tau\right) ds \leq \int_\alpha^t kb^p \epsilon\,ds \cdot \int_\alpha^t ku^p\,ds =$$
$$= -\int_\alpha^t \epsilon'\,ds \cdot \int_\alpha^t ku^p\,ds = (1-\epsilon)\int_\alpha^t ku^p\,ds,$$

and (4.7) implies

$$[1-(1-\epsilon)^{1/p}]\left(\int_\alpha^t ku^p\,ds\right) \leq \left(\int_\alpha^t ka^p\epsilon\,ds\right)^{1/p}.$$

This and (4.6) imply (4.5). □

THEOREM 4.3. *Let u, a, b, k be continuous functions in an interval $J = [\alpha,\beta]$, with b and k nonnegative and $\int_\alpha^\beta k(s)\,ds \leq 1$. Let g be a strictly monotone continuous function of constant sign in an interval I such that $u(J) \subset I$, with either*

i) *g nonnegative, increasing, and concave in I;*

or

ii) *g nonpositive, decreasing, and convex in I.*

Suppose $b(t)(\int_\alpha^t k(s)\,ds)^{-1}$ is integrable over J and

$$u(t) \leq a(t) + b(t)g^{-1}\left(\int_\alpha^t k(s)g(u(s))\,ds\right), \qquad t \in J. \tag{4.8}$$

Then

$$u(t) \leq a(t)+b(t)\left(\int_\alpha^t k(s)\,ds\right)^{-1} \cdot$$
$$\cdot \int_\alpha^t a(s)k(s)\exp\left(\int_s^t b(\tau)k(\tau)\left(\int_\alpha^\tau k(r)\,dr\right)^{-1} d\tau\right) ds \tag{4.9}$$

for all $\alpha \leq t < \beta_1$, where

$$\beta_1 = \sup\left\{v \in J : \left(\int_\alpha^t k(\tau)\,d\tau\right)^{-1}\int_\alpha^t k(s)u(s)\,ds \in I,\ \alpha \leq t \leq v\right\}.$$

If $\leq$ is replaced by $\geq$ in both (4.8) and (4.9), the result still holds, provided the condition i) *or* ii) *is replaced by*

j) *g is nonnegative, decreasing, and concave in I;*

respectively,

jj) *g is nonpositive, increasing, and convex in I.*

Finally, all results remain valid if $\int_\alpha^t$ and $\int_s^t$ are replaced by $\int_t^\beta$ and $\int_t^s$, respectively, and $[\alpha,\beta_1)$ is replaced by $(\alpha_1,\beta]$, where

$$\alpha_1 = \inf\left\{v \in J : \left(\int_t^\beta k(\tau)\,d\tau\right)^{-1}\int_t^\beta k(s)u(s)\,ds \in I,\ v \leq t \leq \beta\right\}.$$

PROOF. For the sake of being specific we consider the case ii). For $t = \alpha$ we interpret $(\int_\alpha^t k(s)\,ds)^{-1}\int_\alpha^t k(s)u(s)\,ds \equiv K(t)$ as the limit value $u(\alpha)$; similarly, the righthand side of (4.9) is understood to mean $a(\alpha)+b(\alpha)a(\alpha)$. If $\alpha \le t \le \beta_1$, Jensen's integral inequality implies

$$\int_\alpha^t k(s)g(u(s))\,ds = \int_\alpha^t k(s)\,ds \frac{\int_\alpha^t k(s)g(u(s))\,ds}{\int_\alpha^t k(s)\,ds} \geq \int_\alpha^t k(s)\,ds \cdot g(K(t)),$$

since g is convex. The nonpositivity of g implies

$$\int_\alpha^t k(s)\,ds \cdot g(K(t)) \geq g(K(t)).$$

Hence

$$\int_\alpha^t k(s)g(u(s))\,ds \geq g(K(t)),$$

and

$$g^{-1}\left(\int_\alpha^t k(s)g(u(s))\,ds\right) \leq K(t)$$

since g is decreasing. From (4.8) and the fact that $b \geq 0$ it follows that

$$u(t) \leq a(t) + b(t)\left(\int_\alpha^t k(s)\,ds\right)^{-1}\int_\alpha^t k(s)u(s)\,ds.$$

An application of theorem 1.5 now gives (4.9).

The proofs for the other cases are similar. □

In the next four theorems we derive from (4.8) various linear integral or differential inequalities, using properties of g. This enables us to apply the results in chapter I to obtain various estimates for $u(t)$.

THEOREM 4.4. (Gollwitzer, 1969) *Let u, a, b, k be continuous functions in an interval $J = [\alpha, \beta]$, with b and k nonnegative. Let g be a nonnegative continuous function in an interval I such that $u(J) \subset I$, with either*

i) *g strictly increasing, convex, and submultiplicative in I;*

or

ii) *g strictly decreasing, concave, and supermultiplicative in I.*

Suppose

$$u(t) \leq a(t) + b(t)g^{-1}\left(\int_\alpha^t k(s)g(u(s))\,ds\right), \qquad t \in J. \tag{4.10}$$

Let p and q be positive continuous functions in J such that $p(t) + q(t) \equiv 1$. If $A(t) = a(t)/p(t)$, $B(t) = b(t)/q(t)$, then

$$\int_\alpha^t k(s)g(u(s))\,ds \;\substack{\leq \\ \geq}\; \int_\alpha^t p(s)g(A(s))k(s)\exp\left(\int_s^t q(\tau)g(B(\tau))k(\tau)\,d\tau\right)ds, \tag{4.11}$$

where $\leq$ *(respectively,* $\geq$*) holds in case* i) *(respectively,* ii)*). In both cases,*

$$u(t) \leq a(t) + b(t) g^{-1} \left(\int_\alpha^t p(s) g(A(s)) k(s) \exp \left(\int_s^t q(\tau) g(B(\tau)) k(\tau)\, d\tau \right) ds \right). \tag{4.12}$$

The inequalities (4.11) *and* (4.12) *hold for* $\alpha \leq t < \beta_1$*, where* $\beta_1 = \min(v_1, v_2, v_3)$*, with*

$$v_1 = \sup\{v \in J : \ A(t) \in I, \ B(t) \in I \ \text{ for } \alpha \leq t \leq v\},$$

$$v_2 = \sup \left\{ v \in J : \ B(t) g^{-1} \left(\int_\alpha^t k(s) g(u(s))\, ds \right) \in I, \ \alpha \leq t \leq v \right\},$$

$$v_3 = \sup \left\{ v \in J : \ \int_\alpha^v p(s) g(A(s)) k(s) \exp \left(\int_s^v q(\tau) g(B(\tau)) k(\tau)\, d\tau \right) ds \in g(I) \right\}.$$

The result still holds if in (4.10)–(4.12) *the inequality signs are reversed, provided that the conditions* i) *and* ii) *are replaced by*

j) *g is strictly decreasing, convex, and submultiplicative in I (for* $\leq$*); respectively,*

jj) *g is strictly increasing, concave, and supermultiplicative in I (for* $\geq$*).*

Finally, all results remain true if $\int_\alpha^t$ *and* $\int_s^t$ *are replaced by* $\int_t^\beta$ *and* $\int_t^s$*, respectively, and* $[\alpha, \beta_1)$ *is replaced by* $(\alpha_1, \beta]$*, where* $\alpha_1 = \max(v_1, v_2, v_3)$*, with*

$$v_1 = \inf\{v \in J : \ A(t) \in I, \ B(t) \in I \ \text{ for } v \leq t \leq \beta\},$$

$$v_2 = \inf \left\{ v \in J : \ B(t) g^{-1} \left(\int_t^\beta k(s) g(u(s))\, ds \right) \in I, \ v \leq t \leq \beta \right\},$$

$$v_3 = \inf \left\{ v \in J : \ \int_v^\beta p(s) g(A(s)) k(s) \exp \left(\int_v^\beta q(\tau) g(B(\tau)) k(\tau)\, d\tau \right) ds \in g(I) \right\}.$$

PROOF. If $\alpha \leq t < \beta_1$, then (4.10) implies (in case i)) that

$$\begin{aligned} g(u(t)) &\leq g \left\{ p(t) A(t) + q(t) \left[B(t) g^{-1} \left(\int_\alpha^t k(s) g(u(s))\, ds \right) \right] \right\} \leq \\ &\leq p(t) g(A(t)) + q(t) g \left[B(t) g^{-1} \left(\int_\alpha^t k(s) g(u(s))\, ds \right) \right] \leq \\ &\leq p(t) g(A(t)) + q(t) g(B(t)) \int_\alpha^t k(s) g(u(s))\, ds. \end{aligned}$$

Then for $v(t) = \int_\alpha^t k(s) g(u(s))\, ds$ we have

$$v'(t) \leq k(t) p(t) g(A(t)) + k(t) q(t) g(B(t)) v(t), \qquad v(\alpha) = 0,$$

and lemma 1.1 implies (4.11), since $g \geq 0$, $k \geq 0$. Hence

$$\begin{aligned} & g^{-1} \left(\int_\alpha^t k(s) g(u(s))\, ds \right) \leq \\ & \leq g^{-1} \left\{ \int_\alpha^t p(s) g(A(s)) k(s) \exp \left(\int_s^t q(\tau) g(B(\tau)) k(\tau)\, d\tau \right) ds \right\}, \end{aligned}$$

and by substituting this into (4.10) the inequality (4.12) follows since $b \geq 0$.

The proofs of the remaining cases are similar. □

THEOREM 4.5. (Gollwitzer, 1969) *Let u, b, k, g, p, and q be functions satisfying the conditions of theorem* 4.4. *Suppose*

$$u(t) \geq u(\tau) - b(t)g^{-1}\left(\int_\tau^t k(s)g(u(s))\,ds\right), \qquad \alpha \leq \tau \leq t \leq \beta. \tag{4.13}$$

Then for $\alpha \leq \tau \leq t \leq \beta_1$ in case i) *or case* ii),

$$u(t) \geq p(t)g^{-1}\left\{\frac{g(u(\tau))}{p(t)} \exp\left[-q(t)g\left(\frac{b(t)}{q(t)}\right)\int_\tau^t k(s)\,ds\right]\right\}, \tag{4.14}$$

where $\beta_1 = \min(v_1, v_2, v_3)$, with

$$v_1 = \sup\left\{v \in J : \frac{u(t)}{p(t)} \in I,\ \frac{b(t)}{q(t)} \in I \ \text{for}\ \alpha \leq t \leq v\right\},$$

$$v_2 = \sup\left\{v \in J : \frac{b(t)}{q(t)}g^{-1}\left(\int_\tau^t k(s)g(u(s))\,ds\right) \in I,\ \alpha \leq \tau \leq t \leq v\right\},$$

$$v_3 = \sup\left\{v \in J : \frac{g(u(\tau))}{p(t)} \exp\left[-q(t)g\left(\frac{b(t)}{q(t)}\right)\int_\tau^t k(s)\,ds\right] \in g(I),\ \alpha \leq \tau \leq t \leq v\right\}.$$

The result still holds if in (4.13) *and* (4.14) *the inequality signs are reversed, provided that the conditions* i) *and* ii) *are replaced by* j) *and* jj).

Finally, all results remain true if $[\alpha, \beta_1)$ is replaced by $(\alpha_1, \beta]$, where $\alpha_1 = \max(v_1', v_2', v_3')$ is defined in the obvious way.

The PROOF is in essence the same as that of theorem 4.4: first we write (4.13) as

$$u(\tau) \leq p(t)\frac{u(t)}{p(t)} + q(t)\frac{b(t)}{q(t)}g^{-1}\left(\int_\tau^t k(s)g(u(s))\,ds\right),$$

and then we use corollary 1.12 instead of lemma 1.1. □

THEOREM 4.6. (Gollwitzer, 1969) *Let u and k be continuous functions, with $k \geq 0$, in an interval $J = [\alpha, \beta]$, and let g be a function of constant sign having a continuous derivative which vanishes nowhere in an interval I such that $u(J) \subset I$. Let a be a nonzero constant. Suppose that either*

i) *g is concave and g and a have the same sign;*

or

ii) *g is convex and g and a have opposite signs.*

If $a + g^{-1}(0) \in I$ and

$$u(t) \leq a + g^{-1}\left(\int_\alpha^t k(s)g(u(s))\,ds\right), \qquad t \in J, \tag{4.15}$$

then

$$u(t) \leq g^{-1}\left\{g(a + g^{-1}(0)) \exp\left(\int_\alpha^t k(s)\,ds\right)\right\}, \qquad \alpha \leq t \leq \beta_1, \tag{4.16}$$

where $\beta_1 = \min(v_1, v_2)$, *with*

$$v_1 = \sup\left\{v \in J:\ a + g^{-1}\left(\int_\alpha^t k(s)g(u(s))\,ds\right) \in I,\ \alpha \le t \le v\right\},$$

$$v_2 = \sup\left\{v \in J:\ g(a + g^{-1}(0))\exp\left(\int_\alpha^t k(s)\,ds\right) \in g(I),\ \alpha \le t \le v\right\}.$$

If i) *(respectively,* ii)*) is replaced by*

j) *g is concave and g and a have opposite signs*

(respectively,

jj) *g is convex and g and a have the same sign),*

then the result remains valid, provided $\le$ is replaced by $\ge$ in (4.15) *and* (4.16). *Finally, all inequalities remain valid if $\int_\alpha^t$ is replaced by $\int_t^\beta$, and $[\alpha, \beta_1)$ is replaced by $(\alpha_1, \beta]$, where $\alpha_1 = \max(\mu_1, \mu_2)$, with*

$$\mu_1 = \inf\left\{\mu \in J:\ a + g^{-1}\left(\int_t^\beta k(s)g(u(s))\,ds\right) \in I,\ \mu \le t \le \beta\right\},$$

$$\mu_2 = \inf\left\{\mu \in J:\ g(a + g^{-1}(0))\exp\left(\int_t^\beta k(s)\,ds\right) \in g(I),\ \mu \le t \le \beta\right\}.$$

PROOF. For $\alpha \le t < \beta_1$, (4.15) implies

$$z(t) \equiv g(u(t)) \lesseqgtr g\left[a + g^{-1}\left(\int_\alpha^t k(s)z(s)\,ds\right)\right] \equiv \phi(t), \tag{4.17}$$

where $\le$ (respectively, $\ge$) holds in accordance with $g' > 0$ (respectively, $g' < 0$). Hence

$$\phi'(t) = \frac{g'\left[a + g^{-1}\left(\int_\alpha^t kz\,ds\right)\right]}{g'\left[g^{-1}\left(\int_\alpha^t kz\,ds\right)\right]}k(t)z(t) \lesseqgtr \frac{g'\left[a + g^{-1}\left(\int_\alpha^t kz\,ds\right)\right]}{g'\left[g^{-1}\left(\int_\alpha^t kz\,ds\right)\right]}k(t)\phi(t), \tag{4.18}$$

where $\le$ (respectively, $\ge$) holds in accordance with $kg' \ge 0$ (respectively, $kg' \le 0$).

Let us assume that if the function g is convex (i.e. g' is increasing), then the parameter σ takes the value 1; while if g is concave (i.e. g' is decreasing), $\sigma = -1$. Then

$$\frac{g'(a + g^{-1}(v))}{g'(g^{-1}(v))} \lesseqgtr 1 \qquad \left(v = \int_\alpha^t kz\,ds\right),$$

where $\le$ (respectively, $\ge$) holds in accordance with $g'\sigma a < 0$ (respectively, $g'\sigma a > 0$).

Taking into account that $k\phi \ge 0$ if and only if $kg \ge 0$, we find

$$\frac{g'(a + g^{-1}(v))}{g'(g^{-1}(v))}k\phi \lesseqgtr k\phi, \tag{4.19}$$

where $\le$ (respectively, $\ge$) holds in accordance with $kgg'\sigma a \le 0$ (respectively, $kgg'\sigma a \ge 0$). It follows from (4.18) and (4.19) that

$$\phi'(t) \lesseqgtr k(t)\phi(t), \tag{4.20}$$

where $\leq$ (respectively, $\geq$) holds in accordance with

a_1) $kgg'\sigma a \leq 0$ and $kg' \geq 0$

(respectively,

b_1) $kgg'\sigma a \geq 0$ and $kg' \leq 0$).

By lemma 1.1, (4.20) implies

$$\phi(t) \lesseqgtr g(a + g^{-1}(0)) \exp\left(\int_\alpha^t k(s)\, ds\right), \tag{4.21}$$

where $\leq$ (respectively, $\geq$) holds in accordance with

a_2) $g\sigma a \leq 0$ and $kg' \geq 0$

(respectively,

b_2) $g\sigma a \leq 0$ and $kg' \leq 0$).

Thus, (4.17) and (4.21) imply

$$g(u(t)) = z(t) \lesseqgtr \phi(t) \lesseqgtr g(a + g^{-1}(0)) \exp\left(\int_\alpha^t k(s)\, ds\right), \tag{4.22}$$

where $\leq$ (respectively, $\geq$) holds in accordance with

a_3) $g\sigma a \leq 0$ and $kg' \geq 0$ and $g' > 0$

(respectively,

b_3) $g\sigma a \leq 0$ and $kg' \leq 0$ and $g' < 0$).

Analysis of all these conditions reveals that all require $k \geq 0$. Then, in each set of alternatives, $g' > 0$ and $g' < 0$, respectively. If $g\sigma a \leq 0$ (i.e. if i) or ii) holds), (4.22) implies (4.16).

The proofs of the other cases are similar. □

The following theorem can be proved in a similar way.

THEOREM 4.7. *Let u, k, and g be as in theorem 4.6, and suppose that u is of fixed sign in J. If either*

i) *g is concave and g and u have the same sign;*

or

ii) *g is convex and g and u have opposite signs,*

while $u(t) + g^{-1}(0) \in I$ for $t \in J$, then the inequality

$$u(t) \geq u(\tau) - g^{-1}\left(\int_\tau^t k(s) g(u(s))\, ds\right), \qquad \alpha \leq \tau \leq t \leq \beta, \tag{4.23}$$

implies

$$u(t) \geq g^{-1}\left[g(u(\tau)) \exp\left(-\int_\tau^t k(s)\, ds\right)\right] - g^{-1}(0), \qquad \alpha \leq \tau \leq t < \beta_1, \tag{4.24}$$

where $\beta_1 = \min(v_1, v_2)$, *with*

$$v_1 = \sup\left\{v \in J : \ u(t) + g^{-1}\left(\int_\tau^t k(s)g(u(s))\,ds\right) \in I,\ \alpha \le \tau \le t \le v\right\},$$

$$v_2 = \sup\left\{v \in J : \ g(u(t) + g^{-1}(0))\exp\left(\int_\tau^t k(s)\,ds\right) \in g(I),\ \alpha \le \tau \le t \le v\right\}.$$

If i) *(respectively,* ii)*) is replaced by*

j) g *is concave and* g *and* u *have opposite signs*

(respectively,

jj) g *is convex and* g *and* u *have the same sign),*

then the result remains valid, provided $\geq$ *is replaced by* $\leq$ *in* (4.23) *and* (4.24). *Finally, all inequalities remain valid if* $[\alpha, \beta_1)$ *is replaced by* $(\alpha_1, \beta]$, *where* $\alpha_1 = \max(\mu_1, \mu_2)$, *with*

$$\mu_1 = \inf\left\{\mu \in J : \ u(t) + g^{-1}\left(\int_\tau^t k(s)g(u(s))\,ds\right) \in I,\ \mu \le \tau \le t \le \beta\right\},$$

$$\mu_2 = \inf\left\{\mu \in J : \ g(u(t) + g^{-1}(0))\exp\left(\int_\tau^t k(s)\,ds\right) \in g(I),\ \mu \le \tau \le t \le \beta\right\}. \quad \square$$

The following theorem is based on lemma 1.1.

THEOREM 4.8. (Stachurska, 1971) *Let* u, a, b, *and* k *be nonnegative continuous functions in* $J = [\alpha, \beta]$, *and let* $p > 1$ *be a constant. Suppose* a/b *is nondecreasing in* J *and*

$$u(t) \le a(t) + b(t)\int_\alpha^t k(s)u^p(s)\,ds, \qquad t \in J. \tag{4.25}$$

Then

$$u(t) \le a(t)\left\{1 - (p-1)\int_\alpha^t k(s)b(s)a^{p-1}(s)\,ds\right\}^{1/(1-p)}, \qquad \alpha \le t \le \beta_p, \tag{4.26}$$

where $\beta_p = \sup\{t \in J\colon\ (p-1)\int_\alpha^t k(s)b(s)a^{p-1}(s)\,ds < 1\}$.

PROOF. Set $v(t) = \int_\alpha^t k(s)u^p(s)\,ds$. Then for $\alpha \le t \le \tau < \beta_p$, (4.25) implies $v(\alpha) = 0$,

$$u(t) \le a(t) + b(t)v(t), \tag{4.27}$$

and

$$\begin{aligned} v'(t) &= k(t)u^p(t) \le k(t)[a(t) + b(t)v(t)]^p = \\ &= k(t)[a(t) + b(t)v(t)]^{p-1}[a(t) + b(t)v(t)] \le \\ &\le k(t)b(t)[a(t) + b(t)v(t)]^{p-1}\left[\frac{a(\tau)}{b(\tau)} + v(t)\right], \end{aligned}$$

i.e.

$$v'(t) \le R(t)\left[\frac{a(\tau)}{b(\tau)} + v(t)\right], \tag{4.28}$$

where

$$R(t) = k(t)b(t)[a(t) + b(t)v(t)]^{p-1}.$$

Lemma 1.1 and (4.28) imply

$$v(t) + \frac{a(\tau)}{b(\tau)} \leq \frac{a(\tau)}{b(\tau)} \exp\left(\int_\alpha^t R(s)\,ds\right), \qquad \alpha \leq t \leq \tau.$$

Thus, for $t = \tau$,

$$a(t) + b(t)v(t) \leq a(t) \exp\left(\int_\alpha^t R(s)\,ds\right). \tag{4.29}$$

From (4.29) we successively obtain

$$[a(t) + b(t)v(t)]^{p-1} \leq a^{p-1}(t) \exp\left(\int_\alpha^t (p-1)R(s)\,ds\right),$$

$$R(t) \leq k(t)b(t)a^{p-1}(t) \exp\left(\int_\alpha^t (p-1)R(s)\,ds\right),$$

$$Z(t) = (p-1)R(t) \leq (p-1)k(t)b(t)a^{p-1}(t) \exp\left(\int_\alpha^t Z(s)\,ds\right).$$

Consequently,

$$Z(t) \exp\left(-\int_\alpha^t Z(s)\,ds\right) \leq (p-1)k(t)b(t)a^{p-1}(t),$$

or

$$\left[-\exp\left(-\int_\alpha^t Z(s)\,ds\right)\right]' \leq (p-1)k(t)b(t)a^{p-1}(t).$$

Integrating this from α to t yields

$$1 - \exp\left(-\int_\alpha^t Z(s)\,ds\right) \leq \int_\alpha^t (p-1)k(s)b(s)a^{p-1}(s)\,ds,$$

from which we conclude that

$$\exp\left(\int_\alpha^t R(s)\,ds\right) \leq \left[1 - (p-1)\int_\alpha^t k(s)b(s)a^{p-1}(s)\,ds\right]^{1/(1-p)}.$$

This, together with (4.27) and (4.29), implies (4.26). □

We will prove the following two theorems by comparison with a differential equation of Bernoulli type. To this end we need the following lemma.

LEMMA 4.1. *Let $v(t)$ be a positive differentiable function satisfying the inequality*

$$v'(t) \leq b(t)v(t) + k(t)v^p(t), \qquad t \in J = [\alpha, \beta], \tag{4.30}$$

where the functions b and k are continuous in J, and $p \geq 0$, $p \neq 1$, is a constant. Then

$$v(t) \leq \exp\left(\int_\alpha^t b(s)\,ds\right)\left[v^q(\alpha) + q\int_\alpha^t k(s)\exp\left(-q\int_\alpha^s b(\tau)\,d\tau\right)ds\right]^{1/q}, \tag{4.31}$$

for $t \in [\alpha, \beta_1)$, where $q = 1 - p$ and β_1 is chosen so that the expression between $[\dots]$ is positive in the subinterval $[\alpha, \beta_1)$.

PROOF. According to (4.30), the function $z = v^q/q$ satisfies

$$z' = v^{q-1} \cdot v' \le v^{q-1}(bv + kv^p) = qbz + k.$$

By lemma 1.1,

$$z(t) \le \frac{v^q(\alpha)}{q} \exp\left(\int_\alpha^t qb(s)\,ds\right) + \int_\alpha^t k(s) \exp\left(\int_s^t qb(\tau)\,d\tau\right) ds,$$

or

$$v^q(t) \lesseqgtr \exp\left(\int_\alpha^t qb(s)\,ds\right)\left[v^q(\alpha) + \int_\alpha^t qk(s)\exp\left(-\int_\alpha^s qb(\tau)\,d\tau\right) ds\right],$$

where $\le$ (respectively, $\ge$) holds for $q > 0$ (respectively, $q < 0$). In both cases this estimate implies (4.31). □

THEOREM 4.9. (Willett and Wong, 1965) *Let u, b, and k be nonnegative continuous functions in $J = [\alpha, \beta]$, and let $p \ge 0$, $p \ne 1$, and $a > 0$ be constants. Suppose that*

$$u(t) \le a + \int_\alpha^t b(s)u(s)\,ds + \int_\alpha^t k(s)u^p(s)\,ds, \qquad t \in J. \tag{4.32}$$

Then

$$u(t) \le \exp\left(\int_\alpha^t b(s)\,ds\right)\left[a^q + q\int_\alpha^t k(s)\exp\left(-q\int_\alpha^s b(\tau)\,d\tau\right) ds\right]^{1/q} \tag{4.33}$$

for $t \in [\alpha, \beta_1)$, where $q = 1 - p$ and β_1 is chosen so that the expression between $[\ldots]$ is positive in the subinterval $[\alpha, \beta_1)$ ($\beta_1 = \beta$ if $q > 0$).

PROOF. Let $v(t)$ denote the righthand side of (4.32). Then $v(\alpha) = a$, $u(t) \le v(t)$,

$$v'(t) = b(t)u(t) + k(t)u^p(t) \le b(t)v(t) + k(t)v^p(t).$$

Lemma 4.1 now implies (4.3). □

THEOREM 4.10. (Gamidov, 1969) *Let u and k be nonnegative continuous functions in $J = [\alpha, \beta]$, and let $0 < p < 1$, $a_1 \ge 0$, $a_2 \ge 0$, and $a_3 > 0$ be constants. Suppose*

$$u(t) \le a_1 + a_2\int_\alpha^t k(s)u^p(s)\,ds + a_3\int_\alpha^\beta k(s)u^p(s)\,ds, \qquad t \in J. \tag{4.34}$$

Then

$$u(t) \le \left(x_0^q + a_2 q\int_\alpha^t k(s)\,ds\right)^{1/q}, \qquad t \in J, \tag{4.35}$$

where $q = 1 - p$ and x_0 is the unique positive root of the equation

$$\left[\frac{a_2 + a_3}{a_3}x - \frac{a_1 a_2}{a_3}\right]^q - x^q - a_2 q\int_\alpha^\beta k(s)\,ds = 0. \tag{4.36}$$

PROOF. Let $v(t)$ denote the righthand side of (4.34). Then $u(t) \le v(t)$,

$$v'(t) = a_2 k(t) u^p(t) \le a_2 k(t) v^p(t), \qquad t \in J,$$

and lemma 4.1 (with $b(s) \equiv 0$) implies that

$$v^q(t) \le v^q(\alpha) + q a_2 \int_\alpha^t k(s)\, ds, \qquad t \in J. \tag{4.37}$$

For $t = \beta$ we thus find

$$v^q(\beta) \le v^q(\alpha) + q a_2 \int_\alpha^\beta k(s)\, ds, \tag{4.38}$$

and the definition of $v(t)$ implies

$$v(\beta) = a_1 + (a_2 + a_3) \int_\alpha^\beta k(s) u^p(s)\, ds,$$
$$v(\alpha) = a_1 + a_3 \int_\alpha^\beta k(s) u^p(s)\, ds.$$

Hence,

$$v(\beta) = \frac{a_2 + a_3}{a_3} v(\alpha) - \frac{a_1 a_2}{a_3},$$

and by (4.38) the following inequality holds for $v(\alpha)$:

$$\left[\frac{a_2 + a_3}{a_3} v(\alpha) - \frac{a_1 a_2}{a_3}\right]^q \le v^q(\alpha) + q a_2 \int_\alpha^\beta k(s)\, ds.$$

Analysis of this inequality shows that

$$\frac{a_1 a_2}{a_2 + a_3} \le v(\alpha) \le x_0,$$

where x_0 is the unique positive root of equation (4.36). Thus (4.37) implies (4.35). □

5. Bihari-type inequalities

THEOREM 5.1. (Bihari, 1956) *Let g be a monotone continuous function in an interval I, containing a point u_0, which vanishes nowhere in I. Let u and k be continuous functions in an interval $J = [\alpha, \beta]$ such that $u(J) \subset I$, and suppose that k is of fixed sign in J. Let $a \in I$. Suppose that*

$$u(t) \le a + \int_\alpha^t k(s) g(u(s))\, ds, \qquad t \in J. \tag{5.1}$$

If either

i) *g is nondecreasing and k is nonnegative;*

or

ii) *g is nonincreasing and k is nonpositive,*

then

$$u(t) \geq G^{-1}\left(G(a) + \int_\alpha^t k(s)\,ds\right), \qquad \alpha \leq t \leq \beta_1, \tag{5.2}$$

where $G(u) = \int_{u_0}^u dx/g(x)$, $u \in I$, *and* $\beta_1 = \min(v_1, v_2)$, *with*

$$v_1 = \sup\left\{v \in J : \, a + \int_\alpha^t k(s)g(u(s))\,ds \in I, \, \alpha \leq t \leq v\right\},$$
$$v_2 = \sup\left\{v \in J : \, G(a) + \int_\alpha^t k(s)\,ds \in G(I), \, \alpha \leq t \leq v\right\}.$$

If equality holds in (5.1) *for* $\alpha \leq t \leq \beta'$ $(< \beta_1)$, *then equality holds in* (5.2) *for these* t, *while if strict inequality holds in* (5.1) *for* $t = t_1 < \beta_1$, *then strict inequality holds in* (5.2) *also for* $t = t_1$. *Moreover, the result remains valid if* $\leq$ *is replaced by* $\geq$ *in both* (5.1) *and* (5.2).

PROOF. First we assume that $g > 0$. Let

$$v(t) \equiv a + \int_\alpha^t k(s)g(u(s))\,ds, \qquad t \in J.$$

Then $v(t) \in I$ for $\alpha \leq t < \beta_1$, and since $u(t) \leq v(t)$, (5.1) implies

$$\frac{d}{ds}G(v(s)) = \frac{g(u(s))}{g(v(s))}k(s) \leq k(s),$$

where either $k \geq 0$ and g is nondecreasing, or $k \leq 0$ and g is nonincreasing. By integration over $\alpha \leq s \leq t$ $(< \beta_1)$ we obtain

$$G(v(t)) \leq G(a) + \int_\alpha^t k(s)\,ds$$

in both cases. Since $t < \beta_1$ implies $t < v_2$ and since G^{-1} is increasing (with G) for $g > 0$, it follows that in both cases

$$v(t) \leq G^{-1}\left(G(a) + \int_\alpha^t k(s)\,ds\right). \tag{5.3}$$

Now (5.1) and (5.3) imply (5.2). The assertions concerning equality and strict inequality are obvious.

If $u(t) \geq v(t)$, all inequality signs in the proof must be replaced by their opposites.

Now we assume $g < 0$ in I. Set $g_1 = -g$, $k_1 = -k$. It is readily verified that the functions u, g_1, k_1 satisfy the conditions of theorem 5.1 with $g_1 > 0$ and $G_1(u) = \int_{u_0}^u dx/g_1(x)$. Thus, $G_1(u) = -G(u)$, $G_1^{-1}(v) = G^{-1}(-v)$, and

$$u(t) \leq G_1^{-1}\left(G_1(a) + \int_\alpha^t k_1(s)\,ds\right) = G^{-1}\left(G(a) + \int_\alpha^t k(s)\,ds\right). \quad \square$$

REMARK 5.1. The choice of the point $u_0 \in I$ does not affect the final result (5.2). Indeed, if $\overline{u}_0 \in I$ and $\overline{G}(u) = \int_{\overline{u}_0}^{u} dy/g(y)$, $u \in I$, then $\overline{G}(u) = G(u) - G(\overline{u}_0)$, and thus $\overline{G}^{-1}(v) = G^{-1}(v + G(\overline{u}_0))$. Consequently,

$$\overline{G}^{-1}\left(\overline{G}(a) + \int_{\alpha}^{t} k(s)\,ds\right) = G^{-1}\left(G(a) + \int_{\alpha}^{t} k(s)\,ds\right). \quad \Box$$

REMARK 5.2. Sometimes $g(u_1) = 0$ at an endpoint of the interval I. For instance, let $I = (u_1, \infty)$, let g be continuous in $[u_1, \infty)$ and $g > 0$ in I. For the sake of being specific, suppose that g is nondecreasing in I and $k \geq 0$ in J. Then theorem 5.1 is valid for continuous functions u in J for which $u(J) \subset [u_1, \infty)$, even if $u(t_1) = u_1$ for some $t_1 \in J$. To show this, let $\epsilon > 0$ and $u_\epsilon(t) = u(t) + \epsilon > u_1$ for all $t \in J$. Then

$$u_\epsilon(t) \leq (a + \epsilon) + \int_{\alpha}^{t} k(s)g(u_\epsilon(s) - \epsilon)\,ds \leq (a + \epsilon) + \int_{\alpha}^{t} k(s)g(u_\epsilon(s))\,ds.$$

Since $u_\epsilon(J) \subset I$, it follows that

$$u_\epsilon(t) \leq G^{-1}\left(G(a + \epsilon) + \int_{\alpha}^{t} k(s)\,ds\right).$$

Letting $\epsilon \to 0$ we arrive at (5.2). $\Box$

REMARK 5.3. Bihari's version of theorem 5.1 treats the case $I = J = (0, \infty)$ and g nondecreasing and positive in I. Taking $g(u) = u$, $G(u) = \log u$, we obtain Bellman's inequality in theorem 1.1. $\Box$

The following results are consequences of theorem 5.1.

COROLLARY 5.1. *Under the conditions of theorem* 5.1, *but with* $J = (\alpha, \beta]$, *the inequality*

$$u(t) \leq a + \int_{t}^{\beta} k(s)g(u(s))\,ds, \qquad t \in J, \tag{5.4}$$

implies

$$u(t) \leq G^{-1}\left(G(a) + \int_{t}^{\beta} k(s)\,ds\right), \qquad \alpha_1 < t \leq \beta, \tag{5.5}$$

where $G(u) = \int_{u_0}^{u} dx/g(x)$ *and* $\alpha_1 = \max(\mu_1, \mu_2)$, *with*

$$\mu_1 = \inf\left\{\mu \in J : \, a + \int_{t}^{\beta} k(s)g(u(s))\,ds \in I, \; \mu \leq t \leq \beta\right\},$$

$$\mu_2 = \inf\left\{\mu \in J : \, G(a) + \int_{t}^{\beta} k(s)\,ds \in G(I), \; \mu \leq t \leq \beta\right\}.$$

The result remains valid if in (5.4) *and* (5.5) *the sign* $\leq$ *is replaced by* $\geq$. $\Box$

COROLLARY 5.2. *Let g be nonzero, monotone, and continuous in an interval I containing u_0. Let u and k be continuous functions in an interval $J = [\alpha, \beta)$ such that $u(J) \subset I$, with k of constant sign. Let $a \in I$. Suppose that either g is nonincreasing and $k \geq 0$, or g is nondecreasing and $k \leq 0$. If*

$$u(t) \geq a - \int_\alpha^t k(s)g(u(s))\,ds, \qquad t \in J, \tag{5.6}$$

then

$$u(t) \geq G^{-1}\left(G(a) - \int_\alpha^t k(s)\,ds\right), \qquad \alpha \leq t < \beta_1, \tag{5.7}$$

where $G(u) = \int_{u_0}^u dx/g(x)$ and $\beta_1 = \min(v_1, v_2)$, with

$$v_1 = \sup\left\{v \in J : a - \int_\alpha^t k(s)g(u(s))\,ds \in I,\ \alpha \leq t \leq v\right\},$$
$$v_2 = \sup\left\{v \in J : G(a) - \int_\alpha^t k(s)\,ds \in G(I),\ \alpha \leq t \leq v\right\}.$$

The same result holds if J is replaced by $(\alpha, \beta]$, $\int_\alpha^t$ by $\int_t^\beta$, and β_1 by α_1. □

REMARK 5.4. In applications to differential equations (see § 9), inequalities of the form (5.6) usually appear with $k \geq 0$ and $g > 0$ but nondecreasing. The simple example

$$u(t) \geq 1 - \int_0^t u(s)\,ds, \qquad 0 \leq t \leq 1,$$

which holds for $u(t) = 2-2t$, so that $u(t) \not\geq e^{-t}$ for t near 1, shows that, in general, the implication (5.6)⇒(5.7) is false. At the same time Langenhop has obtained certain positive results in case $k \geq 0$, $g > 0$, and g nondecreasing. □

THEOREM 5.2. (Langenhop, 1960) *Let g be a nonzero monotone continuous function in an interval I containing a point u_0. Let u and k be continuous functions in $J = [\alpha, \beta]$ such that $u(J) \subset I$ and k of fixed sign in J. Suppose*

$$u(t) \geq u(\tau) - \int_\tau^t k(s)g(u(s))\,ds, \qquad \alpha \leq \tau \leq t \leq \beta. \tag{5.8}$$

If either g is nondecreasing and $k \geq 0$, or g is nonincreasing and $k \leq 0$, then

$$u(t) \geq G^{-1}\left(G(u(\tau)) - \int_\tau^t k(s)\,ds\right), \qquad \alpha \leq \tau \leq t < \beta_1, \tag{5.9}$$

where $G(u) = \int_{u_0}^u dx/g(x)$, $u \in I$, and $\beta_1 = \min(v_1, v_2)$, with

$$v_1 = \sup\left\{v \in J : u(\tau) - \int_\tau^t k(s)g(u(s))\,ds \in I,\ \alpha \leq \tau \leq t \leq v\right\},$$
$$v_2 = \sup\left\{v \in J : G(u(\tau)) - \int_\tau^t k(s)\,ds \in G(I),\ \alpha \leq \tau \leq t \leq v\right\}.$$

The result holds also if $\geq$ *is replaced by* $\leq$ *in* (5.8) *and* (5.9). *Moreover, both results hold if* $[\alpha, \beta_1)$ *is replaced by* $(\alpha_1, \beta]$, *where* $\alpha_1 = \max(\mu_1, \mu_2)$, *with*

$$\mu_1 = \inf\left\{\mu \in J : \ u(\tau) - \int_\tau^t k(s)g(u(s))\,ds \in I, \ \mu \leq \tau \leq t \leq \beta\right\},$$

$$\mu_2 = \inf\left\{\mu \in J : \ G(u(\tau)) - \int_\tau^t k(s)\,ds \in G(I), \ \mu \leq \tau \leq t \leq \beta\right\}.$$

PROOF. We write (5.8) in the form

$$u(\tau) \leq u(t) + \int_\tau^t k(s)g(u(s))\,ds, \qquad \alpha \leq \tau \leq t \leq \beta,$$

and fix a $t \in [\alpha, \beta]$. Continuing with the function $u(\tau)$, $\tau \in [\alpha, \beta]$, we proceed as in the proof of theorem 5.1. We omit the details. □

By theorem 5.1, the righthand side of (5.2) is a solution of the integral equation

$$v(t) = a + \int_\alpha^t k(s)g(v(s))\,ds,$$

which is equivalent to the initial value problem

$$v'(t) = k(t)g(v(t)), \qquad v(\alpha) = a.$$

In fact, in the proof of theorem 5.1 the integral inequality (5.1) is reduced to the differential inequality

$$v'(t) \leq k(t)g(v(t)), \qquad v(\alpha) = a,$$

and we have implicitly used the following comparison result.

LEMMA 5.1. *Let* $v(t)$ *be a differentiable function in* $J = (\alpha, \beta)$ *such that*

$$v'(t) \leq k(t)g(v(t)), \qquad t \in J, \tag{5.10}$$

where g *is a nonzero continuous function in* $I = (v_1, v_2)$ *and* k *is a continuous function in* J. *Let* $t_0 \in J$ *and* $v(t_0) = a \in I$. *Then*

$$v(t) \leq G^{-1}\left(G(a) + \int_{t_0}^t k(s)\,ds\right), \quad t \in [t_0, \beta_1) = J_1, \tag{5.11}$$

where $G(v) = \int_{v_0}^v dx/g(x)$, $v, v_0 \in I$, *and*

$$\beta_1 = \sup\left\{v \in [t_0, \beta) : \ G(a) + \int_{t_0}^t k(s)\,ds \in G(I), \ t_0 \leq t \leq v\right\}.$$

The result still holds if $\leq$ *is replaced by* $\geq$ *in* (5.10) *and* (5.11).
Finally, the result remains valid if $t \in J' = (\alpha_1, t_0]$, *with*

$$\alpha_1 = \inf\left\{\mu \in [\alpha, t_0] : \ G(a) + \int_{t_0}^t k(s)\,ds \in G(I), \ \mu \leq t \leq t_0\right\}, \tag{5.12}$$

and the inequalities (5.10) *and* (5.11) *are of opposite types.*

PROOF. Inequality (5.10) implies

$$\frac{v'(t)}{g(v(t))} \lesseqgtr k(t), \qquad t \in J,$$

where $\leq$ (respectively, $\geq$) holds for $g > 0$ (respectively, $g < 0$). Integration from t_0 to t gives

$$\int_a^{v(t)} \frac{dx}{g(x)} \lesseqgtr \int_{t_0}^t k(s)\,ds,$$

or, taking into account the definition of G,

$$G(v(t)) \lesseqgtr G(a) + \int_{t_0}^t k(s)\,ds.$$

Since G is increasing if $g > 0$ and decreasing if $g < 0$, the last inequality implies in both cases

$$v(t) \leq G^{-1}\left(G(a) + \int_{t_0}^t k(s)\,ds\right),$$

while $G(a) + \int_{t_0}^t k(s)\,ds \in G(I)$.

The proofs of the remaining cases are similar. □

REMARK 5.5. The condition $G(a) + \int_{t_0}^t k(s)\,ds \in G(I)$, from which we determine the interval of variation of t in (5.11) or (5.12), is equivalent to

$$\int_a^{v_1} \frac{dx}{g(x)} < \int_{t_0}^t k(s)\,ds < \int_a^{v_2} \frac{dx}{g(x)} \qquad \text{if } g > 0,$$

$$\int_a^{v_2} \frac{dx}{g(x)} < \int_{t_0}^t k(s)\,ds < \int_a^{v_2} \frac{dx}{g(x)} \qquad \text{if } g < 0. \tag{5.13}$$

□

REMARK 5.6. If $g > 0$, $k \geq 0$, v_2 is finite, and $\int_a^{v_2} dx/g(x) = \infty$, then $\beta = \beta_1$ and $v(t) < v_2$ for $t \in [t_0, \beta)$. Indeed, (5.13) implies that $\beta_1 = \beta$. Assuming $v(t) < v_2$ for $t_0 \leq t < \beta_1 < \beta$ and $v(\beta_1) = v_2$ we arrive at a contradiction from

$$\int_a^{v(t)} \frac{dx}{g(x)} \leq \int_{t_0}^t k(s)\,ds$$

by letting $t \uparrow \beta_1$. Hence $v(t) < v_2$ for $t \in [t_0, \beta)$. □

REMARK 5.7. If $g > 0$, $k \geq 0$, and $\int_a^{v_2} dx/g(x) < \infty$, then either $\beta = \beta_1$ or β_1 is determined by the equality

$$\int_a^{v_2} \frac{dx}{g(x)} = \int_{t_0}^{\beta_1} k(s)\,ds.$$

If, moreover, there is an $a_0 \in (v_1, v_2)$ such that

$$\int_{a_0}^{v_2} \frac{dx}{g(x)} \geq \int_{t_0}^{\beta} k(s)\,ds,$$

then (5.11) holds for $t \in [t_0, \beta)$ if $a \in (v_1, a_0)$. The first assertion follows from remark 5.5, and the second from the fact that

$$\int_a^{v_2} \frac{dx}{g(x)} > \int_{a_0}^{v_2} \frac{dx}{g(x)} \geq \int_{t_0}^{\beta} k(s)\,ds \geq \int_{t_0}^{t} k(s)\,ds$$

for $a \in (v_1, a_0)$ and $t \in [t_0, \beta)$. □

The following two theorems are based on lemma 5.1.

THEOREM 5.3. *Let $u(t)$ be a nonnegative continuous function for $t \geq \alpha$, and suppose that*

$$u(t) \leq a + \sigma \int_\alpha^t e^{-n(t-s)} g(u(s))\,ds, \qquad t \geq \alpha, \tag{5.14}$$

where a and σ are positive constants and $g(u)$ is a nonnegative nondecreasing continuous function for $u \geq 0$. Suppose there is a number $\lambda_0 > 0$ such that

$$\sigma g(a + \lambda_0) - n\lambda_0 = 0, \qquad \sigma g(a+u) - nu > 0 \quad \text{for } u \in [0, \lambda_0). \tag{5.15}$$

Then

$$u(t) \leq a + F^{-1}(t - \alpha) \leq a + \lambda_0 \tag{5.16}$$

for $t \geq \alpha$, where $F(u) = \int_0^u dx/(\sigma g(a+x) - nx)$, $u > 0$.

PROOF. Let $v(t) = \sigma \int_\alpha^t e^{-n(t-s)} g(u(s))\,ds$. Then $v(\alpha) = 0$, and (5.14) implies

$$u(t) \leq a + v(t), \tag{5.17}$$

$$v'(t) = -nv(t) + \sigma g(u(t)) \leq -nv(t) + \sigma g(a + v(t)) \equiv f[v(t)],$$

for $t \geq \alpha$. Since $f(v) = \sigma g(a+v) - nv > 0$, $v \in [0, \lambda_0)$, lemma 5.1 implies

$$v(t) \leq F^{-1}(F(v(\alpha)) + t - \alpha) = F^{-1}(t - \alpha), \qquad t \in [\alpha, \beta_1). \tag{5.18}$$

Since $g(a+v)$ is nondecreasing in $[0, \lambda_0)$, (5.15) implies $\sigma g(a+v) - nv \leq n(\lambda_0 - v)$ for $v \in [0, \lambda_0)$, so that $\int_0^{\lambda_0} dx/(\sigma g(a+x) - nx) = \infty$.

Consequently, by remark 5.6, $\beta_1 = \infty$ and $v(t) \leq \lambda_0$ for $t \geq \alpha$. This and (5.17)–(5.18) imply (5.16). □

THEOREM 5.4. *Let $u(t)$ and $b(t)$ be nonnegative continuous functions for $t \geq \alpha$, let $f(t,u)$ be a nonnegative continuous function for $t \geq \alpha$, $u \geq 0$, and suppose*

$$u(t) \leq a + \int_\alpha^t [b(s)u(s) + f(s, u(s))]\,ds, \qquad t \geq \alpha, \tag{5.19}$$

where $a > 0$ is a constant. Suppose also that

$$f(t,u) \leq k(t) g\left(u \exp\left(-\int_\alpha^t b(s)\,ds\right)\right) \exp\left(\int_\alpha^t b(s)\,ds\right), \qquad t \geq \alpha,\ u \geq 0, \tag{5.20}$$

where $k(t)$ is a nonnegative continuous function for $t \geq \alpha$ and $g(v)$ is a nondecreasing continuous function for $v \geq 0$ which is positive for $v > 0$. Then

$$u(t) \leq G^{-1}\left[G(a) + \int_\alpha^t k(s)\,ds\right] \exp\left(\int_\alpha^t b(s)\,ds\right) \tag{5.21}$$

for $t \in [\alpha, \beta)$, *where* $G(u) = \int_{u_0}^{u} dx/g(x)$, $u > 0$, $u_0 > 0$, *and either* $\beta = \infty$ *or* β *is defined by the equality*

$$\int_{\alpha}^{\beta} k(s)\,ds = \int_{a}^{\infty} \frac{dx}{g(x)}.$$

If, moreover, there is an $a_0 > 0$ *such that*

$$\int_{\alpha}^{\infty} k(s)\,ds \le \int_{a_0}^{\infty} \frac{dx}{g(x)},$$

then (5.21) *holds for* $t \ge \alpha$ *if* $a \in (0, a_0)$.

PROOF. Let $v(t)$ denote the righthand side of (5.19). Then $v(\alpha) = a$, and (5.19) and (5.20) imply

$$\begin{aligned} u(t) &\le v(t), \\ v'(t) = b(t)u(t) + f(t, u(t)) &\le \\ \le b(t)v(t) + k(t)g &\left[v(t) \exp\left(-\int_{\alpha}^{t} b(s)\,ds\right)\right] \exp\left(\int_{\alpha}^{t} b(s)\,ds\right). \end{aligned}$$

Hence $z(t) = v(t) \exp\left(-\int_{\alpha}^{t} b(s)\,ds\right)$ satisfies the inequality

$$z'(t) \le k(t)g(z(t)), \qquad t \ge \alpha,$$

and by lemma 5.1,

$$z(t) \le G^{-1}\left[G(a) + \int_{\alpha}^{t} k(s)\,ds\right], \quad t \in [\alpha, \beta).$$

This implies (5.21), since $u(t) \le v(t) = z(t)\exp(\int_{\alpha}^{t} b(s)\,ds)$.
The assertions concerning β follow from remark 5.7. □

THEOREM 5.5. (Muldowney and Wong, 1968) *Let* g *be a monotone continuous function which is positive on a halfline* I *containing zero. Let* u, a, *and* k *be continuous functions in* $J = [\alpha, \beta]$ *such that* $u(J) \subset I$, $a(J) \subset I$, *and*

$$u(t) \le a(t) + \int_{\alpha}^{t} k(s)g(u(s))\,ds, \qquad t \in J. \tag{5.22}$$

Then:

i) *if* I *is a right halfline,* g *is nondecreasing and subadditive in* I, *and* k *is nonnegative in* J, *then*

$$u(t) \le a(t) + G^{-1}\left[G\left(\int_{\alpha}^{t} k(s)g(a(s))\,ds\right) + \int_{\alpha}^{t} k(s)\,ds\right], \qquad \alpha \le t < \beta_1; \tag{5.23}$$

ii) *if* I *is a left halfline,* g *is nonincreasing and superadditive in* I, *and* k *is nonpositive in* J, *then*

$$u(t) \le a(t) + G^{-1}\left(\int_{\alpha}^{t} k(s)\,ds\right), \qquad \alpha \le t < \beta_2, \tag{5.24}$$

where $G(u) = \int_0^u dx/g(x)$, $u \in I$, *and*

$$\beta_1 = \sup\left\{v \in J:\ G\left(\int_\alpha^t k(s)g(a(s))\,ds\right) + \int_\alpha^t k(s)\,ds \in G(I),\ \alpha \le t \le v\right\},$$

$$\beta_2 = \sup\left\{v \in J:\ \int_\alpha^t k(s)\,ds \in G(I),\ \alpha \le t \le v\right\}.$$

The results remain valid if $\le$ *is replaced by* $\ge$ *in* (5.22)–(5.24), *provided the hypotheses in* i) *and* ii) *are replaced by:*

j) *I is a left halfline, g is nonincreasing and subadditive in I, and $k \le 0$ in J;*

jj) *I is a right halfline, g is nondecreasing and superadditive in I, and $k \ge 0$ in J.*

Finally, all results remain valid if $\int_\alpha^t$ *is replaced by* $\int_t^\beta$, *and* β_1, β_2 *are replaced by* α_1, α_2, *with*

$$\alpha_1 = \inf\left\{\mu \in J:\ G\left(\int_t^\beta k(s)g(a(s))\,ds\right) + \int_t^\beta k(s)\,ds \in G(I),\ \mu \le t \le \beta\right\},$$

$$\alpha_2 = \inf\left\{\mu \in J:\ \int_t^\beta k(s)\,ds \in G(I),\ \mu \le t \le \beta\right\}.$$

PROOF. First we note that in both cases i) and ii), $v(s) = \int_\alpha^s k(\tau)g(u(\tau))\,d\tau \in I$, $A(s) = \int_\alpha^s k(\tau)g(a(\tau))\,d\tau \in I$, and $a(s) + v(s) \in I$ for $s \in J$. Next (5.22) implies

$$g(u(s)) \mathrel{\substack{\le \\ \ge}} g(a(s) + v(s)) \mathrel{\substack{\le \\ \ge}} g(a(s)) + g(v(s)), \qquad s \in J,$$

with $\le$ (respectively, $\ge$) in case i) (respectively, ii)). After multiplication by $k(s)$ this implies in both cases

$$v'(s) \le k(s)g(a(s)) + k(s)g(v(s)), \qquad s \in J.$$

Integration over $[\alpha, t]$ gives

$$v(t) \le A(t) + \int_\alpha^t k(s)g(v(s))\,ds, \qquad t \in J,$$

so that

$$v(t) \le A(T) + \int_\alpha^t k(s)g(v(s))\,ds \qquad (k \ge 0,\ \alpha \le t \le T < \beta),$$

$$v(t) \le \int_\alpha^t k(s)g(v(s))\,ds \qquad (k \le 0,\ \alpha \le t \le \beta).$$

By theorem 5.1, in case i) (respectively, ii)) we have

$$v(t) \le G^{-1}\left[G(A(T)) + \int_\alpha^t k(s)\,ds\right], \qquad \alpha \le t \le T,\ \alpha \le t \le \beta_1(T)$$

(respectively,

$$v(t) \le G^{-1}\left[\int_\alpha^t k(s)\,ds\right], \qquad \alpha \le t < \beta_2),$$

where

$$\beta_1(T) = \sup\left\{v \in J :\ G(A(T)) + \int_\alpha^t k(s)\,ds \in G(I),\ \alpha \le t \le v\right\}.$$

In case ii) the conclusion (5.24) follows immediately from (5.22) and the definition of $v(t)$. In case i) we have

$$u(T) \le a(T) + G^{-1}\left[G\left(\int_\alpha^T k(s)g(a(s))\,ds\right) + \int_\alpha^T k(s)\,ds\right], \qquad \alpha \le T < \beta_1,$$

provided $T < \beta_1$ implies $T < \beta_1(T)$. The latter, however, follows from the definitions of β_1 and $\beta_1(T)$.

The proofs of the remaining cases are similar. □

THEOREM 5.6. (Deo and Murdeshwar, 1971) *Let u, a, and k be continuous functions in $J = [\alpha, \beta)$ such that k does not change sign in J. Let g be a positive, subadditive, continuous function in an interval I such that $u(J) \subset I$ and $a(J) \subset I$. Suppose also that h is a monotone continuous function in an interval K such that $0 \in K$, $h(K) \subset I$, and either*

i) *h and g are not nondecreasing, and $k \ge 0$;*

or

ii) *h is nonincreasing, g is nondecreasing, and $k \le 0$.*

Then

$$u(t) \le a(t) + h\left(\int_\alpha^t k(s)g(u(s))\,ds\right), \qquad t \in J, \tag{5.25}$$

implies

$$u(t) \le a(t) + h\left\{G^{-1}\left[\int_\alpha^t k(s)\,ds + G\left(\int_\alpha^t k(s)g(a(s))\,ds\right)\right]\right\} \tag{5.26}$$

for $\alpha \le t < \beta_1$, where $G(u) = \int_{u_0}^u dx/g(h(x))$, $u, u_0 \in K$, and $\beta_1 = \min(v_1, v_2, v_3)$, with

$$v_1 = \sup\left\{v \in J :\ a(t) + h\left(\int_\alpha^t k(s)g(u(s))\,ds\right) \in I,\ \alpha \le t \le v\right\},$$

$$v_2 = \sup\left\{v \in J :\ \int_\alpha^t k(s)\left\{g(a(s)) + g \circ h\left(\int_\alpha^s k(\tau)g(u(\tau))\,d\tau\right)\right\} ds \in K,\ \alpha \le t \le v\right\},$$

$$v_3 = \sup\left\{v \in J :\ \int_\alpha^t k(s)\,ds + G\left(\int_\alpha^T k(s)g(a(s))\,ds\right) \in G(K),\ \alpha \le t \le T \le v\right\}.$$

The result remains valid if $\le$ is replaced by $\ge$ in (5.25) and (5.26), and conditions i) and ii) are replaced by

j) *h and g are both nonincreasing, and $k \ge 0$;*

or

jj) *h is nondecreasing, g is nonincreasing, and $k \le 0$.*

Moreover, all results remain valid if $[\alpha,\beta)$, $[\alpha,\beta_1)$, *and* $\int_\alpha^t$ *are replaced by* $(\alpha,\beta]$, $(\alpha_1,\beta]$, *and* $\int_t^\beta$, *respectively, where* $\alpha_1 = \max(\mu_1,\mu_2,\mu_3)$, *with*

$$\mu_1 = \inf\left\{\mu \in J : \ a(t) + h\left(\int_t^\beta k(s)g(u(s))\,ds\right) \in I, \ \mu \le t \le \beta\right\},$$

$$\mu_2 = \inf\left\{\mu \in J : \ \int_t^\beta k(s)\left\{g(a(s)) + g\circ h\left(\int_s^\beta k(\tau)g(u(\tau))\,d\tau\right)\right\} ds \in K, \ \mu \le t \le \beta\right\},$$

$$\mu_3 = \sup\left\{\mu \in J : \ \int_t^\beta k(s)\,ds + G\left(\int_T^\beta k(s)g(a(s))\,ds\right) \in G(K), \ \mu \le T \le t \le \beta\right\}.$$

PROOF. We first assume that i) holds. We set

$$v(t) = \int_\alpha^t k(s)g(u(s))\,ds, \qquad t \in J,$$

and note that (5.25) implies $v(J) \subset K$, and

$$\begin{aligned} g(u(s)) &\le g[a(s) + h(v(s))] \le g(a(s)) + g\circ h(v(s)),\\ v'(s) &= k(s)g(u(s)) \le k(s)g(a(s)) + k(s)g\circ h(v(s)). \end{aligned}$$

Integration over $[\alpha, t]$ gives

$$v(t) \le \int_\alpha^t k(s)g(a(s))\,ds + \int_\alpha^t k(s)g\circ h(v(s))\,ds.$$

Now we fix $T \in (\alpha, \beta_1)$, and set $A(t) = \int_\alpha^t k(s)g(a(s))\,ds$. Then

$$v(t) \le A(T) + \int_\alpha^t k(s)g\circ h(v(s))\,ds, \qquad \alpha \le t \le T, \tag{5.27}$$

and theorem 5.1 implies

$$v(t) \le G^{-1}\left[\int_\alpha^t k(s)\,ds + G(A(T))\right], \qquad \alpha \le t \le T. \tag{5.28}$$

(Note that $\alpha \le t \le T \le \beta_1$ implies

$$0 \le A(T) + \int_\alpha^t k(s)g\circ h(v(s))\,ds \le A(T) + \int_\alpha^T k(s)g\circ h(v(s))\,ds,$$

and hence $A(T) + \int_\alpha^t k(s)g\circ h(v(s))\,ds \in K$.) For $t = T$, (5.28) implies

$$h(v(t)) \le h\left\{G^{-1}\left[\int_\alpha^t k(s)\,ds + G(A(t))\right]\right\}, \qquad \alpha \le t \le \beta_1,$$

which proves (5.26).

In case ii) we set $K_1 = -K = \{u\colon -u \in K\}$, $h_1(u) = h(-u)$, $u \in K_1$, $k_1(s) = -k(s)$. Then h_1, g, and k_1 satisfy condition i) in I, J, and K_1, and

$$u(t) \le a(t) + h_1\left(\int_\alpha^t k_1(s)g(u(s))\,ds\right),$$

so that (5.26) holds with h, k replaced by h_1, k_1, and G by

$$G_1(v) = \int_{-u_0}^{v} \frac{dy}{g(h_1(y))} = -\int_{u_0}^{-v} \frac{dx}{g(h(x))} = -G(-v), \qquad v \in K_1.$$

It is readily verified that this leads to (5.26).

The proofs of the remaining cases are similar. □

REMARK 5.8. If, under the conditions of theorem 5.6, the function g is negative and superadditive, then the conclusion of this theorem remains valid if in case $\leq$ the conditions i) and ii) are replaced by

i') h is nondecreasing, g is nonincreasing, and $k \leq 0$;
ii') h and g are both nonincreasing, and $k \geq 0$,

and in case $\geq$ the conditions j) and jj) are replaced by

j') h is nonincreasing, g is nondecreasing, and $k \leq 0$;
jj') h and g are both nondecreasing, and $k \geq 0$. □

THEOREM 5.7. *Let u, a, b, and k be continuous functions in $J = [\alpha, \beta)$ such that k does not change sign in J and $b \geq 0$ in J. Let g be a positive, subadditive, submultiplicative, continuous function in an interval I such that $u(J) \subset I$, $a(J) \subset I$, and $b(J) \subset I$. Suppose also that h is a monotone continuous function in an interval K such that $0 \in K$, $h(K) \subset I$, and either*

i) h and g are both nondecreasing and $k \geq 0$;

or

ii) h is nonincreasing, g is nondecreasing, and $k \leq 0$.

Then

$$u(t) \leq a(t) + b(t) h\left(\int_{\alpha}^{t} k(s) g(u(s))\, ds\right), \qquad t \in J, \tag{5.29}$$

implies

$$u(t) \leq a(t) + b(t) h\left\{G^{-1}\left[\int_{\alpha}^{t} k(s) g(b(s))\, ds + G\left(\int_{\alpha}^{t} k(s) g(a(s))\, ds\right)\right]\right\} \tag{5.30}$$

for $\alpha \leq t < \beta_1$, where $G(u) = \int_{u_0}^{u} dx/g(h(x))$, $u, u_0 \in K$, and $\beta_1 = \min(v_1, v_2, v_3, v_4)$, *with*

$$v_1 = \sup\{v \in J : a(t) + b(t)h(v(t)) \in I,\ \alpha \leq t \leq v\}, \qquad v(t) = \int_{\alpha}^{t} k(s) g(u(s))\, ds,$$

$$v_2 = \sup\{v \in J : b(t)h(v(t)) \in I,\ \alpha \leq t \leq v\},$$

$$v_3 = \sup\left\{v \in J : \int_{\alpha}^{t} k(s) g(a(s))\, ds + \int_{\alpha}^{t} k(s) g(b(s)) g \circ h(v(s))\, ds \in K,\ \alpha \leq t \leq v\right\},$$

$$v_4 = \sup\left\{v \in J : \int_{\alpha}^{t} k(s) g(b(s))\, ds + G\left(\int_{\alpha}^{T} k(s) g(a(s))\, ds\right) \in g(K),\right.$$
$$\left.\alpha \leq t \leq T \leq v\right\}.$$

The result remains valid if $\leq$ *is replaced by* $\geq$ *in both* (5.29) *and* (5.30), *and conditions* i) *and* ii) *are replaced by*

j) h *and* g *are both nonincreasing and* $k \geq 0$;

or

jj) h *is nondecreasing,* g *is nonincreasing, and* $k \leq 0$.

Finally, all results remain valid if $[\alpha, \beta)$, $[\alpha, \beta_1)$, *and* $\int_\alpha^t$ *are replaced by* $(\alpha, \beta]$, $(\alpha_1, \beta]$, *and* $\int_t^\beta$, *respectively, where* $\alpha_1 = \max(\mu_1, \mu_2, \mu_3, \mu_4)$, *with* μ_i *defined in the obvious manner.*

The PROOF of theorem 5.7 can be given along the lines of the proof of theorem 5.6, and we omit it. We note only that in case i) one has to apply theorem 5.1 to the inequality

$$v(t) \leq A(T) + \int_\alpha^t k_1(s) g \circ h(v(s))\, ds, \qquad \alpha \leq t \leq T,$$

where $A(t) = \int_\alpha^t k(s) g(a(s))\, ds$, $k_1(s) = k(s) g(b(s))$. □

In the proofs of theorems 5.5–5.7 we have used a technique whose essence consists of replacing $A(t)$ by its maximum value in the interval $\alpha \leq t \leq T$. In the next theorem this technique is used in an even more explicit way.

THEOREM 5.8. *Let* u, a, *and* b *be nonnegative continuous functions in* $J = [\alpha, \beta]$, *and let* f *and* g *be nonnegative continuous functions in* $\mathbf{R}_+$, *with* f *strictly increasing and* g *nondecreasing. Suppose* $k(t,s)$ *is a nonnegative continuous function in* $\Delta = \{(t,s)\colon \alpha \leq s \leq t \leq \beta\}$, *and set*

$$A(\tau) = \max_{\alpha \leq s \leq \tau} a(s), \qquad B(\tau) = \max_{\alpha \leq s \leq \tau} b(s), \qquad K(\tau, s) = \max_{s \leq t \leq \tau} k(t,s).$$

If

$$f(u(t)) \leq a(t) + b(t) \int_\alpha^t k(t,s) g(u(s))\, ds, \qquad t \in J, \tag{5.31}$$

then

$$u(t) \leq f^{-1}\left\{a(t) - A(t) + F^{-1}\left[F(A(t)) + B(t) \int_\alpha^t K(t,s)\, ds\right]\right\}, \qquad \alpha \leq t < \beta_1, \tag{5.32}$$

where $F(u) = \int_{u_0}^u dx / g(f^{-1}(x))$, $u_0 > 0$, $u > 0$, *and*

$$\beta_1 = \sup\left\{v \in J : \ F(A(t)) + B(t) \int_\alpha^t K(t,s)\, ds \in F(\mathbf{R}_+),\ \alpha \leq t \leq v\right\}.$$

PROOF. Let $\alpha \leq t \leq \tau < \beta_1$ and $v(t) = \int_\alpha^t K(\tau, s) g(u(s))\, ds$. Then (5.31) implies

$$\begin{gathered} f(u(t)) \leq a(t) + B(t) v(t), \\ f(u(t)) \leq A(\tau) + B(\tau) v(t), \end{gathered} \tag{5.33}$$

$$\begin{aligned} v'(t) &= K(\tau, t) g(u(t)) = K(\tau, t) g \circ f^{-1}(f(u(t)) \leq \\ &\leq K(\tau, t) g \circ f^{-1}(A(\tau) + B(\tau) v(t)). \end{aligned}$$

Thus,

$$(A(\tau)+B(\tau)v(t))' = B(\tau)v'(t) \leq$$
$$\leq B(\tau)K(\tau,t)g\circ f^{-1}(A(\tau)+B(\tau)v(t)),$$

and by lemma 5.1,

$$A(\tau)+B(\tau)v(t) \leq F^{-1}\left[F(A(\tau))+B(\tau)\int_\alpha^t K(\tau,s)\,ds\right].$$

In particular, for $t=\tau$,

$$A(t)+B(t)v(t) \leq F^{-1}\left[F(A(t))+B(t)\int_\alpha^t K(t,s)\,ds\right].$$

Taking into account (5.33) we arrive at (5.32). □

COROLLARY 5.3. (Butler and Rogers, 1971) *Since $a(t) \leq A(t)$, (5.32) implies the weaker estimate*

$$u(t) \leq f^{-1}\left\{F^{-1}\left[F(A(t))+B(t)\int_\alpha^t K(t,s)\,ds\right]\right\}, \qquad \alpha \leq t < \beta_1. \quad \square$$

COROLLARY 5.4. *If $a(t)$ is nondecreasing, then $A(t) \equiv a(t)$, and (5.32) takes the form*

$$u(t) \leq f^{-1}\left\{F^{-1}\left[F(a(t))+B(t)\int_\alpha^t K(t,s)\,ds\right]\right\}, \qquad \alpha \leq t < \beta_1.$$

If $a(t)$ is nonincreasing, then $A(t) \equiv a(\alpha)$, and (5.32) takes the form

$$u(t) \leq f^{-1}\left\{a(t)-a(\alpha)+F^{-1}\left[F(a(\alpha))+B(t)\int_\alpha^t K(t,s)\,ds\right]\right\}, \qquad \alpha \leq t < \beta_1. \quad \square$$

COROLLARY 5.5. (The case $f(u)=u$) *If $G(u)=\int_{u_0}^u dx/g(x)$, $u>0$, $u_0>0$, then the inequality*

$$u(t) \leq a(t)+b(t)\int_\alpha^t k(t,s)g(u(s))\,ds, \qquad t \in J,$$

implies

$$u(t) \leq a(t)-A(t)+G^{-1}\left[G(A(t))+B(t)\int_\alpha^t K(t,s)\,ds\right], \qquad \alpha \leq t < \beta_1. \quad \square$$

6. Comparison results

As already noted, the functions $u(t)$ in the inequalities appearing in theorems 1.1–1.5 and theorem 5.1 satisfy estimates of the form

$$u(t) \leq v(t),$$

where $v(t)$ is the unique continuous solution of the corresponding integral equation, and this estimate is best possible. The following question arises: Are assertions of this type valid for a wider class of inequalities? It turns out that under sufficiently general assumptions this is true. However, in the case when the *comparison integral equation* has more than one solution, the function $u(t)$ can be estimated in terms of the maximal (minimal) solution of this equation. For instance, the following theorem holds.

THEOREM 6.1. *Let $u(t)$ and $a(t)$ be continuous functions in $J = [\alpha, \beta)$, and let $f(t,s,u)$ be a continuous function for $\alpha \le s \le t \le \beta$, $u \in \mathbf{R}$, which is nondecreasing with respect to $u \in \mathbf{R}$. Suppose*

$$u(t) \le a(t) + \int_\alpha^t f(t,s,u(s))\,ds, \qquad t \in J,$$

and let $\psi(t)$ be the maximal solution of the equation

$$\psi(t) = a(t) + \int_\alpha^t f(t,s,\psi(s))\,ds \tag{6.1}$$

in an interval $[\alpha, \beta_1)$ $(\beta_1 \le \beta)$. Then

$$u(t) \le \psi(t), \qquad t \in [\alpha, \beta_1). \quad \square$$

Giving an explicit form of the maximal solution of an equation (6.1) is often impossible. Thus the need arises of obtaining estimates for it. A widely used method for obtaining such estimates consists of transition from the comparison integral equation (6.1) to a *comparison differential equation* (or differential inequality), and subsequent estimation of a solution of the latter. To apply this technique we need some results concerning ordinary differential equations and differential inequalities.

Consider the initial value problem

$$v' = F(t,v), \qquad v(\alpha) = v_0, \tag{6.2}$$

where F is a continuous function in an open set D containing the point (α, v_0). There exist $\beta > \alpha$ and $b > 0$ such that the rectangle

$$Q = \{(t,v) : \alpha \le t \le \beta,\ |v - v_0| \le b\} \subset D.$$

DEFINITION 6.1. A differentiable function v defined in an interval $\alpha \le t \le \beta_v$ is said to be a *solution* of (6.2) if $(t, v(t)) \in D$ for $\alpha \le t \le \beta_v$ and

$$v'(t) = F(t, v(t)), \qquad \alpha \le t \le \beta_v, \qquad v(\alpha) = v_0. \quad \square$$

It is well known that under these assumptions (6.2) has a solution which is defined in an interval containing α. This solution, however, need not be unique in this interval.

EXAMPLE 6.1. Solutions of the initial value problem

$$v' = f(t,v), \qquad v(0) = 0,$$

where

$$f(t,v) = \begin{cases} t - \sqrt{t^2 - 2v}, & t^2 - 2v \ge 0, \quad t \ge 0, \\ t + \sqrt{t^2 - 2v}, & t^2 - 2v \ge 0, \quad t < 0, \\ t, & t^2 - 2v < 0, \end{cases}$$

are the functions of the form

$$v = v_{a,b}(t) = \begin{cases} at - \frac{a^2}{2}, & t \le a \le 0, \\ \frac{t^2}{2}, & a < t < b, \\ bt - \frac{b^2}{2}, & t \ge b \ge 0, \end{cases}$$

as well as the functions $v^-(t) \equiv 0$ and $v^+(t) = t^2/2$. Moreover,

$$v^-(t) \leq v_{a,b}(t) \leq v^+(t), \qquad t \in \mathbf{R}, \quad a \leq 0, \quad b \geq 0. \quad \square$$

DEFINITION 6.2. A solution $r(t)$ is said to be a *maximal* (*minimal*) *solution* of (6.2) in an interval $[\alpha, \beta_r]$ if the inequality

$$v(t) \leq r(t) \qquad (v(t) \geq r(t)), \qquad \alpha \leq t \leq \min(\beta_v, \beta_r),$$

holds for any solution $v(t)$ of (6.2). $\square$

In the sequel the maximal and the minimal solution of (6.2) will be denoted by v^+ and v^-, respectively.

THEOREM 6.2. *If the function F is continuous in the rectangle Q, then equation* (6.2) *has a maximal and a minimal solution. If $|F(t,v)| \leq M$ for $(t,v) \in Q$, then these solutions, as any other solution, are defined in the interval $[\alpha, \beta_1]$, where $\beta_1 - \alpha = \min(\beta - \alpha, bM^{-1})$. If $[\alpha, \beta_r]$ is the largest domain of existence of the maximal (minimal) solution r of* (6.2), *then there is an $\epsilon_0 > 0$ such that for $0 \leq \epsilon \leq \epsilon_0$ the maximal (minimal) solution of the initial value problem*

$$x' = F(t,x) \pm \epsilon, \qquad x(\alpha) = v_0 \pm \epsilon,$$

also exists in $[\alpha, \beta_r]$. Here $+$ (respectively, $-$) holds in the case of the maximal (respectively, minimal) solution. $\square$

This theorem has been proved by various authors (cf., e.g., Hille (1969), Reid (1971), Šzarski (1965), Lakshmikantham and Leela (1969)).

Before proving the main comparison theorem, we will need the following lemmas.

LEMMA 6.1. *Let F be a continuous function in an open set D containing the point (α, v_0), and let $v(t)$ be any solution of* (6.2) *for $t \in [\alpha, \beta_1]$. If a differentiable function $x(t)$ satisfies the inequalities*

$$x'(t) < F(t, x(t)), \qquad \alpha \leq t \leq \beta_1, \quad x(\alpha) < v_0, \tag{6.3}$$

then

$$x(t) < v(t), \qquad \alpha \leq t \leq \beta_1. \tag{6.4}$$

The result remains valid if $<$ is replaced by $>$ in (6.3) *and* (6.4).

PROOF. To prove this (for $<$), we assume that, on the contrary, $v(t) \leq x(t)$ for some $t \in (\alpha, \beta_1)$. Let $\tau = \inf\{t \in (\alpha, \beta_1)\colon v(t) \leq x(t)\}$. Then $\alpha < \tau < \beta_1$, and by continuity $v(\tau) = x(\tau)$, while $x(t) < v(t)$ for $\alpha \leq t < \tau$. Moreover, by (6.2) and (6.3),

$$x'(\tau) < F(\tau, x(\tau)) = F(\tau, v(\tau)) = v'(\tau).$$

Since $x(\tau) = v(\tau)$, this implies $v(t) < x(t)$ on some interval $(\tau - \epsilon, \tau)$, contradicting the definition of τ.

The same proof applies if $<$ is replaced by $>$. $\square$

LEMMA 6.2. *Let F be continuous in an open set D containing the point (α, v_0), and let $\{\epsilon_n\}$ be a strictly monotone sequence with $\lim_{n\to\infty} \epsilon_n = 0$. Consider the initial value problem*

$$y' = F(t, y) + \epsilon_n, \qquad y(\alpha) = v_0 + \epsilon_n. \tag{6.5}$$

Let $y = r_n(t)$ be the maximal (respectively, minimal) solution of (6.5) in $[\alpha, \beta_1]$ if $\epsilon_n > 0$ (respectively, $\epsilon_n < 0$). Then the sequence $\{r_n\}$ converges uniformly on $[\alpha, \beta_1]$ to a function r which is the maximal (respectively, minimal) solution of (6.2).

PROOF. First we note that, by theorem 6.2, if the maximal (minimal) solution of (6.2) exists in $[\alpha, \beta_1]$, then so does the maximal (minimal) solution of (6.5) for all sufficiently large n. To simplify the notation we assume this to be true for $n \geq 1$. In lemma 6.1, take $x(t) = r_{n+1}(t)$. Since

$$\begin{aligned} x'(t) = F(t, x(t)) + \epsilon_{n+1} < F(t, x(t)) + \epsilon_n, \qquad \alpha \leq t \leq \beta_1,\\ x(\alpha) = v_0 + \epsilon_{n+1} < v_0 + \epsilon_n, \end{aligned}$$

it follows that $x(t) = r_{n+1}(t) < r_n(t)$ for $\alpha \leq t \leq \beta_1$, so that the sequence $\{r_n\}$ is monotone decreasing in $[\alpha, \beta_1]$. Moreover, if y is any solution of the initial value problem (6.2), then $y(t) < r_n(t)$ for $\alpha \leq t \leq \beta_1$ (by the same reasoning). Hence the decreasing sequence $\{r_n\}$ is bounded below on $[\alpha, \beta_1]$, and so

$$r(t) = \lim_{n\to\infty} r_n(t) \geq y(t) \tag{6.6}$$

exists in $[\alpha, \beta_1]$. Note that

$$\begin{aligned} r_n(t) &= r_n(\alpha) + \int_\alpha^t F(s, r_n(s))\, ds + \epsilon_n(t - \alpha) = \\ &= v_0 + \int_\alpha^t F(s, r_n(s))\, ds + \epsilon_n(1 + t - \alpha). \end{aligned} \tag{6.7}$$

Hence, for $n < m$,

$$|r_n(t) - r_m(t)| \leq (\epsilon_n - \epsilon_m)(1 + \beta_1 - \alpha) + \int_\alpha^{\beta_1} |F(s, r_n(s)) - F(s, r_m(s))|\, ds. \tag{6.8}$$

Since F is uniformly continuous on the compact set

$$K = \{(t, x) : \alpha \leq t \leq \beta_1,\ y(t) \leq x \leq r_1(t)\},$$

(6.8) implies that the convergence is uniform. Letting $n \to \infty$ in (6.7), we find

$$r(t) = v_0 + \int_\alpha^t F(s, r(s))\, ds.$$

Hence r is a solution of (6.2), and by (6.6) it is maximal.

The proof for the case of a minimal solution is similar. □

THEOREM 6.3. (Comparison theorem) *Let $F(t, v)$ be a continuous function in an open set D containing a point (α, v_0), and suppose that the initial value problem*

$$r'(t) = F(t, r), \qquad r(\alpha) = v_0,$$

has a maximal solution $r(t)$ with domain $\alpha \leq t \leq \beta_1$. Let $v(t)$ be a differentiable function in $[\alpha, \beta_1]$ such that $(t, v(t)) \in D$ for $t \in [\alpha, \beta_1]$ and

$$v'(t) \leq F(t, v(t)), \qquad \alpha \leq t \leq \beta_1, \qquad v(\alpha) = v_0. \tag{6.9}$$

Then

$$v(t) \leq r(t), \qquad \alpha \leq t \leq \beta_1. \tag{6.10}$$

Moreover, the result remains valid if 'maximal' is replaced by 'minimal', and $\leq$ is replaced by $\geq$ in both (6.9) *and* (6.10).

PROOF. In the 'maximal' case, with $\{\epsilon_n\}$ and $\{r_n\}$ as in lemma 6.2, we have

$$v'(t) \leq F(t, v(t)) < F(t, v(t)) + \epsilon_n, \qquad \alpha \leq t \leq \beta_1,$$
$$v(\alpha) < v_0 + \epsilon_n.$$

By lemma 6.1, $v(t) < r_n(t)$ for $\alpha \leq t \leq \beta_1$. Letting $n \to \infty$ and using lemma 6.2 (as well as the continuity at β_1), we find $v(t) \leq r(t)$ for $\alpha \leq t \leq \beta_1$. □

REMARK 6.1. The comparison theorem 6.3 remains valid if $v(t)$ is continuous in $[\alpha, \beta_1]$, $(t, v(t)) \in D$, $t \in [\alpha, \beta_1]$, and

$$Dv(t) \leq F(t, v(t)), \qquad \alpha \leq t \leq \beta_1, \qquad v(\alpha) \leq v_0,$$

where $Dv(t)$ is any *Dini derivative* of $v(t)$. □

We will also need the following lemma.

LEMMA 6.3. *Let F and G be continuous functions of constant sign in a domain D containing a point (α, r_0), and suppose that*

$$r_1'(t) = G(t, r_1), \qquad r_1(\alpha) = r_0,$$

has a maximal (if $G \geq 0$) or minimal (if $G \leq 0$) solution $r_1(t)$ in $[\alpha, \beta_1]$ such that

$$S_+ = \{(t, r) : \alpha \leq t \leq \beta_1, r_0 \leq r \leq r_1(t)\} \subset D \qquad \text{if } G \geq 0,$$
$$S_- = \{(t, r) : \alpha \leq t \leq \beta_1, r_0 \geq r \geq r_1(t)\} \subset D \qquad \text{if } G \leq 0.$$

If $0 \leq \pm F(t, r) \leq \pm G(t, r)$ for $(t, r) \in S_\pm$, then all solutions r of the initial value problem

$$r'(t) = F(t, r), \qquad r(\alpha) = r_0, \tag{6.11}$$

exist in $[\alpha, \beta_1]$. Moreover, $r_0 \leq r(t) \leq r_1(t)$ in $[\alpha, \beta_1]$ if $G \geq 0$, and $r_0 \geq r(t) \geq r_1(t)$ in $[\alpha, \beta_1]$ if $G \leq 0$.

The PROOF follows from the fact that if r is a local solution of (6.11) in a subinterval $[\alpha, \beta_r]$ of $[\alpha, \beta_1]$, then by the comparison theorem,

$$r_0 \le r(t) \le r_1(t) \qquad (\text{or} \quad r_0 \ge r(t) \ge r_1(t)), \qquad t \in [\alpha, \beta_r]$$

in the maximal (minimal) case. That is, the graph of r lies in the set S_+ (S_-), and can hence be extended to all of $[\alpha, \beta_1]$. □

In many cases the solutions of integral and differential inequalities are compared with a solution of an initial value problem

$$r'(t) = k(t)g(r) + \sigma(t), \qquad r(\alpha) = r_0. \tag{6.12}$$

In lemmas 1.1, 4.1, and 5.1 we have considered particular instances of this initial value problem (6.12), and obtained its unique solution in explicit form. In the general case, however, only an estimate of the maximal (minimal) solution of (6.12) can be obtained. This is done in the following theorem, which is due to Lakshmikantham (1973) and was made more precise by Beesack (1975).

THEOREM 6.4. *Let σ and k be continuous functions of constant sign in an interval $J = [\alpha, \beta]$, and let g be a nowhere vanishing monotone continuous function in an interval I containing a point r_0. Suppose that either g is nondecreasing and $k \ge 0$, or g is nonincreasing and $k \le 0$. If $\sigma \ge 0$ (respectively, $\sigma \le 0$), then we assume that the maximal (minimal) solution $r(t)$ of (6.12) exists in the interval $[\alpha, \beta_1]$, and we set*

$$v_1 = \sup\left\{v \in J :\ r_0 + \int_\alpha^v \sigma(s)\,ds \in I\right\},$$
$$v_2 = \sup\left\{v \in J :\ G\left(r_0 + \int_\alpha^t \sigma(s)\,ds\right) + \int_\alpha^t k(s)\,ds \in G(I),\ \alpha \le t \le v\right\},$$

where $G(u) = \int_{u_0}^u dx/g(x)$, $u \in I$ ($u_0 \in I$). Let $\beta_2 = \min(v_1, v_2)$, and $\overline{\beta}_1 = \min(\beta_1, \beta_2)$. Then for $\alpha \le t \le \overline{\beta}_1$ we have

$$r(t) \le G^{-1}\left[\int_\alpha^t k(s)\,ds + G\left(r_0 + \int_\alpha^t \sigma(s)\,ds\right)\right] \qquad \text{if } \sigma \ge 0,$$
$$r(t) \ge G^{-1}\left[\int_\alpha^t k(s)\,ds + G\left(r_0 + \int_\alpha^t \sigma(s)\,ds\right)\right] \qquad \text{if } \sigma \le 0.$$

Moreover, if $\sigma \ge 0$ and k, g have the same sign ($\sigma \le 0$ and k, g have opposite signs), then $\beta_1 \ge \beta_2$ and

$$G^{-1}\left[\int_\alpha^t k(s)\,ds + G(r_0)\right] \le r(t)$$

(respectively,

$$G^{-1}\left[\int_\alpha^t k(s)\,ds + G(r_0)\right] \ge r(t)$$

, $\alpha \le t < \beta_2$). □

This theorem is given without proof.

7. Lakshmikantham's comparison method

To be able to apply the comparison theorem in the investigation of a given integral inequality, it is necessary to first transform this inequality into a differential inequality, and also to find a relationship between the function to be estimated and the maximal (minimal) solution of the comparison differential equation. In this transition the various monotonicity properties of the functions participating in the inequality have to be taken into account. The results obtained by Lakshmikantham (1973) enable us to omit all these intermediate transformations and properties, and to pass directly to the comparison differential equation; only the estimation of the maximal (minimal) solution of this equation is left to the investigator. This method was later developed by Beesack (1975), and was applied to the derivation by a single uniform method of a large number of well-known inequalities. However, the constructive potential of the method has not yet been fully enclosed. This is mainly due to a lack of systematic investigation of the special comparison equations used by the method. Here we present the method, following the works of Lakshmikantham (1973), Beesack (1975), Martynyuk and Gutowski (1979).

We will deal with monotone functions of several variables, and, for the sake of brevity, we will introduce some special notation and terminology. Let $F_1(t,s,u)$ be a function which is defined in $I_1 \times J_1 \times K_1$, where I_1, J_1, K_1 are intervals. If for each $(s,u) \in J_1 \times K_1$ the function $f(t) \equiv F_1(t,s,u)$ is monotone (nondecreasing, increasing, etc.) in I_1, we say that $F_1(t)$ is monotone (nondecreasing, increasing, etc.). We will also write $F_1(t) \in \uparrow$ (respectively, $F_1(t) \in \downarrow$) to denote that $F_1(t)$ is nondecreasing (nonincreasing). Similarly, if a function $F_2(t,s,u)$ is defined in $I_2 \times J_2 \times K_2$, then we say that $F_1(t)$ and $F_2(u)$ have the same parity (respectively, have opposite parity) if $F_1(t)$ and $F_2(u)$ are both nonincreasing or nondecreasing (respectively, if one is nonincreasing and the other is nondecreasing) in the respective variables.

The following theorem includes the important and quite general result of Lakshmikantham (1973).

THEOREM 7.1. *Suppose that the following conditions hold:*

1) *$f(u)$ is a strictly monotone continuous function in an interval I;*
2) *$H(t,v)$ is a continuous function in $J \times K$, with $H(v)$ monotone, where $J = [\alpha,\beta]$ and K is an interval containing $[-b,b]$ for some $b>0$;*
3) *$W(t,s,u)$ is a continuous function of constant sign in $\Delta \times I$, with $W(t)$ monotone in J and $W(u)$ monotone in I. Here $\Delta = \{(t,s)\colon \alpha \le s \le t \le \beta\}$;*
4) *u and a are continuous functions in J with $u(J) \subset I$ and*

$$a(t) + H(t,v) \in f(I) \qquad \text{for } t \in J,\ |v| \le b; \tag{7.1}$$

5)

$$f(u(t)) \le a(t) + H\left(t, \int_\alpha^t W(t,s,u(s))\,ds\right), \qquad t \in J; \tag{7.2}$$

6) *the functions $W(u)$ and f have the same (opposite) parity and $r = r(t;T,\alpha)$ is*

the maximal (minimal) solution of the initial value problem

$$r' = W(T, t, f^{-1}[a(t) + H(t, r)]), \qquad \alpha \le t \le T \le \beta_1 \le \beta, \qquad r(\alpha) = 0, \tag{7.3}$$

where $\beta_1 > \alpha$ is chosen so that the maximal (minimal) solution exists for $\alpha \le t \le \beta_1$.

If $W(t)$ and $H(v)$ have the same parity, then

$$u(t) \lesseqgtr f^{-1}[a(t) + H(t, \tilde{r}(t))], \qquad \alpha \le t \le \beta_1, \tag{7.4}$$

where $\tilde{r}(t) = r(t; t, \alpha)$, and $\le$ (respectively, $\ge$) holds if

i) *f is increasing and $W(u)$, $H(v)$ have the same parity (respectively,*
ii) *f is decreasing and $W(u)$, $H(v)$ have opposite parity).*

PROOF. The function $F(T, t, r) = W(T, t, f^{-1}[a(t) + H(t, r)])$ is, by (7.1), continuous in the compact set $\Delta \times [-b, b]$, so there is a constant $M > 0$ such that $|F(T, t, r)| \le M$ for all such (T, t, r). Theorem 6.2 implies that there is a $\beta_1 > \alpha$ (in fact, $\beta_1 \ge \alpha + \min(\beta - \alpha, bM^{-1})$, independent of T, such that the maximal (minimal) solution of (7.3) exists in $[\alpha, \beta_1]$.

Now we fix $T \in (\alpha, \beta]$ and let $\alpha \le t \le T$. Then

$$v(t, t) \equiv \int_\alpha^t W(t, s, u(s))\, ds \lesseqgtr \int_\alpha^t W(T, s, u(s))\, ds \equiv v(t, T), \tag{7.5}$$

where $\le$ ($\ge$) holds if $W(t) \in \uparrow$ ($W(t) \in \downarrow$). Note that (7.2) implies that $v(\tau, \tau) \in K$ for all $\tau \in J$. Since $0 \in K$, it follows that $v(t, T) \in K$ (set $\tau = t$ or $\tau = T$), in both cases of inequality (7.5), irrespective of whether $W \ge 0$ or $W \le 0$. From (7.2) we obtain

$$u(t) \lesseqgtr f^{-1}[a(t) + H(t, v(t, t))], \tag{7.6}$$

where $\le$ ($\ge$) holds if $f \in \uparrow$ ($f \in \downarrow$). Since $v'(t, T) = W(T, t, u(t))$ for $\alpha \le t \le T \le \beta$, we have

$$v'(t, T) \lesseqgtr W(T, t, f^{-1}[a(t) + H(t, v(t, t))]), \tag{7.7}$$

where $\le$ ($\ge$) holds if $W(u)$ and f have the same (opposite) parity. By (7.5),

$$H(t, v(t, t)) \lesseqgtr H(t, v(t, T)), \qquad \alpha \le t \le T, \tag{7.8}$$

where $\le$ ($\ge$) holds if $W(t)$ and $H(v)$ have a_1) the same parity (respectively, b_1) opposite parity). Thus,

$$f^{-1}[a(t) + H(t, v(t, t))] \lesseqgtr f^{-1}[a(t) + H(t, v(t, T))], \qquad \alpha \le t \le T,$$

where $\le$ ($\ge$) holds if

a_2) $(f \in \uparrow, a_1)) \vee (f \in \downarrow, b_1))$ (respectively,
b_2) $(f \in \uparrow, b_1)) \vee (f \in \downarrow, a_1))$).

In turn, this implies that

$$W(T,t,f^{-1}[a(t)+H(t,v(t,t))]) \lesseqgtr W(T,t,f^{-1}[a(t)+H(t,v(t,T))]),$$

where $\leq$ ($\geq$) holds if

a_3) $(W(u) \in \uparrow, a_2)) \vee (W(u) \in \downarrow, b_2))$ (respectively,
b_3) $(W(u) \in \uparrow, b_2)) \vee (W(u) \in \downarrow, a_2))$).

By combining this with (7.7) we see that if $W(t)$ and $H(v)$ have the same parity, then

$$v'(t,T) \lesseqgtr W(T,t,f^{-1}[a(t)+H(t,v(t,T))]), \qquad \alpha \leq t \leq T \leq \beta,$$

where $\leq$ ($\geq$) holds if $W(u)$ and f have the same (opposite) parity.

Since $v(\alpha,T)=0$, the comparison theorem 6.3 shows that if $W(t)$ and $H(v)$ have the same parity, and $r(t,T,\alpha)$ is the maximal (minimal) solution of (7.3), then

$$v(t,T) \lesseqgtr r(t,T,\alpha) \qquad \text{for } \alpha \leq t \leq T \leq \beta_1.$$

Setting $t=T$ and changing notation, we arrive at

$$v(t,t) \lesseqgtr \tilde{r}(t) \qquad \text{for } \alpha \leq t \leq \beta_1,$$

where $\leq$ or $\geq$ according as

A_1) $W(u)$, f have the same parity; or
B_1) $W(u)$, f have opposite parity.

As in the analysis of (7.5) and (7.8), for $\alpha \leq t \leq \beta_1$ we have

$$H(t,v(t,t)) \lesseqgtr h(t,\tilde{r}(t)),$$

where $\leq$ or $\geq$ according as

A_2) $(H(v) \in \uparrow, A_1) \vee (H(v) \in \downarrow, B_1)$; or
B_2) $(H(v) \in \uparrow, B_1) \vee (H(v) \in \downarrow, A_1)$.

Moreover,

$$f^{-1}[a(t)+H(t,v(t,t))] \lesseqgtr f^{-1}[a(t)+H(t,\tilde{r}(t))],$$

where $\leq$ or $\geq$ according as

A_3) $(f \in \uparrow, A_2) \vee (f \in \downarrow, B_2)$; or
B_3) $(f \in \uparrow, B_2) \vee (f \in \downarrow, A_2)$.

Analysing the various cases, we see that if $W(t)$ and $H(v)$ have the same parity, then

$$f^{-1}[a(t)+H(t,v(t,t))] \lesseqgtr f^{-1}[a(t)+H(t,\tilde{r}(t))], \qquad \alpha \leq t \leq \beta_1, \tag{7.9}$$

where $\leq$ ($\geq$) holds if $W(u)$, $H(v)$ have the same (opposite) parity. In cases i) and ii) the conclusion (7.4) now follows from (7.6) and (7.9)). □

In the same way we can prove

THEOREM 7.2. *Under the conditions* 1)–4) *of theorem* 7.1, *suppose that*

5′) $f(u(t)) \geq a(t) + H(t, \int_\alpha^t W(t,s,u(s))\,ds)$, $t \in J$;

6′) *the functions* $W(u)$, f *have opposite (the same) parity, and* $\tilde{r}(t) = r(t,t,\alpha)$, *where* $r(t,T,\alpha)$ *is the maximal (minimal) solution of* (7.3).

If $W(t)$, $H(v)$ *have opposite parity, then*

$$u(t) \underset{\geq}{\leq} f^{-1}[a(t) + H(t, \tilde{r}(t))], \qquad \alpha \leq t \leq \beta_1,$$

where $\leq$ *(*$\geq$*) according as case* i) *(*ii)*) of theorem* 7.1 *holds.* □

REMARK 7.1. Similar results are true for inequalities of the form

$$f(u(t)) \underset{\geq}{\leq} a(t) + H\left(t, \int_t^\beta W(t,s,u(s))\,ds\right), \qquad t \in J. \tag{7.10}$$

We do not give these results, but note that by the change of variable $t = -\tau$ (7.10) becomes

$$f(u_1(\tau)) \underset{\geq}{\leq} a_1(\tau) + H_1\left(\tau, \int_{-\beta}^{\tau} W_1(\tau,\sigma,u_1(\sigma))\,d\sigma\right), \qquad -\beta \leq \tau \leq -\alpha, \tag{7.11}$$

with $H_1(t,v) = H(-t,v)$, $W_1(t,s,u) = W(-t,-s,u)$, $u_1(s) = u(-s)$, $a_1(t) = a(-t)$. To apply theorem 7.1 to (7.11), we also note that if $\leq$ holds in (7.10), then $W(t)$, $H(v)$ must have opposite parity. □

REMARK 7.2. If $K = [0,B]$, then necessarily $W \geq 0$ (since $v(t,t) \in K$). Now condition (7.1) is:

$$a(t) + H(t,v) \in f(I) \qquad \text{for } t \in J,\ 0 \leq v \leq b.$$

Similarly, if $K = [-B,0]$, then $W \leq 0$, and (7.1) is:

$$a(t) + H(t,v) \in f(I) \qquad \text{for } t \in J,\ -b \leq v \leq 0.$$

Here $0 < b \leq B$ is a constant. The conclusion (7.4) remains unchanged. □

REMARK 7.3. The assumption that W is of constant sign has only been used in (7.5), to ensure that $v(t,T) \in K$ for $\alpha \leq t \leq T \leq \beta$. If the latter requirement is added as a hypothesis, then fixed sign for W need not be assumed. □

COROLLARY 7.1. (Lakshmikantham, 1973) *Let* u, a, b, *and* c *be nonnegative continuous functions in* $J = [\alpha,\beta]$, *and* f *and* h *nonnegative continuous functions in* $\mathbf{R}_+$, *with* f *strictly increasing and* h *nondecreasing. Let* $k(t,s)$ *be a nonnegative continuous function in* $\Delta = \{(t,s)\colon \alpha \leq s \leq t \leq \beta\}$, *and* $w(s,u)$ *a nonnegative continuous function in* $J \times \mathbf{R}_+$, *with* $w(u)$ *nonincreasing in* $\mathbf{R}_+$. *Put*

$$C(t) = \max_{\alpha \leq s \leq t} c(s), \qquad K(t,s) = \max_{s \leq \sigma \leq t} k(\sigma,s), \qquad \alpha \leq s \leq t \leq \beta.$$

If

$$f(u(t)) \leq a(t) + b(t)h\left[c(t) + \int_\alpha^t k(t,s)w(s,u(s))\,ds\right], \qquad t \in J, \tag{7.12}$$

then

$$u(t) \leq f^{-1}[a(t) + b(t)h(C(t) + \tilde{y}(t))] \qquad t \in J_1, \tag{7.13}$$

where $\tilde{y}(t) = y(t,t,\alpha)$ *and* $y = y(t,T,\alpha)$ *is the maximal solution of*

$$y' = K(T,t)w(t, f^{-1}[a(t) + b(t)h(C(T) + y)]), \qquad y(\alpha) = 0, \tag{7.14}$$

in some interval $[\alpha, \beta_1] \subset J$. *It is assumed that* $J_1 \subset J$ *is chosen such that the righthand side of* (7.13) *is defined. Moreover,*

$$u(t) \leq f^{-1}[a(t) + b(t)h(c(t) + \tilde{r}(t))], \qquad t \in J_2, \tag{7.15}$$

where $\tilde{r}(t) = r(t,t,\alpha)$ *and* $r = r(t,T,\alpha)$ *is the maximal solution of*

$$r' = K(T,t)w(t, f^{-1}[a(t) + b(t)h(c(t) + r)]), \qquad r(\alpha) = 0, \tag{7.16}$$

in some interval $[\alpha, \beta_2] \subset J$. *It is assumed that* $J_2 \subset J$ *is chosen such that the righthand side of* (7.15) *is defined.*

PROOF. First we note that (7.12) implies

$$f(u(t)) \leq a(t) + b(t)h\left(c(t) + \int_\alpha^t K(t,s)w(s,u(s))\,ds\right), \qquad t \in J, \tag{7.17}$$

and

$$f(u(t)) \leq a(t) + b(t)h\left(C(T) + \int_\alpha^t K(t,s)w(s,u(s))\,ds\right), \qquad \alpha \leq t \leq T. \tag{7.18}$$

By theorem 7.1, (7.18) implies (7.13), and (7.17) implies (7.15). It is readily seen from (7.16) and (7.14) that the estimate (7.15) is better than (7.13). □

In conclusion of this section we show how many of the results in §§4 and 5 can be derived from theorem 7.1. For this, in each case we write down the inequality, define $f(u)$, $H(t,v)$, $W(t,s,u)$, I, J, and K, write down the initial value problem (7.3), its solution, and the concluding inequality. In some cases we give an additional conclusion.

THEOREM 5.1. (Bihari)

$$u(t) \leq a + \int_\alpha^t k(s)g(u(s))\,ds, \qquad t \in J = [\alpha, \beta),$$

$f(u) = u$, $H(t,v) = v$, $W(t,s,u) = k(s)g(u)$; *I is an interval with* $u(J) \subset I$, $a \in I$, $g \neq 0$ *in* I; $K = \mathbf{R}$;

$$r' = k(t)g(a + r), \qquad r(\alpha) = 0,$$

$$u(t) \leq a + \tilde{r}(t) = G^{-1}\left[G(a) + \int_\alpha^t k(s)\,ds\right], \qquad \alpha \leq t < \beta_1,$$

where $G(u) = \int_{u_0}^{u} dx/g(x)$, $u \in I$ ($u_0 \in I$). □

THEOREM 4.2. (Willett and Wong)

$$u(t) \le a(t) + b(t)\left(\int_\alpha^t k(s)u^p(s)\,ds\right)^{1/p}, \qquad t \in J = [\alpha,\beta), \qquad 1 \le p < \infty,$$

$f(u) = u$, $H(t,v) = b(t)v^{1/p}$, $W(t,s,u) = k(s)u^p$; $I = K = \mathbf{R}_+$;

$$r' = k(t)[a(t) + b(t)r^{1/p}]^p, \qquad r(\alpha) = 0, \tag{7.19}$$

$$u(t) \le a(t) + b(t)r^{1/p}(t).$$

Also, by estimating $r^{1/p}(t)$ as has been done in theorem 4.2, we obtain (4.5). To obtain another estimate, we apply to (7.19) the inequality

$$(A+B)^p \le \begin{cases} A^p + B^p, & 0 < p \le 1, \\ 2^{p-1}(A^p + B^p), & 1 \le p < \infty, \end{cases}$$

which holds for $A, B \ge 0$. Now (7.19) implies

$$r' \le c_p k(t)[a^p(t) + b^p(t)r], \qquad r(\alpha) = 0,$$

where c_p is equal to 1 (respectively, 2^{p-1}) for $0 < p \le 1$ (respectively, $p \ge 1$).
Applying lemma 1.1 gives

$$r(t) \le c_p \int_\alpha^t k(s)a^p(s)\exp\left(c_p \int_s^t k(\tau)b^p(\tau)\,d\tau\right) ds, \qquad t \in J,$$

whence

$$u(t) \le a(t) + b(t)\left[c_p \int_\alpha^t k(s)a^p(s)\exp\left(c_p\int_s^t k(\tau)b^p(\tau)\,d\tau\right) ds\right]^{1/p}, \qquad t \in J. \tag{7.20}$$

Unlike (4.5), this estimate is valid for all $p > 0$, but for $p \ge 1$ it is worse than (4.5). □

THEOREM 4.9. (Willett and Wong)

$$u(t) \le a + \int_\alpha^t [b(s)u(s) + k(s)u^p(s)]\,ds, \qquad t \in J = [\alpha,\beta), \quad p > 0,\ p \ne 1,$$

$f(u) = u$, $H(t,v) = v$, $W(t,s,u) = b(s)u + k(s)u^p$; $I = K = \mathbf{R}_+$;

$$r' = b(t)[a + r] + k(t)[a+r]^p, \qquad r(\alpha) = 0. \tag{7.21}$$

Applying lemma 4.1, (7.21) implies the sharp estimate (4.33). □

THEOREM 4.8. (Stachurska)

$$u(t) \le a(t) + b(t)\int_\alpha^t k(s)u^p(s)\,ds, \qquad t \in J = [\alpha,\beta), \qquad p > 0, \quad p \ne 1,$$

where a/b is a nondecreasing function in J, $f(u) = u$, $H(t,v) = b(t)v$, $W(t,s,u) = k(s)u^p$; $I = K = \mathbf{R}_+$;

$$r' = k(t)[a(t) + b(t)r]^p, \qquad r(\alpha) = 0, \tag{7.22}$$

$$u(t) \le a(t) + b(t)\tilde{r}(t). \tag{7.23}$$

Estimating $a(t) + b(t)\tilde{r}(t)$ as in the proof of theorem 4.8 we arrive at (4.26). To obtain another estimate for the solution of (7.22) we write

$$r' \le k(t)b^p(t)\left[\frac{a(T)}{b(T)} + r\right]^p, \qquad \alpha \le t \le T < \beta.$$

Lemma 5.1 implies

$$a(T) + b(T)r(t) \le a(T)\left\{1 - (p-1)\left[\frac{a(T)}{b(T)}\right]^{p-1}\int_\alpha^t k(s)b^p(s)\,ds\right\}^{-1/(p-1)}.$$

Setting $t = T$ and changing notation, (7.23) implies

$$u(t) \le \frac{a(t)}{\left\{1 - (p-1)\left[\frac{a(t)}{b(t)}\right]^{p-1}\int_\alpha^t k(s)b^p(s)\,ds\right\}^{1/(p-1)}}, \qquad \alpha \le t < \beta_1, \tag{7.24}$$

where $\beta_1 = \beta$ if $0 < p < 1$, and for $p > 1$, β_1 is the smallest value of $t \ge \alpha$ such that the expression between brackets vanishes.

Unlike (4.26), (7.24) is valid for every $p > 0$, $p \ne 1$. For $p > 1$, however, it is worse than Stachurska's result. □

THEOREM 4.4. (Gollwitzer)

$$u(t) \le a(t) + b(t)g^{-1}\left(\int_\alpha^t k(s)g(u((s))\,ds\right), \qquad t \in J = [\alpha, \beta),$$

$f(u) = u$, $H(t,v) = b(t)g^{-1}(v)$, $W(t,s,u) = k(s)g(u)$; $K = g(I)$, with I as in theorem 4.4;

$$r' = k(t)g[a(t) + b(t)g^{-1}(r)], \qquad r(\alpha) = 0, \tag{7.25}$$

$$u(t) \le a(t) + b(t)g^{-1}(r(t)),$$

where $r(t)$ is the maximal (minimal) solution of (7.25) if g satisfies condition i) (ii)) in theorem 4.4. □

THEOREM 4.6. (Gollwitzer)

$$u(t) \le a + g^{-1}\left(\int_\alpha^t k(s)g(u((s))\,ds\right), \qquad t \in J = [\alpha, \beta),$$

$f(u) = u$, $H(t,v) = g^{-1}(v)$, $W(t,s,u) = k(s)g(u)$; $K = g(I)$, with I as in theorem 4.6;

$$r' = k(t)g[a + g^{-1}(r)], \qquad r(\alpha) = 0 \tag{7.26}$$

$$u(t) \le a + g^{-1}(r(t)), \qquad \alpha \le t < \beta_1, \tag{7.27}$$

where $r(t)$ is the unique solution of (7.26) *in* $\alpha \le t < \beta_1$. *Putting*

$$F(u) = \int_0^u \frac{dx}{g[a + g^{-1}(x)]}, \qquad u \in g(I) = K,$$

then (7.27) *implies*

$$u(t) \le a + g^{-1}\left[F^{-1}\left(\int_\alpha^t k(s)\,ds\right)\right], \qquad \alpha \le t < \beta_1, \tag{7.28}$$

where $\beta_1 = \sup\{t \in J\colon \int_\alpha^t k(s)\,ds \in F(K)\}$.

It can be proved that (7.28) is better than (4.16). □

THEOREM 5.7.

$$u(t) \le a(t) + b(t)h\left(\int_\alpha^t k(s)g(u((s))\,ds\right), \qquad t \in J = [\alpha, \beta),$$

$f(u) = u$, $H(t,v) = b(t)h(v)$, $W(t,s,u) = k(s)g(u)$; *I and K are intervals such that* $0 \in K$, $h(K) \subset I$, $u(J) \subset I$, $a(J) \subset I$, $b(J) \subset I$;

$$r' = k(t)g[a(t) + b(t)h(r)], \qquad r(\alpha) = 0 \tag{7.29}$$
$$u(t) \le a(t) + b(t)h(r(t)), \qquad \alpha \le t < \beta_1,$$

where $r(t)$ is the maximal (minimal) solution of (7.29) *in* $[\alpha, \beta_1)$ *if condition* i) *(*ii)*) in theorem* 5.7 *holds.* □

Note that the estimates for the solutions of equations (7.25) and (7.29) are obtained as in the corresponding theorems 4.4 and 5.7.

An important particular case of theorem 5.7 which can be proved using theorem 7.1 is the following corollary

COROLLARY 7.2. *Let u, k be continuous functions in an interval $J = [\alpha, \beta]$, with k nonnegative. Let g be a monotone continuous function in an interval I such that $u(J) \subset I$, which does not vanish in I, except possibly at the endpoints of I. Let h be a monotone continuous function in an interval K such that $0 \in K$, and let a, $b > 0$ be constants such that $a + h(v) \in$* Int *I for $v \in K$, $|v| \le b$. If g and h are both nonincreasing or both nondecreasing, and*

$$u(t) \le a + h\left(\int_\alpha^t k(s)g(u(s))\,ds\right), \qquad t \in J,$$

then

$$u(t) \le a + h\left[F^{-1}\left(\int_\alpha^t k(s)\,ds\right)\right], \qquad \alpha \le t < \beta_1,$$

where $F(u) = \int_0^u dx/g(a+h(x))$, $u \in K$, *and* $\beta_1 = \sup\{t \in J\colon \int_\alpha^t k(s)\,ds \in F(K)\}$. □

8. Inequalities of the type of Deo and Dhongade

In this section we investigate, by Lakshmikantham's method, inequalities which resemble those of Deo and Dhongade in [68], [69]. They used the class $\mathcal{F}$ of positive, nondecreasing, continuous functions g in $\mathbf{R}_+$ satisfying the condition

$$\frac{1}{a}g(u) \le g\left(\frac{u}{a}\right), \qquad u \ge 0, \qquad a > 0. \tag{8.1}$$

Beesack [24] noted that these conditions are satisfied only by linear functions of the form $g(u) = g(1)u$, which essentially restricts the scope of applicability of the results of Deo and Dhongade. To extend the class $\mathcal{F}$, Beesack somewhat weakened (8.1), and gave the following definition.

DEFINITION 8.1. A function $g\colon \mathbf{R}_+ \to \mathbf{R}_+$ is said to belong to the *class* $\mathcal{F}$ if it satisfies the following conditions:

a) g is nondecreasing and continuous in $\mathbf{R}_+$, and $g(u) > 0$ for $u > 0$;
b)

$$\frac{1}{a}g(u) \le g\left(\frac{u}{a}\right), \qquad u \ge 0, \qquad a \ge 1. \qquad \square \tag{8.2}$$

Note that (8.2) is equivalent to

$$g(ku) \le kg(u), \qquad u \ge 0, \qquad k \ge 1. \tag{8.3}$$

Moreover, the functions in $\mathcal{F}$ have the following properties:

i) $g \in \mathcal{F}$ if and only if $g(u)/u$ is nonincreasing for $u > 0$;
ii) $g \in \mathcal{F}$ implies that g is subadditive;
iii) if g satisfies a) and is concave in $\mathbf{R}_+$, then $g \in \mathcal{F}$.

Assertion iii) shows that all nondecreasing, concave, continuous functions in $\mathbf{R}_+$ which are positive in $(0, \infty)$ belong to $\mathcal{F}$. However, these functions alone do not exhaust $\mathcal{F}$. E.g., the function

$$g(u) = \frac{u^2 + 2u}{u + 1}$$

belongs to $\mathcal{F}$ but is not concave.

We will need the following two lemmas.

LEMMA 8.1. *Let $f(u)$ be a strictly increasing continuous function for $u \ge u_0 \ge 0$. If $f(u)/u$ is nondecreasing for $u > u_0$, then*

$$a^{-1}f^{-1}(x) \le f^{-1}(xa^{-1}) \qquad \textit{for } x > af(u_0),\ a \ge 1. \tag{8.4}$$

PROOF. Put $u = f^{-1}(x)$, $v = f^{-1}(xa^{-1})$. We have $u_0 < v \le u$, so $v^{-1}f(v) \le u^{-1}f(u)$, which implies (8.4). $\square$

LEMMA 8.2. *Let $f(u)$ be a strictly increasing continuous function for $u \ge u_0 \ge 0$, and suppose that $f(u) \ge u > 0$ for $u > u_0$. If $y_0 > f(u_0)$ and*

$$F(u) = \int_{y_0}^{u} \frac{dy}{f^{-1}(y)}, \qquad u > f(u_0),$$

then

$$(F \circ f)^{-1}(x + F(a))) \le ae^x \qquad x \ge 0, \qquad a > f(u_0). \tag{8.5}$$

PROOF. Since $ae^x \ge a > f(u_0)$, $f(ae^x)$ is defined, and $f(ae^x) \ge ae^x > f(u_0)$ for $x \ge 0$. Hence $F \circ f(ae^x)$ is defined for $x \ge 0$, and (8.5) is equivalent to

$$F \circ f(ae^x) \ge x + F(a), \qquad x \ge 0, \qquad a > f(u_0).$$

Since F is increasing, it suffices to prove that

$$\phi(x) \equiv F(ae^x) - x - F(a) \ge 0, \qquad x \ge 0.$$

However, this follows from the fact that $\phi(0) = 0$ and $\phi'(x) = \{ae^x/f^{-1}(ae^x)\} - 1 \ge 0$, since $f^{-1}(ae^x) \le ae^x$. □

THEOREM 8.1. *Let u, a, b, k be nonnegative continuous functions in $J = [\alpha, \beta)$, with a nondecreasing in J. Let g, h be nondecreasing continuous functions in $\mathbf{R}_+$, with g positive, subadditive, and submultiplicative in $(0, \infty)$, $h(u) > 0$ for $u > 0$, and $a(0) + h(0) > 0$. Let f be a strictly increasing continuous function in $\mathbf{R}_+$, with $f(u) \ge u$ for $u \ge 0$, $f(0) = 0$. If*

$$f(u(t)) \le a(t) + h\left(\int_\alpha^t k(s)g(u(s))\,ds\right) + \int_\alpha^t b(s)u(s)\,ds, \qquad t \in J, \tag{8.6}$$

then

$$\begin{aligned} u(t) \le & (f^{-1} \circ F^{-1})\left\{\int_\alpha^t b(s)\,ds + \right. \\ & \left. + F\left[a(t) + (h \circ G^{-1})\left\{\int_\alpha^t k(s)g(E(s))\,ds + G\left(\int_\alpha^t k(s)g(a(s)E(s))\,ds\right)\right\}\right]\right\} \end{aligned} \tag{8.7}$$

for $\alpha \le t < \beta_1$, where $E(t) = \exp\left(\int_\alpha^t b(s)\,ds\right)$, $F(u) = \int_{y_0}^u dy/f^{-1}(y)$, $u > 0$ $(y_0 > 0)$, and $G(u) = \int_{u_0}^u dy/g(h(y))$, $u > 0$ $(u_0 > 0)$. Here,

$$\beta_1 = \sup\left\{t \in J :\, G\left(\int_\alpha^t k(s)g(a(s)E(s))\,ds\right) + \int_\alpha^t k(s)g(E(s))\,ds \in G(\mathbf{R}_+)\right\}.$$

PROOF. Let $a_1(t)$ denote the first two terms on the righthand side of (8.6). We apply theorem 7.1 with

$$H(t, v) = v, \qquad W(t, s, u) = b(s)u, \qquad I = K = \mathbf{R}_+,$$

noting that $f(I) = I$. The comparison equation (7.3) is

$$r' = b(t)f^{-1}[a_1(t) + r], \qquad r(\alpha) = 0.$$

If $r = r(t)$ denotes the maximal solution of this equation, then for t in the domain of r we have

$$u(t) \le f^{-1}[a_1(t) + r(t)]. \tag{8.8}$$

Since $a_1(t)$ is nondecreasing,

$$r' \le b(t)f^{-1}[a_1(T) + r(t)], \qquad \alpha \le t \le T < \beta.$$

Noting that $F(\mathbf{R}_+) \supset \mathbf{R}_+$, lemma 5.1 implies

$$a_1(T) + r(t) \le F^{-1}\left[F(a_1(T)) + \int_\alpha^t b(s)\,ds\right], \qquad \alpha \le t \le T < \beta.$$

Setting $t = T$ and changing notation, (8.8) yields

$$u(t) \le (f^{-1} \circ F^{-1})\left[F(a_1(t)) + \int_\alpha^t b(s)\,ds\right], \qquad \alpha \le t < \beta,$$

or

$$(F \circ f)(u(t)) \le \int_\alpha^t b(s)\,ds + F\left[a(t) + h\left(\int_\alpha^t k(s)g(u(s))\,ds\right)\right], \qquad t \in J. \quad (8.9)$$

Again we apply theorem 7.1, with $F \circ f$ instead of f,

$$H(t, v) = F[a(t) + h(v)], \qquad W(t, s, u) = k(s)g(u), \qquad I = K = \mathbf{R}_+.$$

(Note that $\int_\alpha^t b(s)\,ds + F[a(t) + h(v)] \in F(f(I)) = F(\mathbf{R}_+)$ for $t \in J$, $v \in K$.) The comparison equation (7.3) is:

$$r' = k(t)g \circ (F \circ f)^{-1}\left\{\int_\alpha^t b(s)\,ds + F[a(t) + h(r)]\right\}, \qquad r(\alpha) = 0. \quad (8.10)$$

For t in the domain of r we find

$$(F \circ f)(u(t)) \le \int_\alpha^t b(s)\,ds + F[a(t) + h(r(t))], \quad (8.11)$$

where $r = r(t)$ is the maximal solution of (8.10). By lemma 8.2, with $u_0 = 0$, $x = \int_\alpha^t b(s)\,ds \ge 0$, $a = a(t) + h(r) \ge a(0) + h(0) > 0$, we have

$$(F \circ f)^{-1}\left[\int_\alpha^t b(s)\,ds + F[a(t) + h(r)]\right] \le E(t)[a(t) + h(r)], \qquad t \in J.$$

Since g is subadditive and submultiplicative, (8.10) implies that

$$r' \le k(t)g(a(t)E(t)) + k(t)g(E(t))g \circ h(r).$$

By theorem 6.4, the initial value problem

$$\psi' = k(t)g(E(t))g \circ h(\psi) + k(t)g(a(t)E(t)), \qquad \psi(\alpha) = 0,$$

has a maximal solution on $[\alpha, \beta_1)$ satisfying

$$\psi(t) \le G^{-1}\left\{\int_\alpha^t k(s)g(E(s))\,ds + G\left(\int_\alpha^t k(s)g(a(s)E(s))\,ds\right)\right\}. \quad (8.12)$$

By lemma 6.3, $r(t) \le \psi(t)$ for $t \in [\alpha, \beta_1)$, so that (8.7) follows from (8.11) and (8.12). □

COROLLARY 8.1. *If $a(t) \equiv a$ and $a + h(0) > 0$, then the assumption that g be subadditive is not necessary. In this case,*

$$u(t) \le (f^{-1} \circ F^{-1}) \left\{ \int_\alpha^t b(s)\, ds + F\left[a + h \circ G_a^{-1} \left(\int_\alpha^t k(s) g(E(s))\, ds \right) \right] \right\} \quad (8.13)$$

for $\alpha \le t < \beta_2$, where $G_a(u) = \int_0^u dy/g[a + h(y)]$, $u \ge 0$, and

$$\beta_2 = \sup \left\{ t \in J : \int_\alpha^t k(s) g(E(s))\, ds \in G_a(\mathbf{R}_+) \right\}.$$

PROOF. The proof is similar to the one above (for nonconstant $a(t)$), but now the second comparison equation is:

$$\psi' = k(t) g(E(t)) g[a + h(\psi)], \qquad \psi(\alpha) = 0.$$

This equation has the unique solution

$$\psi(t) = G_a^{-1} \left(\int_\alpha^t k(s) g(E(s))\, ds \right), \qquad \alpha \le t < \beta_2,$$

and the estimate (8.13) follows from (8.11) and lemma 6.3. □

THEOREM 8.2. *Let u, a, b, k be nonnegative continuous functions in $J = [\alpha, \beta)$, with a nondecreasing and ≥ 1. Let g, g_1, h be nondecreasing continuous functions in $\mathbf{R}_+$, with g positive, subadditive, and submultiplicative in $(0, \infty)$, g_1 positive in $(0, \infty)$ and satisfying (8.2), and $h(u) > 0$ for $u > 0$. Let f be a strictly increasing continuous function in $\mathbf{R}_+$ with $f(u)/u$ nondecreasing for $u > 0$, $f(0) = 0$, and suppose that*

$$f(u(t)) \le a(t) + h\left(\int_\alpha^t k(s) g(u(s))\, ds \right) + \int_\alpha^t b(s) g_1(u(s))\, ds, \qquad t \in J. \quad (8.14)$$

Then

$$u(t) \le$$

$$\le A(t) \left\{ a(t) + h \circ G^{-1} \left[\int_\alpha^t k(s) g(A(s))\, ds + G\left(\int_\alpha^t k(s) g(a(s) A(s))\, ds \right) \right] \right\} \quad (8.15)$$

for $\alpha \le t < \beta_1$, where $A(t) = (f^{-1} \circ F_1^{-1})[F_1(1) + \int_\alpha^t b(s)\, ds]$, $G(u) = \int_{u_0}^u dy/g(h(y))$, $u > 0$, $F_1(u) = \int_{u_0}^u dy/g_1(f^{-1}(y))$, $u > 0$ ($u_0 > 0$), and $\beta_1 = \sup\{t \in J$: $\int_\alpha^t k(s) g(A(s))\, ds + G(\int_\alpha^t k(s) g(a(s) A(s))\, ds) \in G(\mathbf{R}_+)\}$.

PROOF. First we write (8.14) as

$$u(t) \le f^{-1} \left[a_1(t) + \int_\alpha^t b(s) g_1(u(s))\, ds \right], \qquad t \in J,$$

where $a_1(t) = a(t) + h(\int_\alpha^t k(s) g(u(s))\, ds)$. The conditions imposed on f imply that $f(\mathbf{R}_+) = \mathbf{R}_+$. Since $a_1 \ge 1$ and a_1 is nondecreasing, while (by lemma 8.1) g_1 and f^{-1} satisfy (8.2), it follows that

$$v(t) \equiv \frac{u(t)}{a_1(t)} \le f^{-1} \left\{ 1 + \int_\alpha^t b(s) g_1\left(\frac{u(s)}{a_1(s)} \right) ds \right\} = f^{-1} \left\{ 1 + \int_\alpha^t b(s) g_1(v(s))\, ds \right\},$$

or

$$f(v(t)) \le 1 + \int_\alpha^t b(s)g_1(v(s))\,ds, \qquad t \in J. \tag{8.16}$$

Applying to this inequality theorem 7.1, with

$$H(t,v) = v, \qquad W(t,s,u) = b(s)g_1(u), \qquad I = K = \mathbf{R}_+,$$

we obtain

$$v(t) \le f^{-1}[1 + r(t)], \tag{8.17}$$

where $r = r(t)$ is the maximal solution of

$$r' = b(t)g_1 \circ f^{-1}(1+r), \qquad r(\alpha) = 0.$$

This initial value problem has the unique solution

$$r(t) = F_1^{-1}\left[\int_\alpha^t b(s)\,ds + F_1(1)\right] - 1, \qquad t \in J.$$

(Note that $F_1(\mathbf{R}_+) \supset \mathbf{R}_+$, since $g_1(u) \le g_1(1)u$ and $f^{-1}(u) \le f^{-1}(1)u$ by (8.3).) By (8.17),

$$\frac{u(t)}{a_1(t)} \le f^{-1} \circ F_1^{-1}\left[\int_\alpha^t b(s)\,ds + F_1(1)\right] \equiv A(t),$$

or

$$u(t) \le A(t)a(t) + A(t)h\left(\int_\alpha^t k(s)g(u(s))\,ds\right), \qquad t \in J. \tag{8.18}$$

Again we apply theorem 7.1, with

$$f(u) = u, \qquad H(t,v) = A(t)h(v), \qquad W(t,s,u) = k(s)g(u), \qquad I = K = \mathbf{R}_+,$$

and obtain

$$u(t) \le A(t)a(t) + A(t)h(r(t)) \tag{8.19}$$

for t in the domain of r, where $r = r(t)$ is the maximal solution of

$$r' = k(t)g[A(t)a(t) + A(t)h(r)], \qquad r(\alpha) = 0. \tag{8.20}$$

Since g is subadditive and submultiplicative, and $k \ge 0$,

$$r' \le k(t)g(A(t))g \circ h(r) + k(t)g(A(t)a(t)).$$

Applying theorem 6.4 and lemma 6.3, as in the proof of theorem 8.1, we obtain

$$r(t) \le G^{-1}\left\{\int_\alpha^t k(s)g(A(s))\,ds + G\left(\int_\alpha^t k(s)g(a(s)A(s))\,ds\right)\right\}$$

for $\alpha \le t < \beta_1$. Hence (8.15) follows from (8.19). □

COROLLARY 8.2. *For $a(t) \equiv a \geq 1$ the proof is literally as above up to* (8.20). *Then, using that g is submultiplicative, we have*

$$u(t) \leq A(t)\left\{a + h \circ G_a^{-1}\left(\int_\alpha^t k(s)g(A(s))\,ds\right)\right\}, \qquad \alpha \leq t < \beta_2, \tag{8.21}$$

where $G_a(u) = \int_{u_0}^u dy/g[a+h(y)]$, $u \geq 0$, and

$$\beta_2 = \sup\left\{t \in J : \int_\alpha^t k(s)g(A(s))\,ds \in G_a(\mathbf{R}_+)\right\}.$$

PROOF. In this case (8.20) implies

$$r' \leq k(t)g(A(t))g[a + h(r)].$$

By lemma 5.1,

$$r(t) \leq G_a^{-1}\left(\int_\alpha^t k(s)g(A(s))\,ds\right), \qquad \alpha \leq t < \beta_2,$$

and (8.21) follows from this and (8.19) with $a(t) \equiv a$. □

It is possible to obtain a result similar to theorem 8.2 without requiring that $a \geq 1$ and that g_1 satisfies (8.2). To this end we proceed as in the proof of theorem 8.1, and use the following minor modification of lemma 8.2.

LEMMA 8.3. *Let f be a strictly increasing continuous function in $\mathbf{R}_+$, with $f(u) \geq u$ for $u > 0$, and let g_1 be a continuous function in $\mathbf{R}_+$, with $0 < g_1(u) \leq f(u)$ for $u > 0$. If $u_0 > f(0)$ and*

$$F_1(u) = \int_{u_0}^u \frac{dy}{g_1(f^{-1}(y))}, \qquad u > f(0),$$

then

$$(F_1 \circ f)^{-1}[x + F_1(a)] \leq ae^x \qquad \text{for } x \geq 0,\ a > f(0). \ \square$$

THEOREM 8.3. *Let u, b, k, g, h be functions satisfying the hypotheses of theorem 8.2. Let a be a nonnegative, nondecreasing, continuous function in $J = [\alpha, \beta)$, with $a(0) + h(0) > 0$, and let f and g_1 satisfy the hypotheses of lemma 8.3, with $f(0) = 0$. If*

$$f(u(t)) \leq a(t) + h\left(\int_\alpha^t k(s)g(u(s))\,ds\right) + \int_\alpha^t b(s)g_1(u(s))\,ds, \qquad t \in J, \tag{8.22}$$

then

$$u(t) \leq (F_1 \circ f)^{-1}\left(\int_\alpha^t b(s)\,ds + F_1\left\{a(t) + \right.\right.$$
$$\left.\left. + h \circ G^{-1}\left[\int_\alpha^t k(s)g(E(s))\,ds + G\left(\int_\alpha^t k(s)g(a(s)E(s))\,ds\right)\right]\right\}\right) \tag{8.23}$$

for $\alpha \leq t < \beta_1$, where E, G, β_1 are defined in theorem 8.1 and F_1 is defined in theorem 8.2 (or lemma 8.3).

If $a(t) \equiv a > 0$, the assumption that g be subadditive is superfluous, and

$$u(t) \leq (F_1 \circ f)^{-1} \left\{ \int_\alpha^t b(s)\, ds + F_1 \left[a + h \circ G_a^{-1} \left(\int_\alpha^t k(s) g(E(s))\, ds \right) \right] \right\} \tag{8.24}$$

for $\alpha \leq t < \beta_2$, where G_a and β_2 are defined in corollary 8.2.

SKETCH OF THE PROOF. The arguments are the same as in the proof of theorem 8.1, but the first comparison equation is:

$$r' = b(t) g_1 \circ f^{-1}[a_1(t) + r], \qquad r(\alpha) = 0,$$

and its maximal solution $r = r(t)$ satisfies

$$a_1(t) + r(t) \leq F_1^{-1} \left\{ \int_\alpha^t b(s)\, ds + F_1(a_1(t)) \right\}, \qquad t \in J.$$

Instead of (8.9), now we have

$$u(t) \leq (F_1 \circ f)^{-1} \left\{ \int_\alpha^t b(s)\, ds + F_1(a_1(t)) \right\}, \qquad t \in J. \tag{8.25}$$

The remainder of the proof is identical to the proof of theorem 8.1, with F replaced by F_1 and the use of lemma 8.3 instead of lemma 8.2. □

REMARK 8.1. Applying lemma 8.3 to (8.25) we obtain

$$u(t) \leq E(t) a_1(t) = E(t) a(t) + E(t) h \left(\int_\alpha^t k(s) g(u(s))\, ds \right), \qquad t \in J,$$

which is (8.18), with A replaced by E. As in the proof of theorem 8.2 we now obtain

$$u(t) \leq$$

$$E(t) \left\{ a(t) + h \circ G^{-1} \left[\int_\alpha^t k(s) g(E(s))\, ds + G \left(\int_\alpha^t k(s) g(a(s) E(s))\, ds \right) \right] \right\}. \tag{8.26}$$

Using lemma 8.3 it is now easy to verify that (8.23) is better than (8.26). □

REMARK 8.2. When $g_1(u) = u$, we have $F_1 = F$, so that theorem 8.3 includes theorem 8.1 as a particular case. □

REMARK 8.3. If f and g_1 satisfy the hypotheses of theorem 8.2 and theorem 8.3, then lemma 8.3 implies that $A(t) \leq E(t)$, and the estimates (8.15) and (8.23) in them are better than (8.26). In general, (8.15) and (8.23) are incomparable, but when $f(u) = u$, $g_1(u) = \lambda u$ with $0 < \lambda < 1$, (8.15) always gives a better bound than (8.23). In this case we have

$$F_1(u) = \frac{1}{\lambda} \log u, \qquad F_1^{-1}(z) = e^{\lambda z}, \qquad A(t) = e^{\lambda \int_\alpha^t b(s)\, ds} = [E(t)]^\lambda,$$

and (8.15) becomes

$$u(t) \leq [E(t)]^\lambda \left\{ a(t) + h \circ G^{-1} \left[\int_\alpha^t k(s) g(A(s))\, ds + G \left(\int_\alpha^t k(s) g(a(s) A(s))\, ds \right) \right] \right\}.$$

Also, (8.23) becomes

$$u(t) \le [E(t)]^\lambda \left\{a(t) + h \circ G^{-1}\left[\int_\alpha^t k(s)g(E(s))\,ds + G\left(\int_\alpha^t k(s)g(a(s)E(s))\,ds\right)\right]\right\}. \quad \square$$

REMARK 8.4. Applying theorem 7.1 to the inequality

$$u(t) \le a + \int_\alpha^t [k(s)g(u(s)) + b(s)u(s)]\,ds, \qquad t \in J, \tag{8.27}$$

(8.13) (with $f(u) = u$, $h(v) = v$, $a(t) = a > 0$, $W(t,s,u) = k(s)g(u) + b(s)u$) implies that

$$u(t) \le E(t)G^{-1}\left[G(a) + \int_\alpha^t k(s)g(E(s))\,ds\right], \qquad \alpha \le t < \beta_2, \tag{8.28}$$

where $G(u) = \int_{u_0}^u dy/g(y)$, $u > 0$ ($u_0 > 0$), $E(t) = \exp(\int_\alpha^t b(s)\,ds)$, and

$$\beta_2 = \sup\left\{t \in J:\ G(a) + \int_\alpha^t k(s)g(E(s))\,ds \in G(\mathbf{R}_+)\right\}.$$

By direct study of (8.27) we can sometimes obtain, under weaker assumptions, an estimate better than (8.28). $\square$

THEOREM 8.4. *Let u, b, k be nonnegative continuous functions in an interval $J = [\alpha, \beta)$, and let g be a nondecreasing continuous function in $\mathbf{R}_+$ with $g(u) > 0$ for $u > 0$. If u satisfies* (8.27) *for some $a > 0$, then*

$$u(t) \le G^{-1}\left[G(aE(t)) + E(t)\int_\alpha^t \frac{k(s)}{E(s)}\,ds\right], \qquad \alpha \le t < \bar\beta, \tag{8.29}$$

where $\bar\beta = \sup\{t \in J\colon G(aE(t)) + E(t)\int_\alpha^t \frac{k(s)}{E(s)}\,ds \in G(\mathbf{R}_+)\}$.

PROOF. In theorem 7.1 we take

$$f(u) = u, \qquad H(t,v) = v, \qquad W(t,s,u) = k(s)g(u) + b(s)u,$$
$$a(t) = a, \qquad K = \mathbf{R}_+.$$

The comparison equation is:

$$r' = k(t)g(a+r) + b(t)(a+r), \qquad r(\alpha) = 0. \tag{8.30}$$

Moreover, by (7.4),

$$u(t) \le a + r(t) \tag{8.31}$$

for t in the domain of the maximal solution $r = r(t)$ of (8.30). The function $z(t) = (a + r(t))/E(t)$ is the maximal solution of the initial value problem

$$z' = \frac{k(t)}{E(t)}g(E(t)z), \qquad z(\alpha) = a.$$

Hence, since g and E are monotone,

$$E(T)z'(t) \le \frac{E(T)k(t)}{E(t)}g(E(T)z(t)), \qquad z(\alpha) = a, \qquad \alpha \le t \le T.$$

By lemma 5.1,

$$E(T)z(t) \le G^{-1}\left[G(aE(T)) + \int_\alpha^t \frac{E(T)}{E(s)}k(s)\,ds\right], \qquad \alpha \le t \le T < \bar{\beta}.$$

This inequality (for $t = T$) and (8.31) imply (8.29). □

THEOREM 8.5. *Let u, a, b, k be nonnegative continuous functions in $J = [\alpha, \beta)$, with $a \ge 1$, $b \ge 1$, and a nondecreasing in J. Let g and h be nonnegative, nondecreasing, continuous functions in $\mathbf{R}_+$, with $g(u) > 0$ for $u > 0$. Suppose g, h both satisfy (8.2) in $\mathbf{R}_+$, and let g be submultiplicative. If*

$$u(t) \le a(t) + b(t)h\left(\int_\alpha^t k(s)g(u(s))\,ds\right), \qquad t \in J, \tag{8.32}$$

then

$$u(t) \le a(t) + a(t)b(t)h \circ G_1^{-1}\left(\int_\alpha^t k(s)g(b(s))\,ds\right), \qquad \alpha \le t < \beta_1, \tag{8.33}$$

where $G_1(u) = \int_0^u dy/g(1 + h(y))$, $u \ge 0$, and

$$\beta_1 = \sup\left\{t \in J : \int_\alpha^t k(s)g(b(s))\,ds \in G_1(\mathbf{R}_+)\right\}.$$

Moreover, if $b(t) \equiv 1$, the assumption that g be submultiplicative is superfluous, and

$$u(t) \le a(t)\left\{1 + h \circ G_1^{-1}\left(\int_\alpha^t k(s)\,ds\right)\right\}, \qquad \alpha \le t < \beta_2, \tag{8.34}$$

where $\beta_2 = \sup\{t \in J\colon \int_\alpha^t k(s)\,ds \in G_1(\mathbf{R}_+)\}$.

PROOF. The hypotheses on a, g, h, and (8.32) imply that

$$v(t) \equiv \frac{u(t)}{a(t)} \le 1 + b(t)h\left(\int_\alpha^t k(s)g(v(s))\,ds\right), \qquad t \in J.$$

In theorem 7.1 we take

$$f(v) = v, \qquad H(t,v) = b(t)h(v), \qquad W(t,s,u) = k(s)g(u), \qquad I = K = \mathbf{R}_+,$$

and obtain

$$v(t) \le 1 + b(t)h(r(t)) \tag{8.35}$$

for t in the domain of r, where $r = r(t)$ is the maximal solution of

$$r' = k(t)g[1 + b(t)h(r)], \qquad r(\alpha) = 0. \tag{8.36}$$

Since $b \ge 1$ and g is submultiplicative,

$$k(t)g[1 + b(t)h(r)] \le k(t)g[b(t) + b(t)h(r)] \le k(t)g(b(t))g[1 + h(r)],$$

and (8.36) implies

$$r' \le k(t)g(b(t))g[1 + h(r)], \qquad r(\alpha) = 0.$$

By lemma 5.1,

$$r(t) \le G_1^{-1}\left(\int_\alpha^t k(s)g(b(s))\,ds\right), \qquad \alpha \le t < \beta_1.$$

This and (8.35) implies (8.34).

If $b(t) \equiv 1$, the comparison equation (8.36) can be solved explicitly, and gives

$$r(t) = G_1^{-1}\left(\int_\alpha^t k(s)\,ds\right), \qquad \alpha \le t < \beta_2.$$

(8.34) now follows from (8.35). □

We can obtain other results by directly applying theorem 7.1 to (8.32).

THEOREM 8.6. *Let u, a, b, k be nonnegative continuous functions in $J = [\alpha, \beta)$. Let g and h be nonnegative, nondecreasing, continuous functions in $\mathbf{R}_+$, and suppose*

$$u(t) \le a(t) + b(t)h\left(\int_\alpha^t k(s)g(u(s))\,ds\right), \qquad t \in J.$$

If $a \ge 1$, $b \ge 1$, and g is submultiplicative, then

$$u(t) \le a(t) + b(t)h \circ G_1^{-1}\left(\int_\alpha^t k(s)g(a(s))g(b(s))\,ds\right), \qquad \alpha \le t < \bar\beta_1,$$

where $G_1(u) = \int_0^u dy/g(1+h(y))$, $u \ge 0$, and

$$\bar\beta_1 = \sup\left\{t \in J : \int_\alpha^t k(s)g(a(s))g(b(s))\,ds \in G_1(\mathbf{R}_+)\right\}.$$

If $b \ge 1$ and g is subadditive and satisfies (8.2), then

$$u(t) \le a(t) + b(t)h \circ G_1^{-1}\left[\int_\alpha^t k(s)b(s)\,ds + G\left(\int_\alpha^t k(s)b(s)g\left(\frac{a(s)}{b(s)}\right)\,ds\right)\right]$$

for $\alpha \le t < \bar\beta_2$, where $G(u) = \int_{u_0}^u dy/g(h(y))$, $u > 0$ ($u_0 > 0$), and

$$\bar\beta_2 = \sup\left\{t \in J : \int_\alpha^t k(s)b(s)\,ds + G\left(\int_\alpha^t k(s)b(s)g\left(\frac{a(s)}{b(s)}\right)\,ds\right) \in G(\mathbf{R}_+)\right\}. \quad \square$$

9. Applications

9.1. Two-sided estimates. Consider the differential equation

$$x' = f(t, x). \tag{9.1}$$

Suppose that the following condition holds:

H9.1) The function f is continuous in $\mathbf{R}_+ \times \mathbf{R}^n$ and satisfies the condition

$$|f(t,x)| \le \omega(t, |x|), \qquad t \in \mathbf{R}_+, \quad x \in \mathbf{R}^n, \tag{9.2}$$

where $\omega(t,u)$ is a nonnegative continuous function in $\mathbf{R}_+^2$ and is nondecreasing with respect to $u \in \mathbf{R}_+$ for fixed $t \in \mathbf{R}_+$.

Let $x(t)$ be a solution of (9.1) which is defined for $t \in J = (\alpha, \beta) \subset \mathbf{R}_+$. Then the relation

$$x(t) = x(t_0) + \int_{t_0}^{t} f(s, x(s))\, ds, \qquad t, t_0 \in J, \tag{9.3}$$

and (9.2) imply the estimate

$$|x(t)| \le |x(t_0)| + \int_{t_0}^{t} \omega(s, |x(s)|)\, ds, \qquad t_0, t \in J. \tag{9.4}$$

Under additional assumptions regarding ω, (9.4) implies various two-sided estimates for $|x(t)|$. For instance,

i) If $\omega(t, u) = b(t)u$ and $b(t)$ is a nonnegative continuous function in $\mathbf{R}_+$, then, by corollary 1.11,

$$|x(t_0)| \exp\left(-\int_{t_0}^{t} b(s)\, ds\right) \le |x(t)| \le |x(t_0)| \exp\left(\int_{t_0}^{t} b(s)\, ds\right) \tag{9.5}$$

for $t, t_0 \in J$, $t \ge t_0$.

ii) If $\omega(t, u) = k(t)g(u)$, where $k(t)$ is a nonnegative continuous function in $\mathbf{R}_+$ and $g(u)$ is a positive nondecreasing continuous function in $(0, \infty)$, then, by theorems 5.1 and 5.2,

$$G^{-1}\left[G(|x(t_0)|) - \int_{t_0}^{t} k(s)\, ds\right] \le |x(t)| \le G^{-1}\left[G(|x(t_0)|) + \int_{t_0}^{t} k(s)\, ds\right] \tag{9.6}$$

for $t, t_0 \in J$, $t_0 \le t < T$, where T is such that the quantities between $[\dots]$ belong to $\operatorname{Dom} G^{-1}$ and

$$G(u) = \int_{u_0}^{u} \frac{dz}{g(z)}, \qquad u > 0 \quad (u_0 > 0).$$

From estimates of the form (9.5) and (9.6) we can derive various properties of the solutions of (9.1), such as extendability of solutions, boundedness, convergence to a constant, stability and unstability, etc.

Finally we make some remarks concerning estimation from above of solutions of (9.1).

Let $x(t)$ be a solution of (9.1), with $x(t_0) = x_0$, defined for $t \in J_x = [t_0, t_x)$. Given the condition (9.2), the integral equation (9.3) implies

$$|x(t)| \le |x_0| + \int_{t_0}^{t} \omega(t, |x(s)|)\, ds, \qquad t \in J_x. \tag{9.7}$$

Consequently, by theorems 6.1 and 6.3,

$$|x(t)| \le r(t; t_0, u_0), \qquad t \in J_x \cap J_r, \tag{9.8}$$

where $u_0 \ge |x_0|$ and $r(t; t_0, u_0)$ is the maximal solution of the differential equation

$$r' = \omega(t, r), \qquad r(t_0) = u_0, \tag{9.9}$$

with domain of existence $t \in J_r = [t_0, t_r)$.

REMARK 9.1. Note that in theorem 6.1 the condition that ω be nondecreasing in u is essential. However, if instead of the integral equation corresponding to (9.7) we directly use the differential equation (9.1), we find for $u(t) = |x(t)|$,

$$u'_{+}(t) \leq |f(t, x(t))| \leq \omega(t, u(t)), \qquad t \geq t_0, \tag{9.10}$$

where $u'_{+}(t)$ is the right-derivative of $u(t)$. Theorem 6.3 and (9.10) imply the estimate (9.8), without the restriction that ω be nondecreasing in u. □

REMARK 9.2. Note also that condition (9.2) requires that ω be nonnegative, which implies that the solutions $r(t; t_0, u_0)$ of (9.9) are nondecreasing. This unpleasant restriction can be omitted, if we replace (9.2) by

$$[x, f(t, x)]_{+} \equiv \lim_{h \to 0} \frac{1}{h}[|x + hf(t, x)| - |x|] \leq \omega(t, |x|) \tag{9.11}$$

for $(t, x) \in \mathbf{R}_{+} \times \mathbf{R}^n$. This restriction implies the differential inequality

$$Du(t) \leq \omega(t, u(t)), \qquad t \geq t_0, \tag{9.12}$$

for a suitable Dini derivative $Du(t)$. Theorem 6.3 and (9.12) imply (9.8). But in (9.11) the function ω need not be nonnegative, and hence the upper bound obtained from (9.8) is much better.

If $f(t, x) = Ax$, with A an $(n \times n)$-matrix, then it is readily seen that $\omega(t, u) = \mu(A)u$, where $\mu(A)$ is the *logarithmic norm*, defined by

$$\mu(A) = \lim_{h \to 0} \frac{1}{h}[|I + hA| - 1].$$

The value of $\mu(A)$ depends on the concrete vector and matrix norms used . E.g., if $|x|$ is the Euclidean vector norm, then the matrix norm $|A|$ compatible with it is equal to the square root of the largest eigenvalue of A^*A, and $\mu(A)$ is the largest eigenvalue of $\frac{1}{2}(A^* + A)$. On the other hand, if $|x| = \sum_{i=1}^n |x_i|$, then $|A| = \sup_k \sum_{i=1}^n |a_{ik}|$, and $\mu(A) = \sup_k \{\operatorname{Re} a_{kk} + \sum_{i \neq k}^n |a_{ik}|\}$. □

9.2. Global existence. For solutions of the initial value problem

$$x' = f(t, x), \qquad x(t_0) = x_0, \tag{9.13}$$

we have the following theorems.

THEOREM 9.1. *Assume that the condition* H9.1) *holds, and that the maximal solution* $r(t; t_0, u_0)$ *of* (9.9) *exists for* $t \geq t_0$. *Then the largest right interval of existence of any solution* $x(t)$ *of* (9.13), *with* $|x_0| \leq u_0$, *is* $[t_0, \infty)$. *If, in addition,* $r(t; t_0, u_0)$ *is bounded in* $[t_0, \infty)$, *then* $\lim_{t \to \infty} x(t; t_0, u_0)$ *exists.*

PROOF. Let $x(t)$ be a solution of (9.13), with $|x_0| \leq u_0$, defined in $t_0 \leq t < \beta$ for $\beta < \infty$ maximal (i.e., β cannot be taken larger). By (9.9),

$$|x(t)| \leq r(t) \tag{9.14}$$

for $t_0 \le t < \beta$, where $r(t) = r(t; t_0, u_0)$ is the maximal solution of (9.9). For any t_1, t_2 such that $t_0 < t_1 < t_2 < \beta$ we see that (using (9.14) and the nondecrease of ω)

$$|x(t_2) - x(t_1)| \le \int_{t_1}^{t_2} \omega(s, |x(s)|)\, ds \le$$
$$\le \int_{t_1}^{t_2} \omega(s, r(s))\, ds = r(t_2) - r(t_1). \tag{9.15}$$

Since $\lim_{t\to\beta^-} r(t)$ exists and is finite, limit transition as $t_1, t_2 \to \beta^-$, Cauchy's convergence criterion, and (9.15) yield that $\lim_{t\to\beta^-} x(t)$ exists and is finite. Hence $x(t)$ can be continued beyond β, which contradicts our assumption. Therefore, every solution $x(t)$ of (9.13) such that $|x_0| \le u_0$ exists in $[t_0, \infty)$, and (9.14) and (9.15) hold with $\beta = \infty$.

Since $r(t)$ is bounded and nondecreasing in $[t_0, \infty)$, it follows that $\lim_{t\to\infty} r(t)$ exists, which implies, in view of (9.14) and (9.15), that $\lim_{t\to\infty} x(t)$ exists. □

Let $\omega(t, u) = k(t)g(u)$, and let condition (9.2) be replaced by

$$|f(t, x)| \le k(t)g(|x|), \qquad t \in \mathbf{R}_+, \quad x \in \mathbf{R}^n, \tag{9.16}$$

where $k(t)$ is a nonnegative continuous function in $\mathbf{R}_+$ and $g(u)$ is a positive nondecreasing continuous function in $(0, \infty)$.

In this case, for the solution of (9.9) we have

$$r(t) = G^{-1}\left[G(u_0) + \int_{t_0}^{t} k(s)\, ds\right], \qquad t \in [t_0, T), \tag{9.17}$$

where $G(u) = \int_{u_0}^{u} dz/g(z)$, $u > 0$ $(u_0 > 0)$.

Taking into account (9.6), (9.17), theorem 9.1, and remarks 5.5–5.7 about the interval of existence $[t_0, T)$ of $r(t)$, we can prove the following theorems [39].

THEOREM 9.2. (Cooke, 1955) *If condition* (9.16) *holds, then all solutions* $x(t)$ *of* (9.13) *with* $|x(0)| \le u_0$ *exist in the interval* $[0, T)$, *where* $\int_0^T k(s)\, ds = \int_{u_0}^{\infty} dz/g(z)$. *If* $\int_0^{\infty} k(s)\, ds \le \int_{u_0}^{\infty} dz/g(z)$, *these solutions exist in* $\mathbf{R}_+$, *while if* $\int_0^{\infty} k(s)\, ds < \int_{u_0}^{\infty} dz/g(z)$, *they are bounded in* $\mathbf{R}_+$.

THEOREM 9.3. (Wintner, 1946) *If condition* (9.16) *holds and* $\int_{u_0}^{\infty} dz/g(z) = \infty$, *then all solutions of* (9.13) *exist in* $\mathbf{R}_+$. *If, in addition,* $\int_0^{\infty} k(s)\, ds < \infty$, *they are bounded in* $\mathbf{R}_+$.

THEOREM 9.4. (Bihari, 1956) *If condition* (9.16) *holds, then all solutions* $x(t)$ *of* (9.13) *with* $|x(0)| \le u_0$ *satisfy* $|x(t)| \le G^{-1}[G(u_0) + \int_0^t k(s)\, ds]$ *for all* $t \ge 0$ *for which* $\int_0^t k(s)\, ds \le \int_{u_0}^{\infty} dz/g(z)$.

THEOREM 9.5. (Langenhop, 1960) *If condition* (9.16) *holds, then all solutions* $x(t)$ *of* (9.13) *with* $|x(0)| \ge u_0$ *satisfy* $|x(t)| \ge G^{-1}[G(u_0) - \int_0^t k(s)\, ds]$ *for all* $t \ge 0$ *for which* $\int_0^t k(s)\, ds \le \int_0^{u_0} dz/g(z)$.

9.3. Terminal value problem

THEOREM 9.6. *Suppose condition* H9.1) *holds. Assume that for each* $(t_0, u_0) \in \mathbf{R}_+^2$ *the maximal solution of* (9.9) *exists for* $t \geq t_0$ *and is bounded in* $[t_0, \infty)$. *Then the terminal value problem*

$$x' = f(t, x), \qquad x(\infty) = x_\infty, \tag{9.18}$$

has a solution for every $x_\infty \in \mathbf{R}^n$.

PROOF. First we note that for each $(t_0, \lambda) \in \mathbf{R}_+^2$,

$$\int_{t_0}^{\infty} \omega(s, \lambda)\, ds < \infty. \tag{9.19}$$

To see this, let $r(t) = r(t; t_0, \lambda)$ be the maximal solution of (9.9) passing through (t_0, λ). Since $\lim_{t\to\infty} r(t) = r_\infty$ exists, we have

$$r_\infty \geq r(t) = \lambda + \int_{t_0}^{t} \omega(s, r(s))\, ds \geq \lambda + \int_{t_0}^{t} \omega(s, \lambda)\, ds.$$

Now we consider the maximal solution $r(t; 0, |x_\infty|)$ of (9.9). Setting $R = \lim_{t\to\infty} r(t; 0, |x_\infty|)$, by (9.19) we can choose $t_0 \geq 0$ so large that

$$\int_{t_0}^{\infty} \omega(s, 2R)\, ds < R.$$

On the set Ω of bounded continuous functions $x\colon [t_0, \infty) \to \mathbf{R}^n$ with norm $\|x\| = \sup_{t\geq t_0} |x(t)| \leq 2R$ we define the operator

$$Sx(t) = x_\infty - \int_{t}^{\infty} f(s, x(s))\, ds.$$

The set Ω is closed, bounded, and convex. By the continuity of f and (9.19), S is easily shown to be continuous. Moreover, $S(\Omega) \subset \Omega$, since

$$|Sx(t)| \leq |x_\infty| + \int_{t}^{\infty} |f(s, x(s))|\, ds \leq R + \int_{t_0}^{\infty} \omega(s, 2R)\, ds < 2R. \tag{9.20}$$

This inequality also shows that the set $\mathcal{F} = \{Sx(t)\colon x \in \Omega\}$ is uniformly bounded. It is readily verified that $\mathcal{F}$ is equicontinuous on any interval $[t_0, T]$, and that $\lim_{t\to\infty} Sx(t) = x_\infty$ uniformly in $x \in \Omega$. Consequently $\mathcal{F}$ is compact, and Schauder's fixed point theorem implies that S has a fixed point $x \in \Omega$, i.e. there is a bounded continuous function $x(t)$ such that

$$x(t) = x_\infty - \int_{t}^{\infty} f(s, x(s))\, ds, \qquad t \in [t_0, \infty).$$

Direct verification shows that $x(t)$ is a solution of the terminal value problem (9.18). □

THEOREM 9.7. (Nguyen The Hoan, 1981) *Let $f(t,x)$ be a continuous function in $\mathbf{R}_+ \times \mathbf{R}^n$, and suppose*

$$|f(t,x)| \leq b(t)|x| + k(t)|x|^m, \qquad t \in \mathbf{R}_+, \quad x \in \mathbf{R}^n,$$

where $m > 1$ and $b(t)$, $k(t)$ are nonnegative continuous functions with

$$\int_0^\infty b(t)\,dt = b < \infty, \qquad \int_0^\infty k(t)\,dt = k < \infty.$$

Then

i) *for any $x \in \mathbf{R}^n$ such that*

$$|x_0| < r_0 = [(m-1)ke^{(m-1)b}]^{1/(1-m)},$$

all solutions of the initial value problem

$$x' = f(t,x), \qquad x(0) = x_0,$$

have a finite limit as $t \to \infty$.

ii) *For any $x_\infty \in \mathbf{R}^n$ the terminal value problem* (9.18) *has a solution.*

PROOF. Assertions i) and ii) are consequences of theorems 9.1 and 9.6, in which $\omega(t,u) = b(t)u + k(t)u^m$. To complete the proof it suffices to show that the unique solution $u(t)$ of the differential equation

$$u' = b(t)u + k(t)u^m, \qquad u(0) = u_0,$$

is bounded if $u_0 < r_0$. Indeed, since

$$u(t) = \left\{u_0^{1-m}e^{(1-m)\int_0^t b(\tau)\,d\tau} + (1-m)\int_0^t k(s)e^{\int_s^t (1-m)b(\tau)\,d\tau}\,ds\right\}^{1/(1-m)},$$

for $u_0 < r_0$ we have

$$u(t) \leq \left\{u_0^{1-m}e^{(1-m)b} + (1-m)k\right\}^{1/(1-m)} < \infty. \qquad \square$$

9.4. Stability by linear transformation. In the investigation of the stability of a given solution $x = \gamma(t)$ of equation (9.1), the following change of variables is often used:

$$x = \gamma(t) + y;$$

under it, (9.1) becomes

$$y' = A(t)y + P(t,y), \tag{9.21}$$

where $A(t) = f_x(t,\gamma(t))$, $P(t,y) = f(t,\gamma(t)+y) - f(t,\gamma(t)) - f_x(t,\gamma(t))y$. Thus, the problem of the stability of the solution $x = \gamma(t)$ of (9.1) is reduced to the problem of the stability of the solution $y \equiv 0$ of (9.21).

In many actual problems it is natural to assume that

$$|P(t,y)| \leq k(t)|y|^m, \qquad t \in \mathbf{R}_+, \quad y \in \mathbf{R}^n, \quad |y| \leq H, \tag{9.22}$$

$$|Y(t,s)| \leq \phi(t)\psi(s), \qquad 0 \leq s \leq t, \tag{9.23}$$

where $H > 0$ and $m > 1$ are constants, $k(t)$, $\phi(t)$, $\psi(t)$ are positive continuous functions in $\mathbf{R}_+$, and $Y(t,s)$ is the Cauchy matrix of the linear equation

$$y' = A(t)y. \tag{9.24}$$

We will investigate the stability of the solution $y \equiv 0$ of (9.21) under the conditions (9.22)–(9.23).

Let $y(t) = y(t; t_0, y_0)$ be a solution of (9.21) defined for $t \in J = [t_0, T)$. By the formula for variation of parameters, for $t \in J$ we have

$$y(t) = Y(t,t_0)y_0 + \int_{t_0}^{t} Y(t,s)P(s,y(s))\,ds, \tag{9.25}$$

whence

$$|y(t)| \le \phi(t)\psi(t_0)|y_0| + \int_{t_0}^{t} \phi(t)\psi(s)k(s)|y(s)|^m\,ds.$$

Consequently, the function $u(t) = |y(t)|/\phi(t)$ satisfies the inequality

$$u(t) \le \psi(t_0)|y_0| + \int_{t_0}^{t} \phi^m(s)\psi(s)k(s)u^m(s)\,ds.$$

By theorem 4.9 we obtain

$$|y(t)| \le \phi(t)\psi(t_0)|y_0|\{1 - (m-1)[\psi(t_0)|y_0|]^{m-1}D(t_0,t)\}^{-1/(m-1)} \tag{9.26}$$

for all $t \ge t_0$ such that

$$(m-1)[\psi(t_0)|y_0|]^{m-1}D(t_0,t) < 1,$$

where

$$D(t_0,t) = \int_{t_0}^{t} \phi^m(s)\psi(s)k(s)\,ds.$$

The estimate (9.26) readily implies

THEOREM 9.8. (Oziraner, 1977) *Let $A\colon \mathbf{R}_+ \to \mathbf{R}^{n\times n}$ and $P\colon \mathbf{R}_+ \times \mathbf{R}^n \to \mathbf{R}^n$ be continuous functions in their domains of definition, satisfying* (9.22)–(9.23). *Then the solution $y \equiv 0$ of* (9.21) *is:*

i) *stable if $D(0,\infty) < \infty$, and for any $t_0 \ge 0$ there is an $N = N(t_0) > 0$ such that $\phi(t) < N$ for $t \ge t_0$;*
ii) *uniformly stable if there is an $N > 0$ such that*

$$\phi(t)\psi(t_0) < N, \qquad \psi^{m-1}(t_0)D(t_0,\infty) < N \qquad \text{for } 0 \le t_0 \le t;$$

iii) *asymptotically stable if $D(0,\infty) < \infty$ and $\lim_{t\to\infty}\phi(t) = 0$;*
iv) *uniformly asymptotically stable if there is an $N > 0$ such that*

$$\phi(t)\psi(t_0) < N, \qquad \psi^{m-1}(t_0)D(t_0,\infty) < N \qquad \text{for } 0 \le t_0 \le t$$

and

$$\lim_{t\to\infty}\phi(t)\psi(t_0) = 0 \qquad \text{uniformly on } t_0 \ge 0. \qquad \square$$

We consider another approach to obtaining an estimate of solutions of (9.21). Let $A(t) = A_0(t) + B(t)$, i.e., (9.21) has the form

$$y' = A_0(t)y + B(t)y + P(t, y). \tag{9.27}$$

Such a representation may be useful if the equation

$$x' = A_0(t)x \tag{9.28}$$

is easier to investigate than equation (9.24).

Let $A_0, B\colon \mathbf{R}_+ \to \mathbf{R}^{n\times n}$ be continuous functions in $\mathbf{R}_+$, and let $X(t)$ be the fundamental matrix of (9.28) for which $X(0) = I$. By the change of variables $y = X(t)z$, (9.27) becomes

$$z' = X^{-1}(t)B(t)X(t)z + X^{-1}(t)P(t, X(t)z). \tag{9.29}$$

Suppose that for $t \in \mathbf{R}_+$ and $z \in \mathbf{R}^n$,

$$|X^{-1}(t)B(t)X(t)| \le b(t), \tag{9.30}$$

$$|X^{-1}(t)P(t, X(t)z)| \le w(t, |z|), \tag{9.31}$$

where $b(t)$ is a nonnegative continuous function in $\mathbf{R}_+$ and $w(t, u)$ is a nonnegative continuous function in $\mathbf{R}_+^2$ which is nondecreasing for $u \in \mathbf{R}_+$ for fixed $t \in \mathbf{R}_+$.

Since for the solution $z(t) = z(t; t_0, z_0)$ of (9.29) we have

$$z(t) = z_0 + \int_{t_0}^{t} [X^{-1}(s)B(s)X(s)z(s) + X^{-1}(s)P(s, X(s)z(s))]\, ds, \tag{9.32}$$

from (9.30) and (9.31) we obtain

$$|z(t)| \le |z_0| + \int_{t_0}^{t} [b(s)|z(s)| + w(s, |z(s)|)]\, ds.$$

Suppose further that

$$w(t, u) \le k(t) g\left(u e^{-\int_{t_0}^{t} b(s)\, ds}\right) e^{\int_{t_0}^{t} b(s)\, ds}$$

for $t \ge t_0$, $u \ge 0$, where $k(t)$ is a nonnegative continuous function for $t \ge 0$, and $g(v)$ is a nondecreasing continuous function for $v \ge 0$ which is positive for $v > 0$. By theorem 5.4,

$$|z(t)| \le G^{-1}\left[G(|z_0|) + \int_{t_0}^{t} k(s)\, ds\right] \exp\left(\int_{t_0}^{t} b(s)\, ds\right) \tag{9.33}$$

for $t \in [t_0, \beta)$, where $G(u) = \int_{u_0}^{u} dx/g(x)$, $u > 0$, $u_0 > 0$, and either $\beta = \infty$ or β is determined by the condition

$$\int_{t_0}^{\infty} k(s)\, ds = \int_{|z_0|}^{\infty} \frac{dx}{g(x)}.$$

If there is an $a_0 > 0$ such that

$$\int_{t_0}^{\infty} k(s)\, ds \le \int_{a_0}^{\infty} \frac{dx}{g(x)},$$

then (9.33) holds for $t \geq t_0$ if $0 < |z_0| < a_0$.

The estimate (9.33) and remarks 5.5–5.7 imply that there is a constant $M = M(t_0, |z_0|)$ such that

$$|z(t)| \leq M(t_0, |z_0|) \qquad \text{for } t \geq t_0,\ 0 < |z_0| < a_0, \tag{9.34}$$

provided one of the following conditions holds:

H9.2) $\int_{t_0}^{\infty} b(t)\,dt < \infty$, $\int^{\infty} \frac{dx}{g(x)} < \infty$, and there is an $a_0 > 0$ such that $\int_{t_0}^{\infty} k(t)\,dt = \int_{a_0}^{\infty} \frac{dx}{g(x)}$;

H9.3) $\int_{t_0}^{\infty} [b(t) + k(t)]\,dt < \infty$, $\int_0 \frac{dx}{g(x)} < \int^{\infty} \frac{dx}{g(x)} = \infty$;

H9.4) $\int_{t_0}^{\infty} [b(t) + k(t)]\,dt < \infty$, $\int^{\infty} \frac{dx}{g(x)} < \int_0 \frac{dx}{g(x)} = \infty$, and there is an $a_0 > 0$ such that $\int_{t_0}^{\infty} k(t)\,dt = \int_{a_0}^{\infty} \frac{dx}{g(x)}$;

H9.5) $\int_{t_0}^{\infty} [b(t) + k(t)]\,dt < \infty$, $\int_0 \frac{dx}{g(x)} = \int^{\infty} \frac{dx}{g(x)} = \infty$.

Moreover, in the cases H9.3) and H9.5) we can choose $a_0 = \infty$, and in the cases H9.4) and H9.5),

$$M(t_0, |z_0|) \to 0 \qquad \text{as } |z_0| \to 0 \text{ for fixed } t_0 \geq 0.$$

If (9.34) holds, the integral

$$\int_{t_0}^{\infty} [X^{-1}(s)B(s)X(s)z(s) + X^{-1}(s)P(s, X(s)z(s))]\,ds$$

is absolutely convergent. Then the integral equation (9.32) can be written as

$$z(t) = b + \int_{\infty}^{t} [X^{-1}(s)B(s)X(s)z(s) + X^{-1}(s)P(s, X(s)z(s))]\,ds, \tag{9.35}$$

where

$$b = z(t_0) + \int_{t_0}^{\infty} [X^{-1}(s)B(s)X(s)z(s) + X^{-1}(s)P(s, X(s)z(s))]\,ds.$$

By (9.35),

$$z(t) = b + o(1) \qquad \text{as } t \to \infty,$$

and since $y(t) = X(t)z(t)$, we have

$$y(t) = X(t)[b + o(1)] \qquad \text{as } t \to \infty. \tag{9.36}$$

Thus, (9.36) implies that if conditions hold such that the estimate (9.34) is valid, while all solutions of (9.28) are bounded (tend to zero as $t \to \infty$), then all solutions of (9.27) are bounded (tend to zero as $t \to \infty$).

9.5. Perturbed nonlinear equations. Consider the nonlinear equation

$$x' = f(t, x), \tag{9.37}$$

and a corresponding perturbed equation

$$y' = f(t, y) + P(t, y). \tag{9.38}$$

While between solutions of the linear equation (9.24) and the corresponding perturbed equation (9.21) there is a relation expressed by the formula for variation of parameters (9.25), the relation between solutions of the nonlinear equation (9.37) and the corresponding perturbed equation (9.38) is given by Alekseev's formula [11] (or some generalization of it, [119]). To formulate this relation, we need some facts about equations in variations.

Let $x(t; t_0, x_0)$ and $y(t; t_0, y_0)$ be the respective solutions of (9.37) and (9.38) for which $x(t_0; t_0, x_0) = x_0$ and $y(t_0; t_0, y_0) = y_0$, and let

$$\frac{dz}{dt} = f_x(t, x(t; t_0, x_0))z \tag{9.39}$$

be the *equation in variations* for (9.37). Let $\Phi(t, t_0, x_0)$ be the fundamental matrix of (9.39) for which $\Phi(t_0, t_0, x_0) = I$ (the unit matrix).

THEOREM 9.9. *Suppose that f has continuous partial derivatives in $\mathbf{R}_+ \times \mathbf{R}^n$. For any $(t_0, x_0) \in \mathbf{R}_+ \times \mathbf{R}^n$, let $x(t; t_0, x_0)$ be the unique solution of (9.37), with $x(t_0; t_0, x_0) = x_0$, defined for $t \geq t_0$. Then the partial derivatives $\frac{\partial x}{\partial x_0}(t; t_0, x_0)$ and $\frac{\partial x}{\partial t_0}(t; t_0, x_0)$ exist and satisfy the relations*

$$\frac{\partial x}{\partial x_0}(t; t_0, x_0) = \Phi(t, t_0, x_0), \qquad t \geq t_0,$$

$$\frac{\partial x}{\partial t_0}(t; t_0, x_0) = -\Phi(t, t_0, x_0)P(t_0, x_0), \qquad t \geq t_0. \qquad \square$$

Under these conditions *Alekseev's formula* is valid:

$$y(t; t_0, x_0) = x(t; t_0, x_0) + \int_{t_0}^{t} \Phi(t, s, y(s; t_0, x_0))P(s, y(s; t_0, x_0))\, ds, \qquad t \geq t_0. \tag{9.40}$$

Moreover, it is readily verified that for $t \geq t_0$,

$$x(t; t_0, y_0) = x(t; t_0, x_0) + \int_0^1 \Phi(t, t_0, sy_0 + (1 - s)x_0)\, ds \cdot (y_0 - x_0) \tag{9.41}$$

and

$$y(t; t_0, y_0) = x(t; t_0, x_0) + \int_0^1 \Phi(t, t_0, sy_0 + (1 - s)x_0)\, ds \cdot (y_0 - x_0) + \\ + \int_{t_0}^{t} \Phi(t, s, y(s; t_0, y_0))P(s, y(s; t_0, y_0))\, ds. \tag{9.42}$$

From formulas (9.40)–(9.42) we can derive conclusions about the behaviour of solutions of the perturbed equation (9.38), given the behaviour of solutions of the equation (9.37), the fundamental matrix Φ, and the perturbation P.

For instance, the following theorem holds.

THEOREM 9.10. *Suppose the conditions of theorem 9.9 hold, as well as:*

i) $f(t,0) = R(t,0) = 0$, $t \in \mathbf{R}_+$;

ii) $|\Phi(t,t_0,x_0)| \le Ke^{-\alpha(t-t_0)}$, $t \ge t_0 \ge 0$, $x_0 \in \mathbf{R}^n$, $|x_0| < H$;

iii) $|P(t,y)| \le \lambda(t)g(|y|)$, $t \in \mathbf{R}_+$, $y \in \mathbf{R}^n$, $|y| < H$, *where* $\alpha > 0$, $H > 0$, $K \ge 1$ *are constants,* $\lambda(t)$ *is a nonnegative continuous function in* $\mathbf{R}_+$, $g(u)$ *is a positive, nondecreasing, continuous function in* $(0,\infty)$ *with* $g(au) \le ag(u)$ *for* $u > 0$, $0 < a \le 1$, *and* $\int_0^\infty \lambda(s)\,ds < \infty$, $G(\infty) = \infty$, *where* $G(u) = \int_{u_0}^u dx/g(x)$, $u > 0$ *(*$u_0 > 0$*).*

Then the solution $y \equiv 0$ *of* (9.38) *is exponentially stable.*

PROOF. Let $t_0 \in \mathbf{R}_+$ and $x_0 \in \mathbf{R}^n$, $0 < |x_0| < \min(1, H, H/M) = \delta$, where

$$M = G^{-1}\left[G(K) + \int_0^\infty K\lambda(s)\,ds\right].$$

Then for the solution $x(t;t_0,x_0)$ of (9.37) it follows from (9.41) that

$$|x(t;t_0,x_0)| \le K|x_0|e^{-\alpha(t-t_0)}. \tag{9.43}$$

If the solution $y(t;t_0,x_0)$ of (9.38) is defined for $t \in [t_0,T)$, then (9.40) and (9.43) imply the estimate

$$|y(t;t_0,x_0)| \le K|x_0|e^{-\alpha(t-t_0)} + \int_{t_0}^t Ke^{-\alpha(t-s)}\lambda(s)g(|y(s;t_0,x_0)|)\,ds \tag{9.44}$$

for $t \in [t_0,T)$. Set $|y(t;t_0,x_0)| = u(t)|x_0|e^{-\alpha(t-t_0)}$. In view of the properties of g, (9.44) implies

$$u(t) \le K + \int_{t_0}^t K\lambda(s)g(u(s))\,ds,$$

so that by theorem 5.1,

$$u(t) \le G^{-1}\left[G(K) + \int_{t_0}^t K\lambda(s)\,ds\right], \qquad t \in [t_0,T).$$

Consequently, $u(t) \le M = G^{-1}[G(K) + \int_0^\infty K\lambda(s)\,ds]$ and

$$|y(t;t_0,x_0)| \le M|x_0|e^{-\alpha(t-t_0)} \tag{9.45}$$

for $t \in [t_0,T)$. Since $M|x_0| < M\delta \le H$, (9.45) implies that the solution can be continued up to every $t \ge t_0$, i.e. the estimate (9.45) holds for all $t \ge 0$, and the solution $y \equiv 0$ is exponentially stable. □

Chapter II: notes and comments

D. Willett [232] was the first to prove the case $k = 1$ of theorem 4.2; this case readily implies theorem 4.2 [234]. Moreover, the hypothesis of continuity has been replaced by that of Lebesgue integrability. Theorem 4.3 has been taken from [23]. Theorems 4.4–4.7 are also taken from [23], and include results of H.E. Gollwitzer [80]. In [213] there is a proof of Stachurska's theorem by induction, assuming n to be an integer. In theorem 4.8 this restriction has been dropped. In 1968, P. Maroni [124] also considered inequality (4.25), not requiring a/b to be nondecreasing, and obtained two estimates: one for $n = 2$ and one for $n = 3$, which are both more complicated than (4.26). Note that theorem 4.9 was proved as early as 1960 by Yue-Sheng Li (see the reference in [118]).

Corollary 5.2 and theorem 5.2 have been taken from C.E. Langenhop [115]. In fact, (5.8) implies (5.19) even if g is not monotone in I. This follows from a general result of W. Walter [228], which was attributed by him to E. Bayada [18] and F. Cafiero [44].

Theorems 5.3 and 5.4 were proved by R. Gutowski and B. Radziszewski in [84], [85]. Theorems 5.5–5.7 have been taken from [23]. Theorem 5.8 is new. In [2], inequality (5.31) with $f(u) = u$ is considered in another manner.

Theorem 6.1 has been proved in [113], Thm. 1.6.3. For similar results see also N.B. Azbelev and Z.B. Tsalyuk [16], L.F. Rakhmatulina [179], and T. Sato [194]. The proof of theorem 6.2 can be found in J. Šzarski [217], and V. Lakshmikantham and S. Leela [110]. Lemmas 6.1–6.3 have been taken from [23]. Theorem 6.3 is a special case of theorem 18.5. Theorem 6.4 was first given by V. Lakshmikantham in [109], and was made specific by P.R. Beesack in [23] (see also [24]). V.P. Rudakov [190] stated and proved theorem 6.4 in integral form.

Theorems 7.1–7.2 and corollary 7.1 have been taken from V. Lakshmikantham [109]. The proof of theorem 7.1 has been taken from [23]. J. Chandra and V. Lakshmikantham [49] considered inequalities of the form (7.12), but with a 'delay', that is, with $w(s, u(s))$ replaced by $w(s, u(\phi(s)))$, with $\phi(s) \leq s$. Note that the first appearance of comparison results is related to the 1919–1920 work of S.A. Čaplygin (see [50], [51]). For various comparison results, see also V.M. Alekseev [12], N.B. Babkin [17], N.N. Luzin [121], M. Rama Mohan Rao [180], R.M. Redheffer [182], J.S.W. Wong [236], A.D. Ziebur [250], B. Viswanatham [227], and T. Wazewski [229].

Theorem 8.1 includes U.D. Dhongade and S.G. Deo [68], Thm. 1 (for $f(u) = u$, $h(v) = v$, $a(t) = t$) and Thm. 2 (for $f(u) = u$). [68], Thm. 4 (for $f(u) = u$, $h(v) = v$, $a(t) = t$) and Thm. 5 (for $f(u) = u$) are included in theorem 8.2. Theorem 8.5 includes U.D. Dhongade and S.G. Deo [68], Thm. 3, and [69], Lemma 2. Theorems 8.3, 8.4, and 8.6 have been taken from [23].

The logarithmic norm $\mu(A)$ was introduced by S.M. Lozinskii (see [56], pp. 41, 58–59). For more details on $\mu(A)$ and the directional derivatives $[x, y]_{\pm}$ in arbitrary Banach spaces see V. Lakshmikantham and S. Leela [110]. Theorems 9.1 and 9.2 have been taken from [113].

For other applications, see also [11], [64], [73], [110], [113], [123], [125], [127], [129], [188], [215].

CHAPTER III

Analogues of Gronwall's inequality

In this chapter we consider various analogues of Gronwall-type inequalities. These are effective tools in the study of equations of various classes.

10. Integral inequalities involving several integrals

Integral inequalities involving several integrals of the same type are usually 'solved' by induction.

Before considering the first integral inequality of this type, we introduce some notation. For a given continuous function $a\colon [\alpha, \beta) \to \mathbf{R}$ we put, for $t \in [\alpha, \beta)$,

$$\hat{a}(t) = \max_{\alpha \le s \le t} a(s), \qquad \overline{a}(t) = \max(1, a(t)), \qquad \tilde{a}(t) = \max(1, \hat{a}(t)).$$

DEFINITION 10.1. A function $g\colon \mathbf{R}_+ \to \mathbf{R}_+$ is said to be of *class* $H_{r,w}$ if

i) $g(u)$ is nondecreasing and continuous for $u \ge 0$ and positive for $u > 0$;
ii) $g(au) \le r(a)w(u)$ for $a > 0$, $u \ge 0$, where $r(a)$ is a nonnegative continuous function in $\mathbf{R}_+$ and $w(u)$ is a nondecreasing continuous function in $\mathbf{R}_+$ which is positive for $u > 0$. □

REMARK 10.1. We note that

i) if $g \in H_{r,g}$, then g belongs to the class H_r introduced by Dannan [61];
ii) if $g \in H_{g,g}$, then g is submultiplicative;
iii) if $r_0(a) = \max(1, a)$ and $g \in H_{r_0,g}$, then $g \in \mathcal{F}$;
iv) if $g \in H_{r_0,w}$, then g belongs to the class $\mathcal{F} = \mathcal{F}_w$ introduced by Dhongade and Deo [68]. □

THEOREM 10.1. (Kong and Zhang, 1989) *Let u, a, b, k, b_i, q_i $(i = 1, \dots, n)$ be nonnegative continuous functions in $J = [\alpha, \beta)$, and let g and h be nonnegative, nondecreasing, continuous functions in $\mathbf{R}_+$. If $g \in \mathcal{F}$ and*

$$u(t) \le a(t) + \sum_{i=1}^{n} q_i(t) \int_\alpha^t b_i(s)u(s)\, ds + $$
$$+ b(t)h(t)\left(\int_\alpha^t k(s)g(u(s))\, ds\right), \qquad t \in J, \qquad (10.1)$$

then

$$u(t) \le A_n(a) + A_n(b) h \circ F^{-1} \left[\int_\alpha^t k\overline{A}_n(b)\, ds + \right.$$
$$\left. + F\left(\int_\alpha^t k\overline{A}_n(b) g\left(\frac{A_n(a)}{A_n(b)} \right) ds \right) \right] \tag{10.2}$$

for $t \in [\alpha, \beta_1)$, *where* $F(u) = \int_{u_0}^u dx/g(h(x))$, $u > 0$ $(u_0 > 0)$,

$$A_0(y) = y,$$
$$A_{k+1}(y) = A_k(y) + A_k(q_{k+1}) \int_\alpha^t b_{k+1} A_k(y) \exp\left(\int_s^t b_{k+1} A_k(q_{k+1})\, d\tau \right) ds,$$
$$\beta_1 = \sup\left\{ t \in J : \int_\alpha^t k\overline{A}_n(b)\, ds + F\left(\int_\alpha^t k\overline{A}_n(b) g\left(\frac{A_n(a)}{\overline{A}_n(b)} \right) ds \right) \in F(\mathbf{R}_+) \right\}.$$

PROOF. Set $v(t) = h(\int_\alpha^t k(s) g(u(s))\, ds)$. Then

$$u(t) \le [a + bv] + \sum_{i=1}^n q_i \int_\alpha^t b_i u\, ds,$$

and by theorem 1.6,

$$u(t) \le A_n(a + bv).$$

Since $v(t)$ is nondecreasing in $t \in J$, lemma 1.2 implies

$$u(t) \le A_n(a) + A_n(b) h\left(\int_\alpha^t k(s) g(u(s))\, ds \right),$$

whence

$$\frac{u(t)}{A_n(b)} \le \frac{A_n(a)}{A_n(b)} + h\left(\int_\alpha^t k(s)\overline{A}_n(b) g\left(\frac{u(s)}{A_n(b)} \right) ds \right).$$

Now theorem 5.6 implies (10.2). □

In the proofs of our subsequent results we will need the following lemma.

LEMMA 10.1. *Let $k(t,s)$ be a nonnegative continuous function for $\alpha \le s \le t \le \beta$ which is nondecreasing in t for any fixed s. Let $g(u)$ be a nondecreasing continuous function in $\mathbf{R}_+$, with $g(u) > 0$ for $u > 0$. Let $u(t)$ and $a(t)$ be nonnegative continuous functions in $J = [\alpha, \beta]$, and suppose*

$$u(t) \le a(t) + \int_\alpha^t k(t,s) g(u(s))\, ds, \qquad t \in J. \tag{10.3}$$

Now,

i) *if $G(u) = \int_{u_0}^u dx/g(x)$, $u > 0$ $(u_0 > 0)$, then*

$$u(t) \le G^{-1}\left[G(\hat{a}(t)) + \int_\alpha^t k(t,s)\, ds \right], \qquad \alpha \le t < \beta_1. \tag{10.4}$$

ii) *if* $a(t)$ *is positive in* J, $g \in H_{r,w}$, *and* $W(u) = \int_{u_0}^{u} dx/w(x)$, $u > 0$ ($u_0 > 0$), *then*

$$u(t) \le \hat{a}(t) W^{-1}\left[W(1) + \frac{r(\hat{a}(t))}{\hat{a}(t)} \int_\alpha^t k(t,s)\, ds\right], \qquad \alpha \le t < \beta_2. \quad (10.5)$$

Moreover, in this case

$$u(t) \le \hat{a}(t) W^{-1}\left[W(1) + \int_\alpha^t k(t,s) \frac{r(\hat{a}(s))}{\hat{a}(s)}\, ds\right], \qquad \alpha \le t < \beta_3. \quad (10.6)$$

In (10.4)–(10.6) *the numbers* β_i, $i = 1, 2, 3$, *are chosen so that the quantity between brackets* $[\cdots]$ *is in the range of* G *(or* W*) for* $\alpha \le t < \beta_i$.

PROOF. i). Since $\hat{a}(t)$ and $k(t,s)$ are nondecreasing in t, (10.3) implies

$$u(t) \le \hat{a}(T) + \int_\alpha^t k(T,s) g(u(s))\, ds, \qquad \alpha \le t \le T < \beta. \quad (10.7)$$

By theorem 5.1,

$$u(t) \le G^{-1}\left[G(\hat{a}(T)) + \int_\alpha^t k(T,s)\, ds\right], \qquad \alpha \le t \le T < \beta_1.$$

Setting $t = T$ and changing notation, (10.4) is obtained.

ii). If $g \in H_{r,w}$, (10.7) implies that

$$\frac{u(t)}{\hat{a}(T)} \le 1 + \int_\alpha^t k(T,s) \frac{r(\hat{a}(T))}{\hat{a}(T)} w\left(\frac{u(s)}{\hat{a}(T)}\right) ds.$$

By theorem 5.1,

$$\frac{u(t)}{\hat{a}(T)} \le W^{-1}\left[W(1) + \int_\alpha^t k(T,s) \frac{r(\hat{a}(T))}{\hat{a}(T)}\, ds\right], \qquad \alpha \le t \le T < \beta_2,$$

which implies (10.5) by setting $t = T$.

In the same situation (10.3) implies

$$\begin{aligned}\frac{u(t)}{\hat{a}(t)} &\le 1 + \int_\alpha^t \frac{k(t,s)}{\hat{a}(s)} g\left(\hat{a}(s)\frac{u(s)}{\hat{a}(s)}\right) ds \le \\ &\le 1 + \int_\alpha^t k(t,s) \frac{r(\hat{a}(s))}{\hat{a}(s)} w\left(\frac{u(s)}{\hat{a}(s)}\right) ds.\end{aligned}$$

Applying i) we arrive at (10.6). □

REMARK 10.2. If the function $g(u)/u$ is nondecreasing, then (10.6) is better than (10.5). If $g(u)/u$ is nonincreasing, then (10.5) is better than (10.6). □

REMARK 10.3. Since $\hat{a}(t) \le \tilde{a}(t)$, $t \in J$, in case ii) we can replace $\hat{a}$ by $\tilde{a}$ in the proof, and hence the estimates (10.5) and (10.6) hold if $\hat{a}$ is replaced by $\tilde{a}$. □

COROLLARY 10.1. i) *If* $g \in \mathcal{F}$, *i.e.* $g(au) \le a g(u)$, $a \ge 1$, $u \ge 0$, *then*

$$u(t) \le \tilde{a}(t) G^{-1}\left[G(1) + \int_\alpha^t k(t,s)\, ds\right], \qquad \alpha \le t < \beta_4. \quad (10.8)$$

ii) *If g is submultiplicative, i.e. $g(au) \le g(a)g(u)$, $a \ge 0$, $u \ge 0$, and $a(t) > 0$ for $t \ge 0$, then*

$$u(t) \le \hat{a}(t)G^{-1}\left[G(1) + \frac{g(\hat{a}(t))}{\hat{a}(t)}\int_\alpha^t k(t,s)\,ds\right], \qquad \alpha \le t < \beta_5. \quad (10.9)$$

Moreover,

$$u(t) \le \hat{a}(t)G^{-1}\left[G(1) + \int_\alpha^t k(t,s)\frac{g(\hat{a}(s))}{\hat{a}(s)}\,ds\right], \qquad \alpha \le t < \beta_6. \quad (10.10)$$

iii) *If $g \in \mathcal{F}_w$, i.e. $g(au) \le aw(u)$, $a \ge 1$, $u \ge 0$, then*

$$u(t) \le \tilde{a}(t)W^{-1}\left[W(1) + \int_\alpha^t k(t,s)\,ds\right], \qquad \alpha \le t < \beta_7. \quad (10.11)$$

In (10.8)–(10.11) the numbers β_i, $i = 4,5,6,7$, are chosen so that the quantity between brackets $[\cdots]$ is in the range of G (or W) for $\alpha \le t < \beta_i$.

Indeed, (10.9) and (10.10) follow from remarks 10.1 and 10.3 and the fact that $r_0(\tilde{a}(t)) \equiv \tilde{a}(t)$. □

Now we will consider inequalities involving several nonlinear integral terms.

THEOREM 10.2. *Suppose the following conditions hold:*

1. *$a(t)$ is a positive continuous function in $J = [\alpha,\beta)$;*
2. *$k_j(t,s)$, $j = 1,\dots,n$, are nonnegative continuous functions for $\alpha \le s \le t < \beta$ which are nondecreasing in t for any fixed s;*
3. *$g_j(u)$, $j = 1,\dots,n$, are nondecreasing continuous functions in $\mathbf{R}_+$, with $g_j(u) > 0$ for $u > 0$;*
4. *$u(t)$ is a nonnegative continuous function in J.*

Suppose also that $g_j \in H_{r_j,w_j}$, $j = 1,\dots,n$, and

$$u(t) \le a(t) + \sum_{j=1}^n \int_\alpha^t k_j(t,s)g_j(u(s))\,ds, \qquad t \in J. \quad (10.12)$$

Then

$$u(t) \le \hat{a}(t)\psi_n(t), \qquad \alpha \le t < \beta_n', \quad (10.13)$$

where $\psi_0(t) \equiv 1$,

$$\psi_j(t) = \psi_{j-1}(t)W_j^{-1}\left[W_j(1) + \frac{r_j(\hat{a}(t)\psi_{j-1}(t))}{\hat{a}(t)}\int_\alpha^t k_j(t,s)\,ds\right],$$

$$W_j(u) = \int_{u_j}^u \frac{dx}{w_j(x)}, \qquad u > 0 \quad (u_j > 0), \qquad j = 1,\dots,n.$$

Moreover, in this case

$$u(t) \le \hat{a}(t)\phi_n(t), \qquad \alpha \le t < \beta_n'', \quad (10.14)$$

where $\phi_0(t) \equiv 1$,

$$\phi_j(t) = \phi_{j-1}(t) W_j^{-1} \left[W_j(1) + \int_\alpha^t k_j(t,s) \frac{r(\hat{a}(s)\phi_{j-1}(s))}{\hat{a}(s)}\, ds \right], \qquad j = 1, \dots, n.$$

Here the numbers β_n' (β_n'') *are so chosen that the functions* ψ_j (ϕ_j), $j = 1, \dots, n$, *are defined for* $\alpha \le t < \beta_n'$ (β_n'').

PROOF. The proof proceeds by induction with respect to n. For $n = 1$ we have

$$u(t) \le a(t) + \int_\alpha^t k_1(t,s) g_1(u(s))\, ds,$$

and lemma 10.1 and the estimate (10.5) imply

$$u(t) \le \hat{a}(t) W_1^{-1} \left[W_1(1) + \frac{r_1(\hat{a}(t))}{\hat{a}(t)} \int_\alpha^t k_1(t,s)\, ds \right] \equiv \hat{a}(t)\psi_1(t).$$

Suppose (10.13) holds for $n = m - 1$. Then for $n = m$ (10.12) takes the form

$$u(t) \le \left[a(t) + \int_\alpha^t k_m(t,s) g_m(u(s))\, ds \right] + \sum_{i=1}^{m-1} \int_\alpha^t k_j(t,s) g_j(u(s))\, ds.$$

It follows from the induction assumption that

$$u(t) \le \left[\hat{a}(t) + \int_\alpha^t k_m(t,s) g_m(u(s))\, ds \right] \psi_{m-1}(t),$$

or

$$u(t) \le \hat{a}(t)\psi_{m-1}(t) + \int_\alpha^t k_m(t,s) \psi_{m-1}(t) g_m(u(s))\, ds.$$

Since $\hat{a}\psi_{m-1}$ and $k_m \psi_{m-1}$ are nondecreasing in t, lemma 10.1 implies

$$u(t) \le \hat{a}(t)\psi_{m-1}(t) W_m^{-1} \left[W_m(1) + \frac{r_m(\hat{a}(t)\psi_{m-1}(t))}{\hat{a}(t)\psi_{m-1}(t)} \int_\alpha^t k_m(t,s)\psi_{m-1}(t)\, ds \right],$$

i.e. $u(t) \le \hat{a}(t)\psi_m(t)$, and (10.13) holds for all $n = 1, 2, \dots$.

The proof of (10.14) is similar, applying inequality (10.6) from lemma 10.1. □

COROLLARY 10.2. *If* $g_j \in \mathcal{F}$, $j = 1, \dots, n$, *and* $G_j(u) = \int_{u_j}^u dx/g_j(x)$, $u > 0$ $(u_j > 0)$, *then*

$$u(t) \le \tilde{a}(t) b_n(t), \tag{10.15}$$

where $b_0(t) \equiv 1$,

$$b_j(t) = b_{j-1}(t) G_j^{-1} \left[G_j(1) + b_{j-1}(t) \int_\alpha^t k_j(t,s)\, ds \right], \qquad j = 1, \dots, n.$$

Moreover,

$$u(t) \le \tilde{a}(t) P_n(t), \tag{10.16}$$

where $P_0(t) \equiv 1$,

$$P_j(t) = P_{j-1}(t) G_j^{-1} \left[G_j(1) + \int_\alpha^t k_j(t,s) P_{j-1}(s)\, ds \right], \qquad j = 1, \dots, n. \quad \square$$

COROLLARY 10.3. *If g_j, $j = 1, \dots, n$, are submultiplicative, and $G_j(u) = \int_{u_j}^u dx/g_j(x)$, $u > 0$ $(u_0 > 0)$, then*

$$u(t) \le \hat{a}(t) Q_n(t), \tag{10.17}$$

where $Q_0(t) \equiv 1$,

$$Q_j(t) = Q_{j-1}(t) G_j^{-1}\left[G_j(1) + \frac{g_j(\hat{a}(t)Q_{j-1}(t))}{\hat{a}(t)} \int_\alpha^t k_j(t,s)\,ds\right], \qquad j = 1, \dots, n.$$

Moreover,

$$u(t) \le \hat{a}(t) R_n(t), \tag{10.18}$$

where $R_0(t) \equiv 1$,

$$R_j(t) = R_{j-1}(t) G_j^{-1}\left[G_j(1) + \int_\alpha^t k_j(t,s) \frac{g_j(\hat{a}(s)R_{j-1}(s))}{\hat{a}(s)}\,ds\right], \qquad j = 1, \dots, n. \qquad \square$$

COROLLARY 10.4. *If g_j, $j = 1, \dots, n$, are submultiplicative, and there are constants $\lambda_j > 0$, $j = 1, \dots, n$, such that $g_j(u) \le \lambda_j u$, $u \ge 0$, then*

$$u(t) \le \hat{a}(t) B_n(t), \tag{10.19}$$

where $B_0(t) \equiv 1$,

$$B_j(t) = B_{j-1}(t) G_j^{-1}\left[G_j(1) + \lambda_j B_{j-1}(t) \int_\alpha^t k_j(t,s)\,ds\right], \qquad j = 1, \dots, n.$$

Moreover,

$$u(t) \le \hat{a}(t) \Pi_n(t), \tag{10.20}$$

where $\Pi_0(t) \equiv 1$,

$$\Pi_j(t) = \Pi_{j-1}(t) G_j^{-1}\left[G_j(1) + \int_\alpha^t \lambda_j k_j(t,s) \Pi_{j-1}(s)\,ds\right], \qquad j = 1, \dots, n. \qquad \square$$

REMARK 10.4. The estimates (10.15) and (10.19) were proved by Beesack, [27], Thm. 1. It is easily seen that the estimates (10.16) and (10.20) are better than (10.15) and (10.19), respectively. $\square$

COROLLARY 10.5. *If g_j, $j = 1, \dots, n$, are such that $g_j(au) \le a w_j(u)$, $a \ge 1$, $u \ge 0$, and $W_j(u) = \int_{u_j}^u dx/w_j(x)$, $u > 0$ $(u_0 > 0)$, then*

$$u(t) \le \tilde{a}(t) T_n(t), \tag{10.21}$$

where $T_0(t) \equiv 1$,

$$T_j(t) = T_{j-1}(t) W_j^{-1}\left[W_j(1) + \int_\alpha^t k_j(t,s) T_{j-1}(s)\,ds\right], \qquad j = 1, \dots, n. \qquad \square$$

Let $A \subset \mathbf{R}$, and let $g_1, g_2 \colon A \to \mathbf{R} \setminus \{0\}$. We put $g_1 \propto g_2$ if g_2/g_1 is nondecreasing in A.

THEOREM 10.3. *Suppose the conditions 1–4 of theorem 10.2 hold. If* $g_1 \propto g_2 \propto \cdots \propto g_n$ *in* $(0, \infty)$, *then the inequality*

$$u(t) \le a(t) + \sum_{j=1}^{n} \int_{\alpha}^{t} k_j(t,s) g_j(u(s))\, ds, \qquad t \in J, \tag{10.22}$$

implies

$$u(t) \le c_n(t), \qquad \alpha \le t < \beta_0, \tag{10.23}$$

where $c_0(t) \equiv \hat{a}(t)$,

$$c_j(t) = G_j^{-1}\left[G_j(c_{j-1}(t)) + \int_{\alpha}^{t} k_j(t,s)\, ds\right], \qquad j = 1, \ldots, n,$$

$$G_j(u) = \int_{u_j}^{u} \frac{dx}{g_j(x)}, \qquad u > 0 \quad (u_0 > 0),$$

and β_0 *is so chosen that the functions* $c_j(t)$, $j = 1, \ldots, n$, *are defined for* $\alpha \le t < \beta_0$.

PROOF. The proof proceeds by induction with respect to n. For $n = 1$ (10.23) holds by lemma 10.1i). Suppose the assertion of the theorem is true for $n = m - 1$. For $n = m$ and a fixed $T \in (\alpha, \beta)$, (10.22) implies

$$u(t) \le \hat{a}(T) + \sum_{j=1}^{m} \int_{\alpha}^{t} k_j(T,s) g_j(u(s))\, ds \equiv v(t), \qquad \alpha \le t \le T < \beta.$$

We now successively obtain $v(\alpha) = \hat{a}(T)$,

$$v'(t) \le k_1(T,t) g_1(v(t)) + \sum_{j=2}^{m} k_j(T,t) g_j(v(t)), \qquad \alpha \le t \le T < \beta,$$

$$\frac{v'(t)}{g_1(v(t))} \le k_1(T,t) + \sum_{j=2}^{m} k_j(T,t) \frac{g_j(v(t))}{g_1(v(t))}, \qquad \alpha \le t \le T < \beta.$$

Integration from α to t gives

$$G_1(v(t)) \le G_1(\hat{a}(T)) + \int_{\alpha}^{t} k_1(T,s)\, ds + \sum_{j=2}^{m} \int_{\alpha}^{t} k_j(T,s) \frac{g_j(v(s))}{g_1(v(s))}\, ds, \qquad \alpha \le t \le T < \beta_0.$$

Setting $t = T$ and changing notation, we arrive at

$$G_1(v(t)) \le G_1(\hat{a}(t)) + \int_{\alpha}^{t} k_1(t,s)\, ds + \sum_{j=2}^{m} \int_{\alpha}^{t} k_j(t,s) \frac{g_j(v(s))}{g_1(v(s))}\, ds, \qquad \alpha \le t < \beta_0.$$

For $z(t) = G_1(v(t))$ we find

$$z(t) \le G_1(\hat{a}(t)) + \int_{\alpha}^{t} k_1(t,s)\, ds + \sum_{j=2}^{m} \int_{\alpha}^{t} k_j(t,s) \frac{g_j(G_1^{-1}(z(s)))}{g_1(G_1^{-1}(z(s)))}\, ds.$$

By the induction hypothesis,

$$z(t) \le \tilde{c}_{m-1}(t), \tag{10.24}$$

where

$$\tilde{c}_0(t) = G_1(\hat{a}(t)) + \int_\alpha^t k_1(t,s)\,ds \equiv G_1(c_1(t)), \tag{10.25}$$

$$\tilde{c}_j(t) = \tilde{G}_j^{-1}\left[\tilde{G}_j(\tilde{c}_{j-1}(t)) + \int_\alpha^t k_{j+1}(t,s)\,ds\right], \tag{10.26}$$

$$\tilde{G}_j(u) = \int_{u_j^*}^u \frac{g_1(G_1^{-1}(y))}{g_{j+1}(G_1^{-1}(y))}\,dy, \qquad u > 0 \quad (u_j^* > 0), \qquad j = 1,\dots,m-1.$$

But $\tilde{G}_j(u) = G_{j+1} \circ G_1^{-1}(u)$, which is why (10.24)–(10.26) and the induction hypothesis imply $G_1^{-1}(\tilde{c}_j(t)) = c_{j+1}(t)$, $j = 0,\dots,m-1$. Now

$$u(t) \le v(t) = G_1^{-1}(z(t)) \le G_1^{-1}(\tilde{c}_{m-1}(t)) = c_m(t),$$

which proves theorem 10.3 for $n = m$. □

THEOREM 10.4. *Suppose the conditions 1–4 of theorem 10.2 hold. Let $g_j \in H_{r_j,w_j}$, $j = 1,\dots,n$, and $w_1 \propto w_2 \propto \cdots \propto w_n$. Then the inequality*

$$u(t) \le a(t) + \sum_{j=1}^n \int_\alpha^t k_j(t,s) g_j(u(s))\,ds, \qquad t \in J, \tag{10.27}$$

implies

$$u(t) \le \hat{a}(t)\lambda_n(t), \qquad \alpha \le t < \gamma_1, \tag{10.28}$$

where $\lambda_0(t) \equiv 1$,

$$\lambda_j(t) = W_j^{-1}\left[W_j(\lambda_{j-1}(t)) + \frac{r_j(\hat{a}(t))}{\hat{a}(t)} \int_\alpha^t k_j(t,s)\,ds\right], \qquad j = 1,\dots,n.$$

Moreover,

$$u(t) \le \hat{a}(t)\mu_n(t), \qquad \alpha \le t < \gamma_2, \tag{10.29}$$

where $\mu_0(t) \equiv 1$,

$$\mu_j(t) = W_j^{-1}\left[W_j(\mu_{j-1}(t)) + \int_\alpha^t k_j(t,s)\frac{r_j(\hat{a}(s))}{\hat{a}(s)}\,ds\right], \qquad j = 1,\dots,n,$$

$$W_j(u) = \int_{u_j}^u \frac{dx}{w_j(x)}, \qquad u > 0 \quad (u_0 > 0),$$

and the numbers γ_1 and γ_2 are so chosen that the functions λ_j (μ_j), $j = 1,\dots,n$, are defined for $\alpha \le t < \gamma_1$ (γ_2).

PROOF. (10.27) implies the inequalities

$$\frac{u(t)}{\hat{a}(T)} \le 1 + \sum_{j=1}^n \int_\alpha^t \frac{k_j(t,s)}{\hat{a}(T)} r_j(\hat{a}(T)) w_j\left(\frac{u(s)}{\hat{a}(T)}\right) ds, \qquad \alpha \le t \le T < \beta, \tag{10.30}$$

and

$$\frac{u(t)}{\hat{a}(t)} \le 1 + \sum_{j=1}^{n} \int_{\alpha}^{t} k_j(t,s) \frac{r_j(\hat{a}(s))}{\hat{a}(s)} w_j \left(\frac{u(s)}{\hat{a}(s)} \right) ds, \qquad \alpha \le t < \beta. \tag{10.31}$$

Applying (10.30), (10.31), and theorem 10.3, we obtain (10.28) and (10.29). □

Soppose that in inequality (10.22) an additional term $\int_\alpha^t k_0(t,s)u(s)\,ds$ appears. Setting $g_0(u) = u$, and assuming $g_0 \propto g_1 \propto \cdots \propto g_n$, to such an inequality we can apply theorem 10.3. However, in the presence of a linear integral term in (10.22) we can obtain another result, such as the following theorem.

THEOREM 10.5. *Suppose the functions u, a, k_j, $j = 0, \ldots, n$, and g_j, $j = 1, \ldots, n$, satisfy the conditions 1–4 of theorem 10.2. Let $g_j \in H_{r_j, w_j}$, $j = 1, \ldots, n$, and $w_1 \propto w_2 \propto \cdots \propto w_n$.*

Then the inequality

$$u(t) \le a(t) + \int_{\alpha}^{t} k_0(t,s)u(s)\,ds + \sum_{j=1}^{n} \int_{\alpha}^{t} k_j(t,s) g_j(u(s))\,ds, \qquad t \in J, \tag{10.32}$$

implies

$$u(t) \le e(t)\nu_n(t), \qquad \alpha \le t < \beta_1, \tag{10.33}$$

where $e(t) = \exp(\int_\alpha^t k_0(t,s)\,ds)$, $\nu_0(t) \equiv \hat{a}(t)$,

$$\nu_j(t) = W_j^{-1} \left[W_j(\nu_{j-1}(t)) + \int_{\alpha}^{t} k_j(t,s) \frac{r_j(e(s))}{e(s)}\,ds \right], \qquad j = 1, \ldots, n,$$

$$W_j(u) = \int_{u_j}^{u} \frac{dx}{w_j(x)}, \qquad u > 0 \quad (u_0 > 0),$$

and the number β_1 is so chosen that the functions ν_j are defined for $\alpha \le t < \beta_1$.

PROOF. For $T \in (\alpha, \beta)$ fixed, (10.32) implies

$$u(t) \le \hat{a}(T) + \int_{\alpha}^{t} k_0(T,s)u(s)\,ds + \sum_{j=1}^{n} \int_{\alpha}^{t} k_j(T,s) g_j(u(s))\,ds \equiv v(t)$$

for $\alpha \le t \le T < \beta$. This implies that $v(\alpha) = \hat{a}(T)$ and

$$v'(t) \le k_0(T,t)v(t) + \sum_{j=1}^{n} k_j(T,t) g_j(v(t)), \qquad \alpha \le t \le T < \beta.$$

Since $e(t) \equiv \exp(\int_\alpha^t k_0(t,s)\,ds) \le \exp(\int_\alpha^t k_0(T,s)\,ds) \equiv E(t)$, we have

$$\left(\frac{v(t)}{E(t)} \right)' = \frac{v'(t)E(t) - E'(t)v(t)}{E^2(t)} = \frac{v'(t) - k_0(T,t)v(t)}{E(t)} \le$$

$$\le \sum_{j=1}^{n} \frac{k_j(T,t)}{e(t)} g_j(v(t)), \qquad \alpha \le t \le T < \beta.$$

By integration from α to t we find

$$\frac{v(t)}{E(t)} - \hat{a}(T) \leq \sum_{j=1}^{n} \int_{\alpha}^{t} \frac{k_j(T,s)}{e(s)} g_j(v(s))\, ds, \qquad \alpha \leq t \leq T < \beta.$$

Setting $t = T$ and changing notation we arrive at

$$\frac{v(t)}{e(t)} \leq \hat{a}(t) + \sum_{j=1}^{n} \int_{\alpha}^{t} \frac{k_j(t,s)}{e(s)} g_j(v(s))\, ds, \qquad \alpha \leq t < \beta.$$

Since $g_j \in H_{r_j,w_j}$,

$$\frac{v(t)}{e(t)} \leq \hat{a}(t) + \sum_{j=1}^{n} \int_{\alpha}^{t} k_j(t,s) \frac{r_j(e(s))}{e(s)} w_j\left(\frac{v(s)}{e(s)}\right) ds.$$

The result now follows from theorem 10.3. □

REMARK 10.5. From theorems 10.4 and 10.5 we can obtain corollaries similar to the corollaries 10.2–10.5 to theorem 10.2. □

REMARK 10.6. Theorems 10.3–10.5 include the corresponding theorems 1, 3, 2 of Pinto [168]. □

REMARK 10.7. In corollaries 10.2, 10.4, and 10.5, and theorems 10.3 and 10.5, the condition that $a(t)$ be positive can be replaced by the weaker one that $a(t)$ be nonnegative. □

11. Integral inequalities with iterated integrals

First we consider some simple inequalities involving iterated integrals.

THEOREM 11.1. (Bykov, 1967) *Let $u(t)$, $b(t)$, $k(t,s)$, and $h(t,s,\sigma)$ be nonnegative continuous functions for $\alpha \leq \sigma \leq s \leq t \leq \beta$, and suppose that*

$$u(t) \leq a + \int_{\alpha}^{t} b(s)u(s)\, ds + \int_{\alpha}^{t} \left(\int_{\alpha}^{s} k(s,\tau)u(\tau)\, d\tau\right) ds + \\ + \int_{\alpha}^{t} \left(\int_{\alpha}^{s} \left(\int_{\alpha}^{\tau} h(s,\tau,\sigma)u(\sigma)\, d\sigma\right) d\tau\right) ds \tag{11.1}$$

for $t \in [\alpha, \beta]$, where $a \geq 0$ is a constant. Then for $t \in [\alpha, \beta]$,

$$u(t) \leq \exp\left\{\int_{\alpha}^{t} b(s)\, ds + \int_{\alpha}^{t} \left(\int_{\alpha}^{s} k(s,\tau)\, d\tau\right) ds + \\ + \int_{\alpha}^{t} \left(\int_{\alpha}^{s} \left(\int_{\alpha}^{\tau} h(s,\tau,\sigma)\, d\sigma\right) d\tau\right) ds\right\}. \tag{11.2}$$

PROOF. We denote the righthand side of (11.1) by $v(t)$. Then $v(\alpha) = a$,

$$v'(t) = b(t)u(t) + \int_{\alpha}^{t} k(t,\tau)u(\tau)\, d\tau + \int_{\alpha}^{t} \left(\int_{\alpha}^{\tau} h(t,\tau,\sigma)u(\sigma)\, d\sigma\right) d\tau \leq \\ \leq \left[b(t) + \int_{\alpha}^{t} k(t,\tau)\, d\tau + \int_{\alpha}^{t} \left(\int_{\alpha}^{\tau} h(t,\tau,\sigma)\, d\sigma\right) d\tau\right] v(t),$$

since $u(t) \le v(t)$ and $v(t)$ is nondecreasing in $[\alpha, \beta]$. Applying lemma 1.1, we arrive at (11.2). $\square$

In the same maner we can prove the following theorem.

THEOREM 11.2. (Ved', 1965) *Let $u(t)$, $b(t)$, $\sigma(t)$, and $k(t,s)$ be nonnegative continuous functions for $\alpha \le s \le t \le \beta$, and suppose that*

$$u(t) \le a_1 + \sigma(t)\left\{a_2 + \int_\alpha^t b(s)u(s)\,ds + \int_\alpha^t \left(\int_\alpha^s k(s,\tau)u(\tau)\,d\tau\right)ds\right\} \quad (11.3)$$

for $t \in [\alpha, \beta]$, where $a_1, a_2 \ge 0$ are constants. Then

$$u(t) \le a_2 e^{\int_\alpha^t B(s)\sigma(s)\,ds} + \int_\alpha^t a_1 B(s) e^{\int_s^t B(\tau)\sigma(\tau)\,d\tau}\,ds, \qquad \alpha \le t \le \beta, \quad (11.4)$$

where $B(s) = b(s) + \int_\alpha^s k(s,\tau)\,d\tau$. $\square$

THEOREM 11.3. (Bykov and Salpagarov, 1962) *Let $u(t)$ be a nonnegative continuous function for $\alpha \le t \le t \le \beta$, and suppose that*

$$u(t) \le a + \int_\alpha^t k(t,\tau)u(\tau)\,d\tau + \int_\alpha^t \left(\int_\alpha^s h(t,s,\sigma)u(\sigma)\,d\sigma\right)ds, \qquad \alpha \le t \le \beta, \quad (11.5)$$

where $a \ge 0$ is a constant and $k(t,s)$ and $h(t,s,\sigma)$ are nonnegative continuous functions for $\alpha \le \sigma \le s \le t \le \beta$. Suppose the partial derivatives $\frac{\partial k}{\partial t}(t,s)$ and $\frac{\partial h}{\partial t}(t,s,\sigma)$ exist and are nonnegative and continuous for $\alpha \le \sigma \le s \le t \le \beta$. Then

$$u(t) \le a \exp\left\{\int_\alpha^t [R(s) + Q(s)]\,ds\right\}, \qquad \alpha \le t \le \beta, \quad (11.6)$$

where

$$R(t) = k(t,t) + \int_\alpha^t h(t,t,\sigma)\,d\sigma,$$

$$Q(t) = \int_\alpha^t \frac{\partial k}{\partial t}(t,\sigma)\,d\sigma + \int_\alpha^t \left(\int_\alpha^s \frac{\partial h}{\partial t}(t,s,\sigma)\,d\sigma\right)ds. \qquad \square$$

REMARK 11.1. The estimate (11.6) can be replaced by

$$u(t) \le a \exp\left\{\int_\alpha^t k(t,s)\,ds + \int_\alpha^t \left(\int_\alpha^s h(t,s,\sigma)\,d\sigma\right)ds\right\}, \qquad \alpha \le t \le \beta, \quad (11.7)$$

since $\int_\alpha^t [R(s)+Q(s)]\,ds \equiv \int_\alpha^t k(t,s)\,ds + \int_\alpha^t (\int_\alpha^s h(t,s,\sigma)\,d\sigma)\,ds$, $\alpha \le t \le \beta$.

Moreover, (11.7) still holds if in theorem 11.3 the condition that the partial derivatives be nonnegative is replaced by the condition that the functions $k(t,s)$ and $h(t,s,\sigma)$ be nondecreasing in t for fixed s, σ, while in (11.5) and (11.7) the constant a is replaced by a nonnegative, nondecreasing function $a(t)$. In this case the proof of the theorem can be given along the lines of that of theorem 1.9. $\square$

Let $\alpha < \beta$, and set $J_i = \{(t_1, \dots, t_i) \in \mathbf{R}^i \colon \alpha \le t_i \le \dots \le t_1 \le \beta\}$, $i = 1, \dots, n$.

THEOREM 11.4. (Ráb, 1979) *Let $u(t)$, $a(t)$, and $b(t)$ be nonnegative continuous functions in $J = [\alpha, \beta]$, and suppose that*

$$u(t) \le a(t) + b(t) \left[\int_\alpha^t k_1(t,t_1)u(t_1)\,dt_1 + \cdots + \right.$$
$$\left. + \int_\alpha^t \left(\int_\alpha^{t_1} \cdots \left(\int_\alpha^{t_{n-1}} k_n(t,t_1,\dots,t_n)u(t_n)\,dt_n \right) \cdots \right) dt_1 \right] \tag{11.8}$$

for $t \in J$, where $k_i(t,t_1,\dots,t_i)$ are nonnegative continuous functions in J_{i+1}, $i = 1,\dots,n$. Suppose the partial derivatives $\frac{\partial k_i}{\partial t}(t,t_1,\dots,t_i)$ exist and are nonnegative and continuous in J_{i+1}, $i = 1,\dots,n$. Then for $t \in J$,

$$u(t) \le a(t) + b(t) \int_\alpha^t (R[a] + Q[a])(s) \exp\left(\int_s^t (R[b] + Q[b])(\tau)\,d\tau \right) ds, \tag{11.9}$$

where, for $t \in J$,

$$R[w](t) = k_1(t,t)w(t) + \int_\alpha^t k_2(t,t,t_2)w(t_2)\,dt_2 +$$
$$+ \sum_{i=3}^n \int_\alpha^t \left(\int_\alpha^{t_2} \cdots \left(\int_\alpha^{t_{i-1}} k_i(t,t,t_2,\dots,t_i)w(t_i)\,dt_i \right) \cdots \right) dt_2,$$

$$Q[w](t) = \int_\alpha^t \frac{\partial k_1}{\partial t}(t,t_1)w(t_1)\,dt_1 +$$
$$+ \sum_{i=2}^n \int_\alpha^t \left(\int_\alpha^{t_1} \cdots \left(\int_\alpha^{t_{i-1}} \frac{\partial k_i}{\partial t}(t,t_1,\dots,t_i)w(t_i)\,dt_i \right) \cdots \right) dt_1,$$

for each continuous function $w(t)$ in J.

PROOF. First we note that $R[w]$ and $Q[w]$ are linear functionals, and

$$R[w_1] \le R[w_2], \qquad Q[w_1] \le Q[w_2]$$

if $w_1(t) \le w_2(t)$, $t \in J$, and

$$R[w_1w_2] \le R[w_1]w_2, \qquad Q[w_1w_2] \le Q[w_1]w_2$$

if $w_1(t)$ is nonnegative in J and $w_2(t)$ is nondecreasing in J. We set

$$v(t) = \int_\alpha^t k_1(t,t_1)u(t_1)\,dt_1 + \cdots + \int_\alpha^t \left(\int_\alpha^{t_1} \cdots \left(\int_\alpha^{t_{n-1}} k_n(t,t_1,\dots,t_n)u(t_n)\,dt_n \right) \cdots \right) dt_1.$$

Since $v(\alpha) = 0$, $u(t) \le a(t) + b(t)v(t)$, and $v(t)$ is nondecreasing and continuous in J, we have

$$v'(t) = R[u](t) + Q[u](t) \le R[a+bv](t) + Q[a+bv](t) \le$$
$$\le (R[a] + Q[a])(t) + (R[b] + Q[b])(t)v(t), \qquad t \in J,$$

and by lemma 1.1 we obtain (11.9). □

THEOREM 11.5. *Let $u(t)$, $a(t)$, and $b(t)$ be nonnegative continuous functions in $J = [\alpha, \beta]$, and suppose*

$$u(t) \le a(t)+b(t) \left[\int_\alpha^t k_1(t,t_1)u(t_1)\,dt_1 + \cdots + \right.$$
$$\left. + \int_\alpha^t \left(\int_\alpha^{t_1} \cdots \left(\int_\alpha^{t_{n-1}} k_n(t,t_1,\ldots,t_n)u(t_n)\,dt_n\right)\ldots\right) dt_1\right] \quad (11.10)$$

for $t \in J$, where $k_i(t,t_1,\ldots,t_i)$ are nonnegative continuous functions in J_{i+1}, $i = 1,\ldots,n$, which are nondecreasing in $t \in J$ for all fixed $(t_1,\ldots,t_i) \in J_i$, $i = 1,\ldots,n$. Then

$$u(t) \le a(t) + b(t)\int_\alpha^t \hat{R}[a](t,s)\exp\left(\int_s^t \hat{R}[b](t,\tau)\,d\tau\right) ds, \qquad t \in J, \quad (11.11)$$

where, for $(t,s) \in J_2$,

$$\hat{R}[w](t,s) = k_1(t,s)w(s) + \int_\alpha^s k_2(t,s,t_2)w(t_2)\,dt_2 +$$
$$+\sum_{i=3}^{n}\int_\alpha^s \left(\int_\alpha^{t_2} \cdots \left(\int_\alpha^{t_{i-1}} k_i(t,s,t_2,\ldots,t_i)w(t_i)\,dt_i\right)\ldots\right) dt_2,$$

for each continuous function $w(t)$ in J.

PROOF. For a fixed $T \in (\alpha,\beta]$ and $\alpha \le t \le T$ we have

$$u(t) \le a(t) + b(t)w(t) \equiv$$
$$\equiv a(t) + b(t)\left[\int_\alpha^t k_1(T,t_1)u(t_1)\,dt_1 + \cdots + \right.$$
$$\left. + \int_\alpha^t \left(\int_\alpha^{t_1} \cdots \left(\int_\alpha^{t_{n-1}} k_n(T,t_1,\ldots,t_n)u(t_n)\,dt_n\right)\ldots\right) dt_1\right].$$

Since $\frac{\partial k_i}{\partial t}(T,t_1,\ldots,t_i) \equiv 0$ for $i = 1,\ldots,n$ and $t \in [\alpha,T]$, we have

$$w'(t) \le \hat{R}[a](T,t) + \hat{R}[b](T,t)w(t),$$

from which, by lemma 1.1,

$$u(t) \le a(t) + b(t)\int_\alpha^t \hat{R}[a](T,s)\exp\left(\int_s^t \hat{R}[b](T,\tau)\,d\tau\right) ds, \qquad \alpha \le t \le T.$$

In particular, for $T = t$ we find (11.11). □

COROLLARY 11.1. *If, under the conditions of theorem 11.5, the functions $a(t)$ and $b(t)$ are also nondecreasing in J, then*

$$u(t) \le a(t)\exp\left\{b(t)\left[\int_\alpha^t k_1(t,t_1)\,dt_1 + \cdots + \right.\right.$$
$$\left.\left. + \int_\alpha^t \left(\int_\alpha^{t_1} \cdots \left(\int_\alpha^{t_{n-1}} k_n(t,t_1,\ldots,t_n)\,dt_n\right)\ldots\right) dt_1\right]\right\} \quad (11.12)$$

for $t \in J$.

Indeed, (11.11) implies that

$$\begin{aligned} u(t) &\le a(t) + b(t)\int_\alpha^t \hat{R}[a](t,s)\exp\left\{\int_s^t \hat{R}[b](t,\tau)\,d\tau\right\}ds \le \\ &\le a(t) + b(t)\int_\alpha^t a(s)\hat{R}[1](t,s)\exp\left\{\int_s^t b(\tau)\hat{R}[1](t,\tau)\,d\tau\right\}ds \le \\ &\le a(t)\left[1+\int_\alpha^t b(t)\hat{R}[1](t,s)\exp\left\{\int_s^t b(t)\hat{R}[1](t,\tau)\,d\tau\right\}ds\right] = \\ &= a(t)\exp\left\{b(t)\int_\alpha^t \hat{R}[1](t,\tau)\,d\tau\right\}. \qquad \square \end{aligned}$$

THEOREM 11.6. (Ráb, 1979) *Let u, $f_1, \dots, f_n$ be nonnegative continuous functions in $J = [\alpha, \beta]$, and suppose*

$$\begin{aligned} u(t) \le a &+ \int_\alpha^t f_1(t_1)u(t_1)\,dt_1 + \cdots + \\ &+ \int_\alpha^t f_1(t_1)\left(\int_\alpha^{t_1} f_2(t_2)\dots\left(\int_\alpha^{t_{n-1}} f_n(t_n)u(t_n)\,dt_n\right)\dots\right)dt_1 \end{aligned} \tag{11.13}$$

for $t \in J$, where $a \ge 0$ is a constant. Then

$$u(t) \le aR_1(t), \qquad t \in J, \tag{11.14}$$

where

$$R_n(t) = \exp\left(\int_\alpha^t f_n(s)\,ds\right), \qquad t \in J,$$

$$R_i(t) = 1 + \int_\alpha^t f_i(t)R_{i+1}(s)\exp\left(\int_\alpha^s f_i(\tau)\,d\tau\right)ds, \qquad t \in J, \qquad i = n-1, \dots, 1.$$

PROOF. We set

$$u_1(t) = a + L_1[u](t), \qquad u_{j+1}(t) = u_j(t) + L_{j+1}[u](t)$$

for $t \in J$, $j = 1, \dots, n-1$, where

$$\begin{aligned} L_k[u](t) = &\int_\alpha^t f_k(t_k)u(t_k)\,dt_k + \cdots + \\ &+ \int_\alpha^t f_k(t_k)\left(\int_\alpha^{t_k} f_{k+1}(t_{k+1})\dots\left(\int_\alpha^{t_{n-1}} f_n(t_n)u(t_n)\,dt_n\right)\dots\right)dt_k \end{aligned}$$

for $t \in J$, $k = 1, \dots, n$. Now (11.13) implies

$$u(t) \le u_1(t). \tag{11.15}$$

Taking into account that

$$u_k(t) \le u_{k+1}(t), \qquad (L_k[u])' = f_k(u + L_{k+1}[u]), \quad k = 1, \dots, n-1, \qquad (L_n[u])' = f_n u,$$

we successively find

$$\begin{aligned} u_1' &\le f_1 u_2, \\ u_k' &\le (f_1 + \cdots + f_{k-1}) u_k + f_k u_{k+1}, \qquad k = 2, \ldots, n-1, \\ u_n' &\le (f_1 + \cdots + f_n) u_n. \end{aligned} \tag{11.16}$$

Since $u_k(\alpha) = a$, $k = 1, \ldots, n$, (11.16) gives, by successive application of lemma 1.1,

$$u_k(t) \le a R_k(t) \exp\left(\int_\alpha^t \sum_{j=1}^{k-1} f_j(s)\, ds \right), \qquad k = n, n-1, \ldots, 1.$$

For $k = 1$ this and (11.15) imply (11.14). □

COROLLARY 11.2. (Pachpatte, 1973) *If u, g, f are nonnegative continuous functions in J, $u_0 \ge 0$ and*

$$u(t) \le u_0 + \int_\alpha^t f(s) \left[u(s) + \int_\alpha^s g(\tau) u(\tau)\, d\tau \right] ds, \qquad t \in J,$$

then

$$u(t) \le u_0 \left[1 + \int_\alpha^t f(s) e^{\int_\alpha^s [f(\tau) + g(\tau)]\, d\tau} ds \right], \qquad t \in J. \qquad \square$$

THEOREM 11.7. (Young, 1985) *Let u, a, f_i, $i = 1, \ldots, n$, be nonnegative continuous functions in $J = [\alpha, \beta]$, and suppose*

$$\begin{aligned} u(t) \le a(t) + &\int_\alpha^t f_1(t_1) u(t_1)\, dt_1 + \cdots + \\ &+ \int_\alpha^t f_1(t_1) \left(\int_\alpha^{t_1} f_2(t_2) \ldots \left(\int_\alpha^{t_{n-1}} f_n(t_n) u(t_n)\, dt_n \right) \ldots \right) dt_1 \end{aligned} \tag{11.17}$$

for $t \in J$. Then

$$u(t) \le a(t) + \int_\alpha^t f_1(s)[a(s) + v_2(s)]\, ds, \tag{11.18}$$

where

$$v_n(t) = \int_\alpha^t (f_1(s) + \cdots + f_n(s)) a(s) e^{\int_s^t (f_1(\tau) + \cdots + f_n(\tau))\, d\tau} ds,$$

$$v_k(t) = \int_\alpha^t [(f_1(s) + \cdots + f_k(s)) a(s) + f_k(s) v_{k+1}(s)] e^{\int_s^t (f_1(\tau) + \cdots + f_{k-1}(\tau))\, d\tau} ds,$$

$$k = n-1, \ldots, 2, \qquad t \in J.$$

PROOF. Let $L_k[u](t)$ be defined as in theorem 11.6, and put

$$\begin{aligned} v_1(t) &= L_1[u](t), \\ v_{k+1}(t) &= v_k(t) + L_k[u](t), \qquad k = 1, \ldots, n-1. \end{aligned}$$

Then $v_k(\alpha) = 0$, $k = 1, \dots, n$, and

$$\begin{aligned} v_1' &\le f_1(a + v_2), \\ v_k' &\le (f_1 + \dots + f_{k-1})v_k + (f_1 + \dots + f_k)a + f_k v_{k+1}, \qquad k = 2, \dots, n-1, \\ v_n' &\le (f_1 + \dots + f_n)v_n + (f_1 + \dots + f_n)a. \end{aligned} \tag{11.19}$$

Solving the system (11.19) 'backwards', and applying lemma 1.1 we arrive at (11.18). □

REMARK 11.2. For $a(t) \equiv a = \text{const}$ the estimates (11.14) and (11.18) coincide. This follows from the fact that if in (11.19) we set $v_k + a = u_k$, we obtain (11.16). □

COROLLARY 11.3. (Pachpatte, 1974) *If u, p, q, h are nonnegative continuous functions in J, $u_0 \ge 0$ is a constant, and*

$$u(t) \le u_0 + \int_\alpha^t [p(s)u(s) + h(s)]\, ds + \int_\alpha^t p(s) \left(\int_\alpha^s q(\tau)u(\tau)\, d\tau \right) ds, \qquad t \in J,$$

then for $t \in J$ we have

$$\begin{aligned} u(t) \le u_0 + &\int_\alpha^t h(s)\, ds + \\ &+ \int_\alpha^t p(s) \left[u_0 e^{\int_\alpha^s (p(\tau)+q(\tau))\, d\tau} + \int_\alpha^s h(\tau) e^{\int_\alpha^s (p(r)+q(r))\, dr}\, d\tau \right] ds. \end{aligned} \tag{11.20}$$

Indeed, (11.20) follows from theorem 11.7 with $f_1 = p$, $f_2 = q$, and $a(t) = u_0 + \int_\alpha^t h(s)\, ds$. □

The estimates obtained in theorems 11.6 and 11.7 are not sharp. Sharp estimates for functions satisfying inequalities of the form (11.17) can be obtained by reduction to an equivalent system of linear differential inequalities, and application of a comparison result.

THEOREM 11.8. *Let u, a, b, f_i, $i = 1, \dots, n$, and g_j, $j = 1, \dots, n-1$, be nonnegative continuous functions in $J = [\alpha, \beta]$, and suppose*

$$\begin{aligned} u(t) \le a(t) + b(t) \Bigg[&\int_\alpha^t f_1(t_1)u(t_1)\, dt_1 + \int_\alpha^t g_1(t_1) \left(\int_\alpha^{t_1} f_2(t_2)u(t_2)\, dt_2 \right) dt_1 + \dots + \\ &+ \int_\alpha^t g_1(t_1) \left(\int_\alpha^{t_1} g_2(t_2) \dots \left(\int_\alpha^{t_{n-2}} g_{n-1}(t_{n-1}) \left(\int_\alpha^{t_{n-1}} f_n(t_n)u(t_n)\, dt_n \right) \dots \right) \right) dt_1 \Bigg] \end{aligned} \tag{11.21}$$

for $t \in J$. Then

$$u(t) \le a(t) + b(t)v_1(t), \qquad t \in J, \tag{11.22}$$

where v_1 is the first component of the n-vector function

$$V(t) = \int_\alpha^t Y(t)Y^{-1}(s)B(s)\, ds$$

and $Y(t)$ is a fundamental matrix of the system

$$Y'(t) = A(t)Y(t), \tag{11.23}$$

and

$$A = \begin{bmatrix} f_1 b & g_1 & 0 & \cdot & 0 \\ f_2 b & 0 & g_2 & \cdot & 0 \\ \cdots & \cdots & \cdots & \cdots & \cdots \\ \cdots & \cdots & \cdots & \cdots & \cdots \\ \cdots & \cdots & \cdots & \cdots & g_{n-1} \\ f_n b & 0 & 0 & \cdot & 0 \end{bmatrix}, \qquad B = \begin{bmatrix} f_1 a \\ f_2 a \\ \vdots \\ f_n a \end{bmatrix}. \tag{11.24}$$

PROOF. We define the functions $z_k(t)$, $k = 1, \dots, n$, by

$$\begin{aligned} z_k(t) = &\int_\alpha^t f_k(t_k)u(t_k)\,dt_k + \int_\alpha^t g_k(t_k)\left(\int_\alpha^{t_k} f_{k+1}(t_{k+1})u(t_{k+1})\,dt_{k+1}\right)dt_k + \cdots + \\ &+ \int_\alpha^t g_k(t_k)\left(\int_\alpha^{t_k} g_{k+1}(t_{k+1})\dots\left(\int_\alpha^{t_{n-2}} g_{n-1}(t_{n-1})\right.\right. \\ &\left.\left.\left(\int_\alpha^{t_{n-1}} f_n(t_n)u(t_n)\,dt_n\right)\dots\right)\right)dt_k. \end{aligned}$$

With this notation, (11.21) is equivalent to the system

$$u \le a + bz_1, \tag{11.25}$$

$$z_k(\alpha) = 0, \qquad k = 1, \dots, n, \tag{11.26}$$

$$z_k' = f_k u + g_k z_{k+1} \le f_k b z_1 + g_k z_{k+1} + f_k a, \qquad k = 1, \dots, n-1,$$

$$z_n' = f_n u \le f_n b z_n + f_n a. \tag{11.27}$$

In matrix notation the relations (11.26)–(11.27) have the form

$$Z'(t) \le A(t)Z(t) + B(t), \qquad Z(\alpha) = 0,$$

where $Z = \mathrm{col}[z_1, \dots, z_n]$ and A, B are defined by (11.24). By corollary 18.1, with $B_k \equiv 0$, $Z(t) \le V(t)$, where $V(t)$ is the solution of the initial value problem

$$V'(t) = A(t)V(t) + B(t), \qquad V(\alpha) = 0, \tag{11.28}$$

i.e.

$$Z(t) \le V(t) = \int_\alpha^t Y(t)Y^{-1}(s)B(s)\,ds,$$

where $Y(t)$ is a fundamental matrix of (11.23). □

REMARK 11.3. Thm. 2 in Young [247] is the particular case of our theorem 11.8 when (11.21) coincides with (11.17), i.e. for $g_k = f_k$, $k = 1, \dots, n-1$, and $b \equiv 1$. □

A result more general than theorem 11.8 is contained in the following theorem.

THEOREM 11.9. *Let $u(t)$ and $a(t)$ be continuous functions in $J = [\alpha, \beta]$, let $b_k(t, s_1, \dots, s_k)$ be nonnegative continuous functions for $\alpha \le s_k \le \dots \le s_1 \le t \le \beta$, and suppose that*

$$u(t) \le a(t) + \sum_{k=1}^{n} \int_\alpha^t \left(\int_\alpha^{s_1} \cdots \left(\int_\alpha^{s_{k-1}} b_k(t, s_1, \dots, s_k) u(s_k)\, ds_k \right) \cdots \right) ds_1 \tag{11.29}$$

for $t \in J$. Then

i)

$$u(t) \le \mu(t), \qquad t \in J, \tag{11.30}$$

where $\mu(t)$ is a solution of the equation

$$u(t) = a(t) + \sum_{k=1}^{n} \int_\alpha^t \left(\int_\alpha^{s_1} \cdots \left(\int_\alpha^{s_{k-1}} b_k(t, s_1, \dots, s_k) u(s_k)\, ds_k \right) \cdots \right) ds_1 \tag{11.31}$$

for $t \in J$;

ii) *the solution $\mu(t)$ of* (11.31) *is unique and can be obtained as the sum of the series*

$$u_0(t) + u_1(t) + \dots + u_m(t) + \dots,$$

where $u_0(t) \equiv a(t)$,

$$u_m(t) = \sum_{k=1}^{n} \int_\alpha^t \left(\int_\alpha^{s_1} \cdots \left(\int_\alpha^{s_{k-1}} b_k(t, s_1, \dots, s_k) u_{m-1}(s_k)\, ds_k \right) \cdots \right) ds_1,$$

$$m = 1, 2, \dots. \qquad \square$$

REMARK 11.4. Finding an exact solution of a linear system of differential equations (11.28) or of an integral equation (11.31) often proves to be very difficult. Therefore such solutions are estimated as, e.g., in the following two theorems. □

THEOREM 11.10. (Agarwal and Thandapani, 1982) *Suppose the conditions of theorem* 11.8 *hold. Then* (11.21) *implies*

$$u(t) \le a(t) + b(t) Q_k(t), \qquad k = 1, \dots, n, \qquad t \in J, \tag{11.32}$$

where

$$Q_k(t) = \int_\alpha^t \left[a(s) \sum_{i=1}^{k} f_i(s) + g_k(s) Q_{k+1}(s) \right] e^{\int_s^t [M_k(\tau) - g_k(\tau)]\, d\tau}\, ds, \qquad k = 1, \dots, n,$$

$$M_k(t) = \max \left[b(t) \sum_{i=1}^{k} f_i(t), g_1(t), \dots, g_{k-1}(t) \right], \qquad k = 2, \dots, n,$$

$$M_1(t) = b(t) f_1(t), \qquad Q_{n+1}(t) \equiv 0, \qquad g_n(t) \equiv 0.$$

PROOF. We start from the relations (11.25)–(11.27), which can be obtained as in theorem 11.8. We set

$$s_k = \sum_{i=1}^{k} z_i, \qquad k = 1, \ldots, n, \qquad z_{n+1} \equiv 0.$$

Adding up the first k inequalities in (11.27) and taking into account that $b\sum_{i=1}^{k} f_i \le M_k$ and $g_j \le M_k$ for $1 \le j \le k-1$, we arrive at

$$\begin{gathered} s_k' \le M_k s_k + a\sum_{i=1}^{k} f_i + g_k z_{k+1}, \qquad k = 1, \ldots, n, \qquad (11.33) \\ s_k(\alpha) = 0, \qquad k = 1, \ldots, n. \end{gathered}$$

The estimates (11.32) are obtained by solving 'backwards' (11.33), and by taking into account that for each $k = n-1, \ldots, 2, 1$,

$$z_{k+1} \le Q_{k+1} - s_k. \qquad \square$$

COROLLARY 11.4. (Pachpatte, 1973) *Let u, p, q, r be nonnegative continuous functions for $t \ge \alpha$, and suppose*

$$u(t) \le u_0 + \int_\alpha^t p(s)u(s)\,ds + \int_\alpha^t q(s)\left[u(s) + \int_\alpha^s r(\tau)u(\tau)\,d\tau\right] ds, \qquad t \ge \alpha.$$

Then

$$u(t) \le u_0 e^{\int_\alpha^t p(s)\,ds}\left[1 + \int_\alpha^t q(s)e^{\int_\alpha^s [q(\tau)+r(\tau)]\,d\tau}\,ds\right], \qquad t \ge \alpha. \qquad (11.34)$$

Indeed, (11.34) follows from theorem 11.10 for $n = 2$, $a(t) \equiv u_0$, $b(t) \equiv 1$, $f_1(t) = p(t) + q(t)$, $f_2(t) = r(t)$, $g_1(t) = q(t)$. $\square$

THEOREM 11.11. *Let $u(t)$ and $a(t)$ be nonnegative continuous functions in $J = [\alpha, \beta]$, with $a(t)$ nondecreasing in J, and let $f_i(t,s)$, $i = 1, \ldots, n$, be nonnegative continuous functions for $\alpha \le s \le t \le \beta$ which are nondecreasing in t for fixed $s \in J$. If*

$$u(t) \le a(t) + \int_\alpha^t f_1(t,t_1)\left(\int_\alpha^{t_1} f_2(t_1,t_2)\ldots\left(\int_\alpha^{t_{n-1}} f_n(t_{n-1},t_n)u(t_n)\,dt_n\right)\ldots\right)dt_1 \qquad (11.35)$$

for $t \in J$, then

$$u(t) \le R_1(t,t), \qquad t \in J, \qquad (11.36)$$

where $R_1(T,t)$ can be successively determined from the formulas

$$R_n(T,t) = a(T)\exp\left(\int_\alpha^t \sum_{i=1}^{n} f_i(T,s)\,ds\right),$$

$$R_k(T,t) = E_k(T,t)\left[a(T) + \int_\alpha^t f_k(T,s)\frac{R_{k+1}(T,s)}{E_k(T,s)}\,ds\right],$$

$$E_k(T,t) = \exp\left(\int_\alpha^t \left[\sum_{i=1}^{k-1} f_i(T,\tau) - f_k(T,\tau)\right] d\tau\right),$$

for $k = n-1, \dots, 1$, $\alpha \le t \le T \le \beta$.

PROOF. Fix $T \in (\alpha, \beta]$. For $\alpha \le t \le T$ we obtain from (11.35),

$$u(t) \le a(T) + \int_\alpha^t f_1(T,t_1)\left(\int_\alpha^{t_1} f_2(T,t_2)\dots\left(\int_\alpha^{t_{n-1}} f_n(T,t_n)u(t_n)\,dt_n\right)\dots\right)dt_1. \tag{11.37}$$

Now we introduce the functions

$$m_1(t) = a(T) + \int_\alpha^t f_1(T,t_1)\left(\int_\alpha^{t_1} f_2(T,t_2)\dots\left(\int_\alpha^{t_{n-1}} f_n(T,t_n)u(t_n)\,dt_n\right)\dots\right)dt_1,$$
$$m_k(t) = m_{k-1}(t) + \int_\alpha^t f_k(T,t_k)\left(\int_\alpha^{t_k}\dots\left(\int_\alpha^{t_{n-1}} f_n(T,t_n)m_{k-1}(t_n)\,dt_n\right)\dots\right)dt_k$$

for $t \in [\alpha, T]$ and $k = 2, \dots, n$. Then (11.37) implies that $m_k(\alpha) = a(T)$, $k = 1, \dots, n$, and

$$u(t) \le m_1(t) \le \dots \le m_n(t), \qquad t \in [\alpha, T].$$

Thus, induction with respect to k gives

$$m_k'(t) \le \left[\sum_{i=1}^{k-1} f_i(T,t) - f_k(T,t)\right] m_k(t) + f_k(T,t)m_{k+1}(t), \tag{11.38}$$

$$m_n'(t) \le \sum_{i=1}^{n} f_i(T,t)m_n(t) \tag{11.39}$$

for $t \in [\alpha, T]$, $k = 1, \dots, n-1$. Lemma 1.1 and (11.39) imply that

$$m_n(t) \le a(T)\exp\left(\int_\alpha^t \sum_{i=1}^n f_i(T,s)\,ds\right) = R_n(T,t), \qquad \alpha \le t \le T.$$

Applying the same lemma to (11.38) for $k = n-1, \dots, 2, 1$, we obtain

$$u(t) \le m_1(t) \le R_1(T,t), \qquad \alpha \le t \le T \le \beta,$$

which implies (11.36) for $T = t$. □

COROLLARY 11.5. (En Hao Yang, 1984) *Under the conditions of theorem* 11.11, (11.35) *implies*

$$u(t) \le a(t)Q_1(t,t), \qquad \alpha \le t \le \beta, \tag{11.40}$$

where $Q_1(T,t)$ *can be successively determined from the formulas*

$$Q_n(T,t) = \exp\left(\int_\alpha^t \sum_{i=1}^n f_i(T,s)\,ds\right),$$
$$Q_k(T,t) = E_k(T,t)\left[1 + \int_\alpha^t f_k(T,s)\frac{Q_{k+1}(T,s)}{E_k(T,s)}\,ds\right],$$

for $k = n-1, \dots, 1$, $\alpha \le t \le T \le \beta$. □

12. Integral inequalities for functions of N variables

One of the first results concerning integral inequalities for functions of several variables is the following, unpublished, theorem, mentioned in Beckenbach and Bellman [21].

THEOREM 12.1. (Wendroff) *Let $u(x,y)$, $k(x,y)$ be nonnegative continuous functions in $\mathbf{R}_+^2$, and suppose*

$$u(x,y) \leq a(x) + b(y) + \int_0^x \int_0^y k(s,t)u(s,t)\,ds\,dt$$

for $x \geq 0$, $y \geq 0$, where $a(x)$, $b(y)$ are positive continuous functions in $\mathbf{R}_+$ having nonnegative derivatives:

$$\frac{da}{dx}(x) \geq 0, \qquad \frac{db}{dx}(x) \geq 0, \qquad x \in \mathbf{R}_+.$$

Then, for $x \geq 0$, $y \geq 0$,

$$u(x,y) \leq \frac{[a(x)+b(0)][a(0)+b(y)]}{a(0)+b(0)} \exp\left[\int_0^x \int_0^y k(t,s)\,ds\,dt\right]. \qquad \square \qquad (12.1)$$

Theorem 12.1 is part of the following more general result.

THEOREM 12.2. *Let $u(x,y)$, $a(x,y)$, $k(x,y)$ be nonnegative continuous functions for $x \geq x_0$, $y \geq y_0$, and let $a(x,y)$ be nondecreasing in each of the variables for $x \geq x_0$, $y \geq y_0$. Suppose that*

$$u(x,y) \leq a(x,y) + \int_{x_0}^x \int_{y_0}^y k(s,t)u(s,t)\,ds\,dt, \qquad x \geq x_0, \quad y \geq y_0. \qquad (12.2)$$

Then

$$u(x,y) \leq a(x,y) \exp\left(\int_{x_0}^x \int_{y_0}^y k(s,t)\,ds\,dt\right), \qquad x \geq x_0, \quad y \geq y_0. \qquad (12.3)$$

PROOF. Let $X > x_0$ and $Y > y_0$ be fixed. Then for $x_0 \leq x \leq X$, $y_0 \leq y \leq Y$ we have

$$u(x,y) \leq a(X,Y) + \int_{x_0}^x \left(\int_{y_0}^y k(s,t)u(s,t)\,dt\right) ds \equiv v(x,y).$$

The function $v(x,y)$ is nondecreasing in each variable x, y, and

$$v(x_0,y) = a(X,Y),$$

$$\frac{\partial v}{\partial x}(x,y) = \int_{y_0}^y k(x,t)u(x,t)\,dt \leq \int_{y_0}^y k(x,t)\,dt v(x,y), \qquad (12.4)$$

since $u(x,t) \leq v(x,t) \leq v(x,y)$. By lemma 1.1, the differential inequality (12.4) implies

$$v(x,y) \leq a(X,Y) \exp\left(\int_{x_0}^x \left(\int_{y_0}^y k(s,t)\,dt\right) ds\right), \qquad x_0 \leq x \leq X, \quad y_0 \leq y \leq Y.$$

Setting $x = X$ and $y = Y$ and changing notation we arive at (12.3). $\square$

COROLLARY 12.1. *If*

$$u(x,y) \leq a + \int_{x_0}^{x} \int_{y_0}^{y} k(s,t)u(s,t)\, ds\, dt, \qquad x \geq x_0, \quad y \geq y_0,$$

then

$$u(x,y) \leq a \exp\left(\int_{x_0}^{x} \int_{y_0}^{y} k(s,t)\, ds\, dt\right), \qquad x \geq x_0, \quad y \geq y_0. \qquad \square \quad (12.5)$$

COROLLARY 12.2. *If, in* (12.2), $a(x,y) = a(x) + b(y)$, *where* $a(x)$, $b(y)$ *are nondecreasing functions for, respectively,* $x \geq x_0$, $y \geq y_0$, *then*

$$u(x,y) \leq [a(x) + b(y)] \exp\left(\int_{x_0}^{x} \int_{y_0}^{y} k(s,t)\, ds\, dt\right), \qquad x \geq x_0, \quad y \geq y_0. \qquad \square \quad (12.6)$$

Note that under the conditions of theorem 12.1, (12.6) with $x_0 = 0$, $y_0 = 0$ is better than (12.1), since

$$a(x) + b(y) \leq \frac{[a(x) + b(0)][a(0) + b(y)]}{a(0) + b(0)}, \qquad x \geq 0, \quad y \geq 0.$$

By a reasoning similar to the proof of theorem 12.2 we can prove the following assertions.

THEOREM 12.3. *Let* $u(x,y)$, $a(x,y)$, $k(x,y)$ *be nonnegative continuous functions in* $\mathbf{R}_+^2$, *and let* $a(x,y)$ *be nonincreasing in each of the variables* x, y. *Suppose that*

$$u(x,y) \leq a(x,y) + \int_{x}^{\infty} \int_{y}^{\infty} k(s,t)u(s,t)\, ds\, dt, \qquad x \geq 0, \quad y \geq 0,$$

and

$$\int_{x}^{\infty} \int_{y}^{\infty} k(s,t)\, ds\, dt < \infty, \qquad x \geq 0, \quad y \geq 0.$$

Then

$$u(x,y) \leq a(x,y) \exp\left(\int_{x}^{\infty} \int_{y}^{\infty} k(s,t)\, ds\, dt\right), \qquad x \geq 0, \quad y \geq 0. \qquad \square \, (12.7)$$

THEOREM 12.4. *Let* $u(x,y)$, $a(x,y)$, $k(x,y)$ *be nonnegative continuous functions in* $\mathbf{R}_+^2$, *and let* $a(x,y)$ *be nondecreasing in* x *and nonincreasing in* y. *Suppose that*

$$u(x,y) \leq a(x,y) + \int_{0}^{x} \int_{y}^{\infty} k(s,t)u(s,t)\, ds\, dt, \qquad x \geq 0, \quad y \geq 0,$$

and

$$\int_{0}^{x} \int_{y}^{\infty} k(s,t)\, ds\, dt < \infty, \qquad x \geq 0, \quad y \geq 0.$$

Then

$$u(x,y) \leq a(x,y) \exp\left(\int_{0}^{x} \int_{y}^{\infty} k(s,t)\, ds\, dt\right), \qquad x \geq 0, \quad y \geq 0. \qquad \square \, (12.8)$$

Note that the proof of theorem 12.2 is carried out according to the following scheme.

i) By a *single* integration with respect to one of the variables (say x) we reduce the integral inequality in question to a differential inequality in which the other variable (in this case y) is treated as a constant.

ii) To the differential inequality obtained we apply certain comparison results, and arrive at the result required.

This simple scheme of the proof has been successfully applied by many authors to investigate integral inequalities for functions of arbitrarily many ($n \geq 2$) variables. Before illustrating this, we introduce some notation and give some definitions to be used in the sequel.

If $x = (x_1, \ldots, x_n)$ and $y = (y_1, \ldots, y_n)$ belong to $\mathbf{R}^n$, we write $x \leq y$ ($x < y$) if and only if $x_i \leq y_i$ ($x_i < y_i$), $i = 1, \ldots, n$. If $x < y$, then $[x; y]$ denotes the n-dimensional interval $\{z \in \mathbf{R}^n \colon x \leq z \leq y\}$. We also adopt the notation

$$\int_x^y f(s)\,ds = \int_{x_1}^{y_1} \cdots \int_{x_n}^{y_n} f(s_1, \ldots, s_n)\,ds_1 \ldots ds_n =$$

$$= \int_{x_1}^{y_1} \left(\int_{x_2}^{y_2} \cdots \left(\int_{x_n}^{y_n} f(s_1, \ldots, s_n)\,ds_n \right) \ldots ds_2 \right) ds_1,$$

$$x = (x_1, x^1), \qquad x^1 = (x_2, \ldots, x_n), \qquad dx^1 = dx_2 \ldots dx_n.$$

If $D \subset \mathbf{R}^n$ and $f \colon D \to \mathbf{R}$, we say that $f(x)$ is a *nondecreasing function* in D if $x, y \in D$ and $x \leq y$ imply $f(x) \leq f(y)$.

THEOREM 12.5. (Zahariev and Bainov, 1981) *Let $\alpha, \beta \in \mathbf{R}^n$, $\alpha < \beta$. Let $u(x)$, $b(x)$ be nonnegative continuous functions for $x \in [\alpha; \beta]$ satisfying the inequality*

$$u(x) \leq a + \int_\alpha^x b(s)u(s)\,ds, \qquad x \in [\alpha; \beta], \tag{12.9}$$

where $a \geq 0$ is a constant. Then

$$u(x) \leq a \exp\left(\int_\alpha^x b(s)\,ds \right), \qquad x \in [\alpha; \beta]. \tag{12.10}$$

PROOF. (12.9) implies

$$u(t) \leq a + \int_{\alpha_1}^{x_1} \left(\int_{\alpha^1}^{x^1} b(s_1, s^1)u(s_1, s^1)\,ds^1 \right) ds_1 \equiv v(x_1, x^1). \tag{12.11}$$

For fixed $x^1 \in [\alpha^1; \beta^1]$ the function $w(x_1) = v(x_1, x^1)$ satisfies the relations

$$w(\alpha_1) = a,$$

$$w'(x_1) = \int_{\alpha^1}^{x^1} b(x_1, s^1)u(x_1, s^1)\,ds^1 \leq$$

$$\leq \int_{\alpha^1}^{x^1} b(x_1, s^1)\,ds^1 \cdot w(x_1), \tag{12.12}$$

since $v(x_1, x^1)$ is nondecreasing in $[\alpha; \beta]$ and $u(x_1, s^1) \le v(x_1, s^1) \le v(x_1, x^1) = w(x_1)$. Lemma 1.1 and (12.12) imply

$$w(x_1) \le a \exp\left(\int_{\alpha_1}^{x_1} \left(\int_{\alpha^1}^{x^1} b(s_1, s^1)\, ds^1 \right) ds_1 \right),$$

which together with (12.11) implies (12.10). □

COROLLARY 12.3. *If $a(x)$ is a nondecreasing function in $[\alpha; \beta] \subset \mathbf{R}^n$ and*

$$u(x) \le a(x) + \int_\alpha^x b(s) u(s)\, ds, \qquad x \in [\alpha; \beta],$$

then

$$u(x) \le a(x) \exp\left(\int_\alpha^x b(s)\, ds \right), \qquad x \in [\alpha; \beta]. \qquad \square \tag{12.13}$$

The following three theorems can be similarly proved.

THEOREM 12.6. *Let $\alpha, \beta \in \mathbf{R}^n$, $\alpha < \beta$. Let $u(x)$, $b(x)$ be nonnegative continuous functions for $x \in [\alpha; \beta]$ satisfying the inequality*

$$u(x) \le a + \int_x^\beta b(s) u(s)\, ds, \qquad x \in [\alpha; \beta],$$

where $a \ge 0$ is a constant. Then

$$u(x) \le a \exp\left(\int_x^\beta b(s)\, ds \right), \qquad x \in [\alpha; \beta]. \qquad \square \tag{12.14}$$

THEOREM 12.7. (Zahariev and Bainov, 1981) *Let $\alpha, \beta \in \mathbf{R}^n$, $\alpha < \beta$. Let $u(x)$, $b(x)$ be nonnegative continuous functions for $x \in [\alpha; \beta]$ satisfying the inequality*

$$u(x) \le u(\tau) + \int_x^\tau b(s) u(s)\, ds$$

for $\alpha \le x \le \tau \le \beta$. Then

$$u(x) \ge u(\alpha) \exp\left(- \int_\alpha^x b(s)\, ds \right), \qquad x \in [\alpha; \beta]. \qquad \square \tag{12.15}$$

THEOREM 12.8. (Hristova and Bainov, 1978) *Let $\alpha, \beta \in \mathbf{R}^n$, $\alpha < \beta$. Let $u(s)$, $b(s)$, $k(s, \tau)$ be nonnegative continuous functions for $\alpha \le \tau \le s \le \beta$ satisfying the inequality*

$$u(x) \le a + \int_\alpha^x \left[b(s) u(s) + \int_\alpha^s k(s, \tau) u(\tau)\, d\tau \right] ds, \qquad x \in [\alpha; \beta],$$

where $a \ge 0$ is a constant. Then

$$u(x) \le a \exp\left(\int_\alpha^x \left[b(s) + \int_\alpha^s k(s, \tau)\, d\tau \right] ds \right), \qquad x \in [\alpha; \beta]. \qquad \square \tag{12.16}$$

THEOREM 12.9. *Let $\alpha, \beta \in \mathbf{R}^n$, $\alpha < \beta$. Let $u(x)$, $a(x)$, $f(x)$, $g(x)$ be nonnegative continuous functions for $x \in [\alpha; \beta]$, with $a(x)$ nondecreasing in $[\alpha; \beta]$. Then the inequality*

$$u(x) \le a(x) + \int_\alpha^x f(s) \left[u(s) + \int_\alpha^s g(\tau) u(\tau)\, d\tau \right] ds, \qquad x \in [\alpha; \beta], \quad (12.17)$$

implies

$$u(x) \le a(x) + \int_\alpha^t a(s) f(s) \exp\left(\int_\alpha^s [f(\tau) + g(\tau)]\, d\tau \right) ds, \qquad x \in [\alpha; \beta]. \quad (12.18)$$

PROOF. We set $r(s) \equiv u(s) + \int_\alpha^s g(\tau) u(\tau)\, d\tau$. Then (12.17) takes the form

$$u(x) \le a(x) + \int_\alpha^x f(s) r(s)\, ds. \quad (12.19)$$

Taking into account that $u(s) \le r(s)$ we obtain

$$r(x) = u(x) + \int_\alpha^x g(s) u(s)\, ds \le$$
$$\le a(x) + \int_\alpha^x f(s) r(s)\, ds + \int_\alpha^x g(s) r(s)\, ds.$$

Consequently, corollary 12.3 implies

$$r(x) \le a(x) \exp\left(\int_\alpha^x [f(\tau) + g(\tau)]\, d\tau \right), \qquad x \in [\alpha; \beta],$$

which together with (12.19) implies (12.18). □

COROLLARY 12.4. *If $a(x)$ is nondecreasing in $[\alpha; \beta]$, (12.18) implies*

$$u(x) \le a(x) \left[1 + \int_\alpha^t f(s) \exp\left(\int_\alpha^s [f(\tau) + g(\tau)]\, d\tau \right) ds \right], \qquad x \in [\alpha; \beta]. \quad □ \quad (12.20)$$

THEOREM 12.10. *Let $\alpha, \beta \in \mathbf{R}^n$, $\alpha < \beta$. Let $u(x)$, $b(x)$ be nonnegative continuous functions for $x \in [\alpha; \beta]$ satisfying the inequality*

$$u(x) \le a + \int_\alpha^x b(s) g(u(s))\, ds, \qquad x \in [\alpha; \beta], \quad (12.21)$$

where $a \ge 0$ is a constant and $g(u)$ is a nondecreasing continuous function for $u \ge 0$ with $g(u) > 0$ for $u > 0$. Then

$$u(x) \le G^{-1} \left[G(a) + \int_\alpha^x b(s)\, ds \right], \qquad x \in D, \quad (12.22)$$

where $G(u) = \int_{u_0}^u dz/g(z)$, $u > 0$ ($u_0 > 0$), and

$$D = \left\{ x \in [\alpha; \beta] : G(a) + \int_\alpha^x b(s)\, ds \in \operatorname{Dom} G^{-1} \right\}.$$

PROOF. (12.21) implies that

$$u(x) \le a + \int_{\alpha_1}^{x_1} \left(\int_{\alpha^1}^{x^1} b(s_1, s^1) g(u(s_1, s^1))\, ds^1 \right) ds_1 \equiv v(x_1, x^1). \qquad (12.23)$$

Let $x^1 \in [\alpha; \beta]$ be fixed. Then the function $w(x_1) = v(x_1, x^1)$ satisfies the relations

$$w(\alpha_1) = a,$$

$$w'(x_1) = \int_{\alpha^1}^{x^1} b(x_1, s^1) g(u(x_1, s^1))\, ds^1 \le$$

$$\le \int_{\alpha^1}^{x^1} b(x_1, s^1)\, ds^1 \cdot g(w(x_1)),$$

since $v(x_1, x^1)$ and $g(u)$ are nondecreasing. Hence, by lemma 5.1,

$$w(x_1) \le G^{-1} \left[G(a) + \int_{\alpha_1}^{x_1} \left(\int_{\alpha^1}^{x^1} b(s_1, s^1)\, ds^1 \right) ds_1 \right],$$

which implies (12.22) in view of (12.23). □

COROLLARY 12.5. *If $a(x)$ is a nondecreasing function in $[\alpha; \beta] \in \mathbf{R}^n$ and*

$$u(x) \le a(x) + \int_\alpha^x b(s) g(u(s))\, ds, \qquad x \in [\alpha; \beta],$$

then

$$u(x) \le G^{-1} \left[G(a(x)) + \int_\alpha^x b(s)\, ds \right] \qquad (12.24)$$

for all $x \in [\alpha; \beta]$ such that $G(a(x)) + \int_\alpha^x b(s)\, ds \in \operatorname{Dom} G^{-1}$. □

THEOREM 12.11. (Young, 1982) *Let $\alpha, \beta \in \mathbf{R}^n$, $\alpha < \beta$. Let $u(x)$, $b(x)$, $f(x)$, $g(x)$ be nonnegative continuous functions for $x \in [\alpha; \beta]$, and let $a(x)$ be a nondecreasing function in $[\alpha; \beta]$. Suppose*

$$u(x) \le a(x) + \int_\alpha^x f(s) \left[u(s) + \int_\alpha^s g(\tau) u(\tau)\, d\tau \right] ds + \int_\alpha^x b(s) h(u(s))\, ds \qquad (12.25)$$

for $x \in [\alpha; \beta]$, where $h(u)$ is a nondecreasing, submultiplicative, continuous function for $u \ge 0$ with $h(u) > 0$ for $u > 0$. Then

$$u(x) \le r(x) H^{-1} \left(H(a(x)) + \int_\alpha^x b(s) h(r(s))\, ds \right), \qquad (12.26)$$

where $H(u) = \int_{u_0}^u dz/h(z)$, $u > 0$ ($u_0 > 0$),

$$r(x) = 1 + \int_\alpha^x f(s) \exp \left(\int_\alpha^s [f(\tau) + g(\tau)]\, d\tau \right) ds,$$

and $x \in [\alpha; \beta]$ is such that $H(a(x)) + \int_\alpha^x b(s) h(r(s))\, ds \in \operatorname{Dom} H^{-1}$.

PROOF. We set $p(x) = a(x) + \int_\alpha^x b(s)h(u(s))\,ds$. Then (12.25) takes the form

$$u(x) \le p(x) + \int_\alpha^x f(s)\left[u(s) + \int_\alpha^s g(\tau)u(\tau)\,d\tau\right] ds,$$

and by corollary 12.4,

$$u(x) \le p(x)\left\{1 + \int_\alpha^x f(s)\exp\left(\int_\alpha^s [f(\tau)+g(\tau)]\,d\tau\right) ds\right\}.$$

Hence,

$$u(x) \le r(x)\left[a(x) + \int_\alpha^x b(s)h(u(s))\,ds\right].$$

Since h is submultiplicative,

$$\frac{u(x)}{r(x)} \le a(x) + \int_\alpha^x b(s)h(r(s))h\left(\frac{u(s)}{r(s)}\right) ds.$$

Applying now corollary 12.5, we obtain

$$\frac{u(x)}{r(x)} \le H^{-1}\left[H(a(x)) + \int_\alpha^x b(s)h(r(s))\,ds\right],$$

which implies (12.26). □

Thus, the proofs of theorems 12.1–12.11 are based on the method of 'preliminary *single* differentiation'. The following question arises: Is it possible to find better estimates than those given in these theorems, if the righthand sides of the inequalities are first differentiated with respect to all n variables? Such an approach is used, e.g., by Gutowski, Corduneanu, Pachpatte, Chen-Chin Yeh, Mau-Hsiang Shin, En Hao Yang, Young (cf. [59], [83], [161], [198], [239], [240], [241], [243], [246]). In these works the crucial inequalities are:

$$u(x) \le a + \int_\alpha^x b(s)u(s)\,ds \equiv v(x), \qquad x \in [\alpha;\beta], \tag{12.27}$$

$$u(x) \le a + \int_x^\beta b(s)u(s)\,ds \equiv v(x), \qquad x \in [\alpha;\beta], \tag{12.28}$$

$$u(x) \le a + \int_\alpha^x b(s)g(u(s))\,ds \equiv v(x), \qquad x \in [\alpha;\beta], \tag{12.29}$$

$$u(x) \le a + \int_x^\beta b(s)g(u(s))\,ds \equiv v(x), \qquad x \in [\alpha;\beta]. \tag{12.30}$$

After differentiation of the righthand sides $v(x)$ with respect to all variables $x_1, \dots, x_n$, these inequalities reduce to the differential inequalities

$$D_1 \dots D_n v(x) \le b(x)v(x), \tag{12.31}$$

$$(-1)^n D_1 \dots D_n v(x) \le b(x)v(x), \tag{12.32}$$

$$D_1 \dots D_n v(x) \le b(x)g(v(x)), \tag{12.33}$$

$$(-1)^n D_1 \dots D_n v(x) \le b(x)g(v(x)), \tag{12.34}$$

respectively, where $x \in [\alpha;\beta]$ and $D_i = \partial/\partial x_i$, $i = 1, \dots, n$.

Below we estimate the function $v(x)$ defined by, respectively (12.27)–(12.30) using the inequalities (12.31)–(12.34).

To demonstrate this estimation procedure, we consider, following Young [245], the differential inequality (12.33) for the function $v(x)$ defined by (12.29).

Thus, let $v(x)$ satisfy (12.29). Then it also satisfies (12.33), hence

$$\frac{D_1 \dots D_n v(x)}{g(v(x))} \le b(x).$$

Since

$$D_n \left[\frac{D_1 \dots D_{n-1} v(x)}{g(v(x))} \right] = \frac{D_1 \dots D_n v(x)}{g(v(x))} - \frac{D_n g(v(x)) \cdot D_1 \dots D_{n-1} v(x)}{g^2(v(x))}$$

and $D_n g(v(x)) = g'(v(x)) D_n v(x) \ge 0$, $D_1 \dots D_{n-1} v(x) \ge 0$, the inequality above implies

$$D_n \left[\frac{D_1 \dots D_{n-1} v(x)}{g(v(x))} \right] \le b(x),$$

provided that $g'(u) \ge 0$ for $u \ge 0$.

Consequently, integration with respect to x_n from α_n to x_n, taking into account the fact that $D_1 \dots D_{n-1} v(x) = 0$ for $x_n = \alpha_n$, gives

$$\frac{D_1 \dots D_{n-1} v(x)}{g(v(x))} \le \int_{\alpha_n}^{x_n} b(x_1, \dots, x_{n-1}, s_n)\, ds_n.$$

Repeating this, we find (after $n-1$ steps)

$$\frac{D_1 v(x)}{g(v(x))} \le \int_{\alpha_2}^{x_2} \dots \left(\int_{\alpha_n}^{x_n} b(x_1, s_2, \dots, s_n)\, ds_n \right) \dots ds_2.$$

For $G(u) = \int_{u_0}^{u} dz/g(z)$, $u > 0$ ($u_0 > 0$) we find $D_1 G(v(x)) = D_1 v(x)/g(v(x))$, so that

$$D_1 G(v(x)) \le \int_{\alpha^1}^{x^1} b(x_1, s^1)\, ds^1.$$

Integration with respect to x_1 from α_1 to x_1 now gives

$$G(v(x_1, \dots, x_n)) - G(v(\alpha_1, x_2, \dots, x_n)) \le \int_{\alpha}^{x} b(s)\, ds,$$

and since $v(\alpha_1, x_2, \dots, x_n) = a$,

$$u(x) \le v(x) \le G^{-1} \left[G(a) + \int_{\alpha}^{x} b(s)\, ds \right]. \tag{12.35}$$

Comparing this estimation procedure with the proof of theorem 12.10, we conclude that

i) both ways of estimation lead to the same results—the estimates (12.22) and (12.35);
ii) the estimation procedure indicated is more involved, and can only be applied under additional restrictions, e.g., the existence of a nonnegative derivative $g'(u)$ is required.

Consequently, the method of single differentiation is to be preferred to the estimation procedure above. However, a sharp estimate for the function $u(x)$ using the inequalities (12.27)–(12.30) cannot be obtained by either method. Finding such estimates is related to the investigation of a corresponding comparison integral equation.

E.g., for $n = 2$ the comparison equation for (12.27) is

$$u(x, y) = a + \int_{x_0}^{x} \int_{y_0}^{y} b(s, t)u(s, t)\, ds\, dt.$$

This equation is closely related to the partial differential equation

$$v_{xy} = b(x, y)v(x, y),$$

and this relationship is reflected in the following theorem.

THEOREM 12.12. (Snow, 1972) *Let $u(x, y)$, $a(x, y)$, $b(x, y)$ be continuous functions in a domain D, with $b \geq 0$ in D. Let $P_0(x_0, y_0)$, $P(x, y)$ be two points in D such that $(x - x_0)(y - y_0) > 0$, and let R be the rectangular region with P and P_0 as two of its opposite vertices. Let $v(s, t; x, y)$ be the solution of the characteristic initial value problem*

$$L[v] = v_{st} - b(s, t)v = 0, \qquad v(x, t; x, y) = v(s, y; x, y) = 1,$$

and let D^+ be a connected subdomain of D, containing P, on which $v > 0$. If $R \subset D^+$ and u satisfies

$$u(x, y) \leq a(x, y) + \int_{x_0}^{x} \int_{y_0}^{y} b(s, t)u(s, t)\, ds\, dt,$$

then u also satisfies

$$u(x, y) \leq a(x, y) + \int_{x_0}^{x} \int_{y_0}^{y} a(s, t)b(s, t)v(s, t; x, y)\, ds\, dt. \qquad \square \tag{12.36}$$

Theorem 12.12 is a particular case of the following theorem of Young, which is based on two lemmas of independent interest.

Let D be a domain in $\mathbf{R}^n$, and let $v_x(x) = D_1 \ldots D_n v(x)$, where $D_i = \partial/\partial x_i$, $i = 1, \ldots, n$.

LEMMA 12.1. *Let $b(s)$ be a continuous function in $D \subset \mathbf{R}^n$. Then the characteristic initial value problem*

$$\begin{aligned} &(-1)^n v_s(s; x) - b(s)v(s; x) = 0 \qquad \text{in } D, \\ &v(s; x) = 1 \qquad \text{on } s_i = x_i,\ i = 1, \ldots, n, \end{aligned} \tag{12.37}$$

has a unique solution $v(s; x)$ for s near x satsifying $\prod_{i=1}^{n}(x_i - s_i) \geq 0$. This solution is continuous; if $b(s)$ is nonnegative, so is $v(s; x)$.

PROOF. The function $v(s; x)$ is the Riemann function relative to the point x. Problem (12.37) is equivalent to the integral equation

$$v(s; x) = 1 + \int_{s}^{x} b(t)v(t; x)\, dt.$$

The existence, uniqueness, and possible nonnegativity of $v(s;x)$ follows by successive approximation arguments, as given in [53], [214]. Since $v(s;x)$ is continuous, and $v = 1$ on $s_i = x_i$, $i = 1, \ldots, n$, there is a domain D^+, containing x, on which $v \geq 0$ even if $b(s)$ is not nonnegative. □

LEMMA 12.2. *Let $b(x)$, $f(x)$ be continuous functions in $D \subset \mathbf{R}^n$. Let $v(s;x)$ be the solution of* (12.37), *and let D^+ be a connected subdomain of D, containing x, on which $v \geq 0$ for all $s \in D^+$. Let $[\alpha;x] \subset D$, $\alpha \leq x$, and let*

$$L[w] = w_x(x) - b(x)w(x) \leq f(x), \tag{12.38}$$

where w vanishes together with all its mixed derivatives up to order $n-1$ on $x_i = \alpha_i$, $i = 1, \ldots, n$. Then

$$w(x) \leq \int_\alpha^x f(t)v(t;x)\,dt. \tag{12.39}$$

PROOF. The proof of lemma 12.2 follows from that of Young's theorem [244].
If ϕ is an n times continuously differentiable function in D, then

$$\phi Lw - wM\phi = \sum_{i=1}^n (-1)^{i-1} D_i[(D_0 D_1 \ldots D_{i-1}\phi)(D_{i+1} \ldots D_n D_{n+1} w)], \tag{12.40}$$

where $M\phi = (-1)^n \phi_x(x) - b(x)\phi(x)$, with $D_0 \equiv D_{n+1} = I$, the identity operation. Integration of (12.40) over $[\alpha;x]$, with t as integration variable, and noting that w vanshes together with all its mixed derivatives up to order $n-1$ on $t_i = \alpha_i$, $i = 1, \ldots, n$, gives

$$\int_\alpha^x (\phi Lw - wM\phi)\,dt = \sum_{i=1}^n (-1)^{i-1} \int_{t_i = x_i} (D_1 \ldots D_{i-1}\phi)(D_{i+1} \ldots D_n w)\,dt^i, \tag{12.41}$$

where $dt^i = dt_1 \ldots dt_{i-1}\,dt_{i+1} \ldots dt_n$.

We set now ϕ equal to the function v satisfying (12.37). Since $v = 1$ on $t_i = x_i$, $i = 1, \ldots, n$, it follows that $D_1 \ldots D_{i-1} v(t;x) = 0$ on $t_i = x_i$ for $i = 2, \ldots, n$. Thus, (12.41) becomes

$$\begin{aligned} \int_\alpha^x v(t;x)L[w](t)\,dt &= \int_{t_1 = x_1} v(t;x) D_2 \ldots D_n w(t)\,dt^1 = \\ &= \int_{\alpha_2}^{x_2} \cdots \int_{\alpha_n}^{x_n} D_2 \ldots D_n w(x_1, t_2, \ldots, t_n)\,dt_2 \ldots dt_n = w(x). \end{aligned} \tag{12.42}$$

By lemma 12.1 there is a domain D^+, containing x, on which $v \geq 0$. Multiplying (12.38) throughout by v and using (12.42), we obtain (12.39). □

REMARK 12.1. Lemma 12.2 still holds if $\leq$ is replaced by $\geq$ in (12.38) and (12.39), or even if the inequalities in (12.38) and (12.39) are replaced by equalities. In this sense (12.39) is the best estimate for a function $w(x)$ satsifying (12.38). □

THEOREM 12.13. (Young, 1973) *Let x, α belong to a domain $D \subset \mathbf{R}^n$, and $\alpha \leq x$. Let $u(x)$, $a(x)$, $b(x) \geq 0$ be continuous functions in D. Let $v(s;x)$ be the solution of the characteristic initial value problem*

$$\begin{aligned} (-1)^n v_s(s;x) - b(s)v(s;x) = 0 \qquad & \text{in } D, \\ v(s;x) = 1 \qquad \text{on } s_i = x_i,\ i = 1, \ldots, n, & \end{aligned}$$

and let D^+ be a connected subdomain of D, containing x, on which $v \geq 0$ for all $s \in D^+$. If $[\alpha; x] \subset D^+$ and

$$u(x) \leq a(x) + \int_\alpha^x b(s)u(s)\,ds,$$

then

$$u(x) \leq a(x) + \int_\alpha^x a(s)b(s)v(s;x)\,ds. \tag{12.43}$$

PROOF. Set $w(x) = \int_\alpha^x b(s)u(s)\,ds$, so that

$$u(x) \leq a(x) + w(x), \tag{12.44}$$

and $w_x(x) = b(x)u(x) \leq a(x)b(x) + b(x)w(x)$, or

$$L[w] = w_x(x) - b(x)w(x) \leq a(x)b(x).$$

Since w vanishes with all its mixed derivatives up to order $n-1$ on $x_i = \alpha_i$, $i = 1, \ldots, n$, lemma 12.2 gives

$$w(x) \leq \int_\alpha^x a(s)b(s)v(s;x)\,ds.$$

Now (12.44) implies (12.43). □

The following theorem can be similarly proved.

THEOREM 12.14. *Let x, α belong to a domain $D \subset \mathbf{R}^n$, and $\alpha \leq x$. Let $u(x)$, $a(x)$, $b(x) \geq 0$, $q(x) \geq 0$ be continuous functions in D. Let $v(s;x)$ be the solution of the characteristic initial value problem*

$$\begin{aligned} (-1)^n v_s(s;x) - q(s)b(s)v(s;x) = 0 \qquad & \text{in } D, \\ v(s;x) = 1 \qquad \text{on } s_i = x_i,\ i = 1, \ldots, n, & \end{aligned}$$

and let D^+ be a connected subdomain of D, containing x, on which $v \geq 0$ for all $s \in D^+$. If $[\alpha; x] \subset D^+$ and

$$u(x) \leq a(x) + q(x)\int_\alpha^x b(s)u(s)\,ds,$$

then

$$u(x) \leq a(x) + q(x)\int_\alpha^x a(s)b(s)v(s;x)\,ds. \qquad \square \tag{12.45}$$

REMARK 12.2. If, in theorems 12.13–12.14, $u(x)$ is a function of one variable, then $v(s;x)$ has the form

$$v(s;x) = \exp\left(\int_s^x b(\tau)\,d\tau\right), \qquad \text{respectively, } v(s;x) = \exp\left(\int_s^x q(\tau)b(\tau)\,d\tau\right),$$

and the estimates (12.43), (12.45) become (1.11) and (1.21), respectively. □

Using lemma 12.2 we can also prove the following theorems.

THEOREM 12.15. (Pachpatte, 1979) *Let x, α belong to a domain $D \subset \mathbf{R}^n$, and $\alpha \le x$. Let $u(x)$, $a(x)$, $b(x)$, $c(x)$, and $\sigma(x)$ be nonnegative, continuous functions in D. Let $v(s;x)$ be the solution of the characteristic initial value problem*

$$(-1)^n v_s(s;x) - [b(s)+c(s)]v(s;x) = 0 \qquad \text{in } D,$$
$$v(s;x) = 1 \qquad \text{on } s_i = x_i,\ i = 1,\dots,n,$$

and let D^+ be a connected subdomain of D, containing x, on which $v \ge 0$ for all $s \in D^+$. Let $[\alpha;x] \subset D^+$ and

$$u(x) \le a(x) + \int_\alpha^x b(s)u(s)\,ds + \int_\alpha^x b(s)\left[\sigma(s) + \int_\alpha^s c(\tau)u(\tau)\,d\tau\right] ds. \tag{12.46}$$

Then

$$u(x) \le a(x) + \int_\alpha^x b(s)\left[a(s)+\sigma(s)+\int_\alpha^s \{a(\tau)c(\tau)+b(\tau)[a(\tau)+\sigma(\tau)]\}v(s;x)\,d\tau\right] ds. \tag{12.47}$$

PROOF. Inequality (12.46) is equivalent to the system

$$u(x) \le a(x) + \phi(x), \tag{12.48}$$
$$\phi(x) = \int_\alpha^x b(s)u(s)\,ds + \int_\alpha^x b(s)[\sigma(s)+\psi(s)]\,ds,$$
$$\psi(s) = \int_\alpha^x c(s)u(s)\,ds.$$

This implies that

$$\phi_x(x) \le b(x)[a(x)+\sigma(x)] + b(x)[\phi(x)+\psi(x)], \tag{12.49}$$
$$\psi_x(x) \le a(x)c(x) + c(x)\phi(x). \tag{12.50}$$

Adding (12.49) and (12.50), we obtain

$$[\phi+\psi]_x(x) \le a(x)c(x) + b(x)[a(x)+\sigma(x)] + [b(x)+c(x)][\phi(x)+\psi(x)],$$

so that, by lemma 12.2,

$$\phi(x)+\psi(x) \le \int_\alpha^x \{a(\tau)c(\tau)+b(\tau)[a(\tau)+\sigma(\tau)]\}v(\tau;x)\,d\tau \equiv R(x). \tag{12.51}$$

From (12.49) and (12.51) we obtain

$$\phi_x(x) \le b(x)[a(x)+\sigma(x)] + b(x)R(x),$$

which implies

$$\phi(x) \le \int_\alpha^x b(s)[a(s) + \sigma(s) + R(s)]\, ds.$$

From (12.48) we thus obtain (12.47). □

THEOREM 12.16. (Pachpatte, 1980) *Let x, α belong to a domain $D \subset \mathbf{R}^n$, and $\alpha \le x$. Let $u(x)$, $a(x)$, $b(x)$, $c(x)$, $p(x)$, $q(s)$ be nonnegative, continuous functions in D. Let $v(s;x)$ and $w(s;x)$ be the solutions of the characteristic initial value problems*

$$\begin{aligned} &(-1)^n v_s(s;x) - [p(s) + b(s)(c(s) + q(s))]v(s;x) = 0 \qquad \text{in } D, \\ &v(s;x) = 1 \qquad \text{on } s_i = x_i,\ i = 1, \dots, n, \end{aligned} \tag{12.52}$$

and

$$\begin{aligned} &(-1)^n w_s(s;x) - [b(s)c(s) - p(s))]w(s;x) = 0 \qquad \text{in } D, \\ &w(s;x) = 1 \qquad \text{on } s_i = x_i,\ i = 1, \dots, n, \end{aligned}$$

respectively, and let D^+ be a connected subdomain of D, containing x, on which $v \ge 0$, $w \ge 0$ for all $s \in D^+$. If $[\alpha; x] \subset D^+$ and

$$\begin{aligned} u(x) \le{} & a(x) + \\ & + b(x)\left[\int_\alpha^x c(s)u(s)\, ds + \int_\alpha^x p(s)\left(\int_\alpha^s q(\tau)u(\tau)\, d\tau\right) ds\right], \end{aligned} \tag{12.53}$$

then

$$\begin{aligned} u(x) \le{} & a(x) + \\ & + b(x)\left[\int_\alpha^x w(s;x)\left\{a(s)c(s) + p(s)\int_\alpha^s a(\tau)[c(\tau) + q(\tau)]v(\tau;s)\, d\tau\right\} ds\right]. \end{aligned} \tag{12.54}$$

PROOF. Inequality (12.53) is equivalent to the system

$$u(x) \le a(x) + b(x)\phi(x), \tag{12.55}$$
$$\phi(x) = \int_\alpha^x c(s)u(s)\, ds + \int_\alpha^x p(s)\psi(s)\, ds,$$
$$\psi(x) = \int_\alpha^x q(s)u(s)\, ds.$$

This implies that

$$\phi_x(x) \le a(x)c(x) + b(x)c(x)\phi(x) + p(x)\psi(x), \tag{12.56}$$
$$\psi_x(x) \le a(x)q(x) + b(x)q(x)\phi(x). \tag{12.57}$$

Adding (12.56) and (12.57), we obtain

$$[\phi + \psi]_x(x) \le a(x)[c(x) + q(x)] + [p(x) + b(x)(c(x) + q(x))][\phi(x) + \psi(x)],$$

so that, by lemma 12.2,

$$\phi(x) + \psi(x) \le \int_\alpha^x a(\tau)[c(\tau) + q(\tau)]v(\tau;x)\, d\tau \equiv Q(x).$$

From $\psi \leq Q - \phi$ and (12.56) we obtain

$$\phi_x(x) \leq a(x)c(x) + p(x)Q(x) + [b(x)c(x) - p(x)]\phi(x).$$

Applying lemma 12.2 again, we find

$$\phi(x) \leq \int_\alpha^x w(s;x)[a(s)c(s) + p(s)Q(s)]\, ds,$$

which together with (12.55) yields (12.54). □

REMARK 12.3. If in (12.56)–(12.57) the functions ϕ, ψ are estimated in another way, we can obtain one more estimate for $u(x)$. Indeed, (12.56)–(12.57) imply that

$$(\phi + \psi)_x(x) \leq a(x)[c(x) + q(x)] + M(x)[\phi(x) + \psi(x)],$$

where $M(x) = \max[b(x)(c(x) + q(x)), p(x)]$. Let $E(s;x)$ be the solution of the characteristic initial value problem

$$\begin{aligned} &(-1)^n E_s(s;x) - M(x)E(s;x) = 0 \qquad \text{in } D, \\ &E(s;x) = 1 \qquad \text{on } s_i = x_i,\ i = 1, \dots, n, \end{aligned} \tag{12.58}$$

and let D^+ be a connected subdomain of D, containing x, on which $E \geq 0$, $w \geq 0$ for all $s \in D^+$. Then by lemma 12.2,

$$\phi(x) + \psi(x) \leq \int_\alpha^x a(\tau)[c(\tau) + q(\tau)]E(\tau;x)\, d\tau \equiv T(x),$$

and hence by (12.56),

$$\phi_x(x) \leq a(x)c(c) + p(x)T(x) + [b(x)c(x) - p(x)]\phi(x).$$

Again we apply lemma 12.2, to obtain

$$\phi(x) \leq \int_\alpha^x w(s;x)[a(s)c(s) + p(s)T(s)]\, ds,$$

or, finally,

$$\begin{aligned} u(x) \leq & a(x) + \\ & + b(x)\left[\int_\alpha^x w(s;x)\left\{a(s)c(s) + p(s)\int_\alpha^s a(\tau)[c(\tau) + q(\tau)]E(\tau;s)\, d\tau\right\} ds\right]. \end{aligned} \tag{12.59}$$

Moreover, since the solutions $v(s;x)$ and $E(s;x)$ of the initial value problems (12.52) and (12.58) satisfy the integral equations

$$v(s;x) = 1 + \int_s^x [p(\tau) + b(\tau)(c(\tau) + q(\tau))]v(\tau;x)\, d\tau, \qquad s \leq x,$$

respectively,

$$E(s;x) = 1 + \int_s^x M(\tau)E(\tau;x)\, d\tau, \qquad s \leq x,$$

and since

$$M(x) \leq p(x) + b(x)[c(x) + q(x)], \qquad \alpha \leq x \leq \beta,$$

we have $E(s;x) \leq v(s;x)$ for $s \in [\alpha;x] \subset D^+$. Consequently, the estimate (12.59) is beter than (12.54). □

THEOREM 12.17. (Pachpatte, 1980) *Let x, α belong to a domain $D \subset \mathbf{R}^n$, and $\alpha \leq x$. Let $u(x)$, $a(x)$, $b(x)$, $q(x)$, $\gamma(x)$, and $\lambda(x)$ be nonnegative continuous functions in D. Let $V(s;x)$ and $W(s;x)$ be the solutions of the characteristic initial value problems*

$$\begin{aligned}(-1)^n V_s(s;x) - b(s)[\gamma(s) + \lambda(s) + q(s)]V(s;x) = 0 \qquad &\text{in } D,\\ V(s;x) = 1 \qquad \text{on } s_i = x_i,\ i = 1,\dots,n,&\end{aligned}$$

and

$$\begin{aligned}(-1)^n W_s(s;x) - b(s)\gamma(s)W(s;x) = 0 \qquad &\text{in } D,\\ W(s;x) = 1 \qquad \text{on } s_i = x_i,\ i = 1,\dots,n,&\end{aligned}$$

respectively, and let D^+ be a connected subdomain of D, containing x, on which $V \geq 0$, $W \geq 0$ for all $s \in D^+$. If $[\alpha;x] \subset D^+$ and

$$\begin{aligned}u(x) \leq & a(x) + \\ &+b(x)\left[\int_\alpha^x \gamma(s)u(s)\,ds + \int_\alpha^x \lambda(s)\left\{u(s) + b(s)\int_\alpha^s q(\tau)u(\tau)\,d\tau\right\}ds\right],\end{aligned} \tag{12.60}$$

then

$$\begin{aligned}u(x) \leq & a(x) + \\ &+ b(x)\left[\int_\alpha^x W(s;x)\left\{a(s)[\gamma(s) + \lambda(s)] + b(s)\lambda(s)\int_\alpha^s a(\tau)\times\right.\right.\\ &\times \left.\left.[\gamma(\tau) + \lambda(\tau) + q(\tau)]V(s;x)\,d\tau\right\}ds\right].\end{aligned} \tag{12.61}$$

PROOF. We write (12.60) as

$$\begin{aligned}u(x) \leq & a(x) + \\ &+b(x)\left[\int_\alpha^x [\gamma(s) + \lambda(s)]u(s)\,ds + \int_\alpha^x \lambda(s)b(s)\left(\int_\alpha^s q(\tau)u(\tau)\,d\tau\right)ds\right].\end{aligned}$$

Then (12.61) follows from (12.59) with $c(x) = \gamma(x) + \lambda(x)$, $p(x) = \lambda(x)b(x)$, and w, E replaced by W and V, respectively, since $M(x) = b(x)[\gamma(x) + \lambda(x) + q(x)]$, $b(x)c(x) - p(x) = b(x)\gamma(x)$, and $c(x) + q(x) = \gamma(x) + \lambda(x) + q(x)$. □

Let $a \vee b = \max(a,b)$.

THEOREM 12.18. (Agarwal, 1982) *Let x, α belong to a domain $D \subset \mathbf{R}^n$, and $\alpha \leq x$. Let $u(x)$, $a(x)$, $b(x)$, $c(x)$, $h(x)$, $p(x)$, and $q(x)$ be nonnegative continuous functions in D. Let $v(s;x)$, $w(s;x)$, and $e(s;x)$ be the solutions of the characteristic initial value problems*

$$(-1)^n v_s(s;x) - [(b(s) + h(s) + q(s)) \vee c(s) \vee p(s)]v(s;x) = 0 \qquad \text{in } D,$$
$$v(s;x) = 1 \qquad \text{on } s_i = x_i,\ i = 1, \dots, n,$$

$$(-1)^n w_s(s;x) - [b(s) + h(s) \vee p(s) - c(s)]w(s;x) = 0 \qquad \text{in } D,$$
$$w(s;x) = 1 \qquad \text{on } s_i = x_i,\ i = 1, \dots, n,$$

$$(-1)^n e_s(s;x) - [b(s) - c(s)]e(s;x) = 0 \qquad \text{in } D,$$
$$e(s;x) = 1 \qquad \text{on } s_i = x_i,\ i = 1, \dots, n,$$

respectively, and let D^+ be a connected subdomain of D, containing x, on which $v \geq 0$, $v \geq 0$, $e \geq 0$ for all $s \in D^+$. If $[\alpha; x] \subset D^+$ and

$$\begin{aligned} u(x) \leq & a(x) + \int_\alpha^x b(x)u(s)\,ds + \int_\alpha^x c(s)\left\{\int_\alpha^s h(\tau)u(\tau)\,d\tau\right\} ds + \\ & + \int_\alpha^x c(s)\left(\int_\alpha^s p(t)\left(\int_\alpha^t q(\tau)u(\tau)\,d\tau\right) dt\right) ds, \end{aligned} \tag{12.62}$$

then

$$\begin{aligned} u(x) \leq & a(x) + \int_\alpha^x e(s;x)\left[a(s)b(s) + c(s)\int_\alpha^s w(t;s) \times \right. \\ & \left. \times \left\{a(t)[b(t) + h(t)] + p(t)\int_\alpha^t v(\tau;t)a(\tau)[b(\tau) + h(\tau) + q(\tau)]\right\} dt\right] ds. \end{aligned} \tag{12.63}$$

PROOF. Inequality (12.62) is equivalent to the system

$$\begin{aligned} u(x) &\leq a(x) + u_1(x), \\ u_1(x) &= \int_\alpha^x [b(s)u(s) + c(s)u_2(s)]\,ds, \\ u_2(x) &= \int_\alpha^x [h(t)u(t) + p(t)u_3(t)]\,dt, \\ u_3(x) &= \int_\alpha^x q(\tau)u(\tau)\,d\tau. \end{aligned} \tag{12.64}$$

This implies that

$$\left.\begin{aligned} u_{1x}(x) \leq & a(x)b(x) + b(x)u_1(x) + c(x)u_2(x), \\ u_{2x}(x) \leq & a(x)h(x) + h(x)u_1(x) + p(x)u_2(x), \\ u_{3x}(x) \leq & a(x)q(x) + q(x)u_1(x). \end{aligned}\right\} \tag{12.65}$$

Adding these inequalities leads to

$$\begin{aligned}(u_1+u_2+u_3)_x(x) \leq & a(x)[b(x)+h(x)+q(x)]+ \\ &+[(b(x)+h(x)+q(x)) \vee c(x) \vee p(x)][u_1(x)+u_2(x)+u_3(x)].\end{aligned}$$

By lemma 12.2,

$$u_1(x)+u_2(x)+u_3(x) \leq \int_\alpha^x v(\tau;x)a(\tau)[b(\tau)+h(\tau)+q(\tau)]\,d\tau \equiv R_3(x). \tag{12.66}$$

Then (12.65) and $u_3 \leq R_3 - u_1 - u_2$ imply that

$$\begin{aligned}(u_1+u_2)_x(x) \leq & a(x)[b(x)+h(x)]+p(x)R_3(x)+ \\ &+[(b(x)+h(x)) \vee c(x) - p(x)][u_1(x)+u_2(x)].\end{aligned}$$

Applying lemma 12.2 once more, we obtain

$$u_1(x)+u_2(x) \leq \int_\alpha^x \{a(\tau)[b(\tau)+h(\tau)]+p(\tau)R_3(\tau)\}w(\tau;x)\,d\tau \equiv R_2(x). \tag{12.67}$$

Consequently, (12.65) and $u_2 \leq R_2 - u_1$ imply that

$$u_{1x}(x) \leq a(x)b(x)+c(x)R_2(x)+[b(x)-c(x)]u_1(x),$$

so that lemma 12.2 gives

$$u_1(x) \leq \int_\alpha^x [a(s)b(s)+c(s)R_2(s)]e(s;x)\,ds. \tag{12.68}$$

Now (12.64), (12.68),(12.67), and (12.66) imply (12.63). □

REMARK 12.4. In the course of the proof we have also obtained the estimates

$$u(x) \leq a(x)+\int_\alpha^x v(\tau;x)a(\tau)[b(\tau)+h(\tau)+q(\tau)]\,d\tau \tag{12.69}$$

and

$$\begin{aligned}u(x) \leq a(x)+&\int_\alpha^x w(t;x)\,\{a(t)[b(t)+h(t)]+ \\ &+p(t)\int_\alpha^t v(\tau;t)a(\tau)[b(\tau)+h(\tau)+q(\tau)]\,d\tau\Big\}\,dt. \qquad \square\end{aligned} \tag{12.70}$$

To apply more effectively the estimates obtained in theorems 12.13–12.18, it is desirable to know the analytic form of the solution $v(s;x)$ of the characteristic initial value problem (12.37). For $n=1$,

$$v(s;x) = \exp\left(\int_s^x b(\tau)\,d\tau\right).$$

However, for $n \geq 2$, such an analytic representation is impossible. Fortunatedly, it turns out that for $n \geq 1$,

$$v(s;x) \leq \exp\left(\int_s^x b(\tau)\,d\tau\right). \tag{12.71}$$

To arrive at (12.71), we first note that the problem (12.37) is equivalent to the integral equation

$$v(s;x) = 1 + \int_s^x b(\tau)v(\tau;x)\,d\tau. \tag{12.72}$$

The solution of (12.72) can be represented as the sum of the series

$$v(s;x) = \sum_{m=0}^{\infty} v_m(s;x), \tag{12.73}$$

where $v_0(s;x) \equiv 1$,

$$v_m(s;x) = \int_s^x b(\tau)v_{m-1}(\tau;x)\,d\tau, \qquad m = 1,2,\dots. \tag{12.74}$$

We can obtain (12.71) from (12.73)–(12.74), but we will first prove some assertions justifying (12.73), and being of independent interest.

Let $U(x_0, r) = \{x \in \mathbf{R}^n\colon |x - x_0| < r\}$.

Let $x, y \in \mathbf{R}^n$, and let $Q(x;y)$ be sets satisfying the condition H12:

H12.1) $Q(x;y)$ is a compact measurable subset of $\mathbf{R}^n$;
H12.2) Measure$[Q(x;y) \setminus Q(x;z)] \cup [Q(x;z) \setminus Q(x;y)] \to 0$ as $y \to z$;
H12.3) $Q(x;y) \subseteq U(x, |x-y|)$;
H12.4) if $z \in Q(x;y)$, then $Q(x;z) \subseteq Q(x;y)$;
H12.5) $Q(x;z) = Q(x;y)$ if and only if $z = y$;
H12.6) if $z \in Q(x;y)$, then $\{w \in Q(x;y)\colon z \in Q(x;w)\} = Q(z;y)$.

Note that, e.g., the n-dimensional intervals

$$I[x;y] = \{z \in \mathbf{R}^n : z_i = \lambda_i x_i + (1-\lambda_i)y_i, \quad 0 \le \lambda_i \le 1, \quad i = 1,\dots,n\}$$

satisfy the condition H12.

THEOREM 12.19. (Simeonov and Bainov, 1983) *Let $Q(x;y)$ be sets satisfying the conditions* H12.1–4. *Let $u, a\colon Q(\alpha;\beta) \to \mathbf{R}$, $G\colon Q(\alpha;\beta) \times Q(\alpha;\beta) \to \mathbf{R}$ be continuous functions, with G nonnegative. For $x \in Q(\alpha;\beta)$, let*

$$u(x) \le a(x) + \int_{Q(\alpha;x)} G(x,s)u(s)\,ds. \tag{12.75}$$

Then

1) *$u(x) \le v(x)$, $x \in Q(\alpha;\beta)$, where $v(x)$ is a solution of the equation*

$$v(x) = a(x) + \int_{Q(\alpha;x)} G(x,s)v(s)\,ds. \tag{12.76}$$

2) *The solution $v(x)$ of (12.76) is unique, and can be represented as the sum of the series*

$$v(x) = v_0(x) + v_1(x) + \dots + v_m(x) + \dots, \tag{12.77}$$

where $v_0(x) \equiv a(x)$,

$$v_m(x) = \int_{Q(\alpha;x)} G(x,s)v_{m-1}(s)\,ds, \qquad m = 1,2,\dots.$$

PROOF. We define on the space C of continuous functions $w\colon Q(\alpha;\beta) \to \mathbf{R}$ with norm $\|w\| = \max_{z\in Q(\alpha;\beta)} |w(z)|$ the operator T by

$$Tw(x) = \int_{Q(\alpha;x)} G(x,s)w(s)\,ds.$$

The conditions H12.1–2 imply that $T(C) \subset C$. The nonnegativity of G and (12.75) successively imply that

$$\begin{aligned} u \le & a + T(a+Tu) = a + Ta + T^2u \le a + Ta + T^2a + T^3u \le \\ \le & a + Ta + T^2a + \dots + T^m a + \dots, \end{aligned}$$

or

$$u(x) \le v(x),$$

where $v(x) = v_0(x) + v_1(x) + \dots + v_m(x) + \dots,$

$$v_0(x) \equiv a(x), \qquad v_m(x) = T^m a(x) = \int_{Q(\alpha;x)} G(x,s)v_{m-1}(s)\,ds.$$

By induction with respect to m we can derive from the conditions H12.3, H12.1, and H12.4 that

$$\|T^m w\| \le \frac{Q^m}{m!}\|w\|, \qquad w \in C, \qquad m = 1,2,\dots, \tag{12.78}$$

where $Q = \frac{|\beta-\alpha|^n}{n!}\cdot \sigma M$, where $M = \max_{Q(\alpha;\beta)\times Q(\alpha;\beta)} |G(x,s)|$ and $\sigma = \int_{S(1)} ds$ is the measure (area) of the unit sphere $S(1) = \{x \in \mathbf{R}^n\colon |x| = 1\}$.

In particular, (12.78) holds for $w = a$. Weierstrass' theorem and (12.78) imply that the series (12.77) is uniformly convergent in $Q(\alpha;\beta)$. Straightforward verification shows that its sum $v(x)$ satisfies (12.76). Since, moreover, (12.78) implies that some power of T is a contraction, $v(x)$ is the unique solution of (12.76). □

LEMMA 12.3. (Simeonov and Bainov, 1983) *Let $Q(x;y)$ be sets satisfying the conditions* H12.1, H12.4–5, *and let $b\colon Q(\alpha;\beta) \to \mathbf{R}$ be a nonnegative continuous function. Then the functions*

$$v_0(x) \equiv 1, \qquad v_m(x) = \int_{Q(\alpha;x)} b(s)v_{m-1}(s)\,ds, \qquad x \in Q(\alpha;\beta), \qquad m = 1,2\dots,$$

satisfy the estimate

$$0 \le v_m(x) \le \frac{1}{m!}\left(\int_{Q(\alpha;x)} b(s)\,ds\right)^m, \qquad x \in Q(\alpha;\beta), \qquad m = 0,1\dots. \tag{12.79}$$

PROOF. Let $\nu = (1,\dots,m)$, and let μ be a permutation of ν, i.e. $\mu = (i_1,\dots,i_m)$. Then we define the following $m!$ sets:

$$\begin{aligned} P_m(\alpha,x;\mu) = \{(s_1,\dots,s_m) \in \mathbf{R}^{mn} : \\ z_{i_1} \in Q(\alpha;x), z_{i_2} \in Q(\alpha;z_{i_1}),\dots, z_{i_m} \in Q(\alpha;z_{i_{m-1}})\}. \end{aligned}$$

In particular, when $\mu = \nu$ we set

$$P_m(\alpha, x) = P_m(\alpha, x; \nu) = \{(s_1, \ldots, s_m) \in \mathbf{R}^{mn} : \\ z_1 \in Q(\alpha; x), z_2 \in Q(\alpha; z_1), \ldots, z_m \in Q(\alpha; z_{m-1})\}.$$

We also introduce the notation

$$D_m(x) = \{(s_1, \ldots, s_m) \in \mathbf{R}^{mn} : \ s_1 \in Q(\alpha; x), \ldots, s_m \in Q(\alpha; x)\}.$$

With this notation the functions $v_m(x)$ take the form

$$v_m(x) = \int \cdots \int_{P_m(\alpha;x)} b(s_1) \ldots b(s_m)\, ds_1 \ldots ds_m. \tag{12.80}$$

We divide the proof in several steps.

I. The equality

$$\int \cdots \int_{P_m(\alpha,x;\mu)} b(s_1) \ldots b(s_m)\, ds_1 \ldots ds_m = \int \cdots \int_{P_m(\alpha;x)} b(s_1) \ldots b(s_m)\, ds_1 \ldots ds_m \tag{12.81}$$

holds, since each integral in (12.81) is equal to the integral

$$\int \cdots \int_{P_m(\alpha,x;\mu)} b(s_{i_1}) \ldots b(s_{i_m})\, ds_{i_1} \ldots ds_{i_m}.$$

II. The definitions of $P_m(\alpha, x; \mu)$ and $D_m(x)$ imply that

$$\cup_\mu P_m(\alpha, x; \mu) \subset D_m(x), \tag{12.82}$$

where the union is over all permutations μ.

III. Let μ and λ be two distinct permutations, in which the numbers k and j form an inversion.

Let $(s_1, \ldots, s_k, \ldots, s_j, \ldots, s_m) \in P_m(\alpha, x; \mu) \cap P_m(\alpha, x; \lambda)$. The definition of the sets $P_m(\alpha, x; \mu)$ implies that $s_k \in Q(\alpha; s_j)$ and $s_j \in Q(\alpha; s_k)$, and H12.4 implies that $Q(\alpha; s_k) \subset Q(\alpha; s_j)$ and $Q(\alpha; s_j) \subset Q(\alpha; s_k)$, i.e. $Q(\alpha; s_j) = Q(\alpha; s_k)$.

Consequently, by H12.5, $s_k = s_j$, so that the set

$$P_m(\alpha, x; \mu) \cap P_m(\alpha, x; \lambda)$$

is contained in the subset $L = \{(s_1, \ldots, s_m) \in \mathbf{R}^{mn}\colon s_k = s_j\}$ of $\mathbf{R}^{mn}$. Since $\dim L < mn$, we have

$$\int \cdots \int_{P_m(\alpha,x;\mu) \cap P_m(\alpha,x;\lambda)} b(s_1) \ldots b(s_m)\, ds_1 \ldots ds_m = 0. \tag{12.83}$$

IV. The following equality holds:

$$\int \cdots \int_{\cup_\mu P_m(\alpha,x;\mu)} b(s_1) \ldots b(s_m)\, ds_1 \ldots ds_m = \\ = \sum_\mu \int \cdots \int_{P_m(\alpha,x;\mu)} b(s_1) \ldots b(s_m)\, ds_1 \ldots ds_m + J, \tag{12.84}$$

where in J we have collected the integrals over all possible intersections of sets of the type $P_m(\alpha,x;\mu)$, $P_m(\alpha,x;\lambda)$ with distinct μ,λ. But (12.83) implies that all such integrals vanish, i.e. $J=0$.

Thus, (12.84), (12.82) and the nonnegativity of b imply

$$\sum_{\mu}\int\cdots\int_{P_m(\alpha,x;\mu)} b(s_1)\dots b(s_m)\,ds_1\dots ds_m \le$$
$$\le \int\cdots\int_{D_m(x)} b(s_1)\dots b(s_m)\,ds_1\dots ds_m = \left(\int_{Q(\alpha;x)} b(s)\,ds\right)^m,$$

and (12.81) and (12.80) imply (12.79). The inequalities $v_m(x)\ge 0$, $m=0,1,\dots$, are obvious. □

COROLLARY 12.6. *If $b(x)\ge 0$ is continuous for $x\in[\alpha;\beta]$, $\alpha<\beta$, then the solution $v(x)$ of (12.37) satisfies the estimate*

$$0\le v(s;x)\le \exp\left(\int_s^x b(\tau)\,d\tau\right),\qquad \alpha\le s\le x\le\beta. \tag{12.85}$$

PROOF. Indeed, theorem 12.19 with $G(x;s)=b(s)$, $a(x)=1$, and $Q(s;x)=[s;x]$ implies the representation (12.73). By lemma 12.3 with $Q(s;x)=[s;x]$ we obtain

$$0\le v(s;x)=\sum_{m=0}^{\infty} v_m(s;x)\le\sum_{m=0}^{\infty}\frac{1}{m!}\left(\int_s^x b(\tau)\,d\tau\right)^m=\exp\left(\int_s^x b(\tau)\,d\tau\right). \qquad \square$$

COROLLARY 12.7. (Simeonov and Bainov, 1983) *If $\alpha,\beta\in\mathbf{R}^n$, $\alpha<\beta$, and $u(x)$, $a(x)$, $b(x)$, $q(x)$ are nonnegative continuous functions in $[\alpha;\beta]$, then the inequality*

$$u(x)\le a(x)+q(x)\int_\alpha^x b(s)u(s)\,ds,\qquad x\in[\alpha;\beta],$$

implies

$$u(x)\le a(x)+q(x)\int_\alpha^x a(s)b(s)\exp\left(\int_s^x q(\tau)b(\tau)\,d\tau\right)ds,\qquad x\in[\alpha;\beta]. \tag{12.86}$$

In particular, if $q(x)=1$ and $a(x)$ is nondecreasing in $[\alpha;\beta]$, then

$$u(x)\le a(x)\left[1+\int_\alpha^x b(s)\exp\left(\int_s^x b(\tau)\,d\tau\right)ds\right],\qquad x\in[\alpha;\beta]. \tag{12.87}$$

PROOF. Indeed, (12.87) follows from the estimate (12.45) in theorem 12.14 and corollary 12.6. Moreover, (12.86) is better than the estimate

$$u(x)\le a(x)\exp\left(\int_\alpha^x b(s)\,ds\right). \qquad \square \tag{12.88}$$

REMARK 12.5. Analysis of the proof of lemma 12.3 shows that (12.79) also holds if condition H12.1 is replaced by the condition

$$\int_{Q(\alpha;\beta)} b(s)\,ds<\infty. \qquad \square$$

To obtain estimates similar to (12.86)–(12.88) with $[\alpha; x]$ replaced by $Q(\alpha; x)$, we will use the following lemma.

LEMMA 12.4. *Let $Q(x; y)$ be sets satisfying the conditions* H12.4–6, *and let a, b, q: $Q(\alpha; \beta) \to \mathbf{R}$ be nonnegative continuous functions, with*

$$\int_{Q(\alpha;\beta)} q(s)b(s)\,ds < \infty, \qquad \int_{Q(\alpha;\beta)} a(s)b(s)\,ds < \infty.$$

If $v_0(x) = a(x)$, then the functions

$$v_m(x) = q(x)\int_{Q(\alpha;x)} b(s)v_{m-1}(s)\,ds, \qquad m = 1, 2, \dots,$$

satisfy the estimate

$$0 \le v_{m+1}(x) \le \frac{q(x)}{m!}\int_{Q(\alpha;x)} a(s)b(s)\left(\int_{Q(s;x)} q(\tau)b(\tau)\,d\tau\right)^m ds, \tag{12.89}$$

$m = 0, 1, \dots$.

PROOF. For the function $v_{m+1}(x)$, $m = 0, 1, \dots$, we have

$$\begin{aligned}
v_{m+1}(x) = &\int_{Q(\alpha;x)} q(x)b(s_1)\left(\int_{Q(\alpha;s_1)} q(s_1)b(s_2)\left(\int_{Q(\alpha;s_2)} \cdots\right.\right.\\
&\cdots\left.\left.\left(\int_{Q(\alpha;s_m)} q(s_m)b(s)a(s)\,ds\right) ds_m\right)\cdots\right) ds_1 =\\
=&\int\cdots\int_{P_{m+1}(\alpha,x)} q(x)q(s_1)b(s_1)\dots q(s_m)b(s_m)a(s)b(s)\,ds_1\dots ds_m\,ds =\\
=&q(x)\int_{Q(\alpha;x)} a(s)b(s)\left(\int\cdots\int_{Q(s)} q(s_1)b(s_1)\dots q(s_m)b(s_m)\,ds_1\dots ds_m\right) ds,
\end{aligned}$$

where $Q(s) = \{(s_1, \dots, s_m) \in \mathbf{R}^{mn}\colon (s_1, \dots, s_m, s) \in P_{m+1}(\alpha, x)\}$. For fixed $s \in Q(\alpha; x)$ we find, using H12.4 and H12.6,

$$\begin{aligned}
Q(s) =&\{(s_1, \dots, s_m) \in \mathbf{R}^{mn} :\\
&\quad s_1 \in Q(\alpha; x), s_2 \in Q(\alpha; s_1), \dots, s_m \in Q(\alpha; s_{m-1}),\\
&\quad s \in Q(\alpha, s_1), \dots, s \in Q(\alpha; s_m)\} =\\
=&\{(s_1, \dots, s_m) \in \mathbf{R}^{mn} :\\
&\quad s_1 \in Q(s; x), s_2 \in Q(s; s_1), \dots, s_m \in Q(s; s_{m-1})\}\\
=&P_m(s, x).
\end{aligned}$$

By lemma 12.3 we obtain

$$\int\cdots\int_{P_m(s,x)} q(s_1)b(s_1)\dots q(s_m)b(s_m)\,ds_1\dots ds_m \le \frac{1}{m!}\left(\int_{Q(s;x)} q(\tau)b(\tau)\,d\tau\right)^m,$$

which implies (12.89). □

THEOREM 12.20. *Let $Q(x;y)$ be sets satisfying the conditions* H12.4–6, *and let $u, a, b, q\colon Q(\alpha;\beta) \to \mathbf{R}$ be nonnegative continuous functions, with*

$$\int_{Q(\alpha;\beta)} u(s)b(s)\,ds < \infty, \qquad \int_{Q(\alpha;\beta)} q(s)b(s)\,ds < \infty, \qquad \int_{Q(\alpha;\beta)} a(s)b(s)\,ds < \infty. \tag{12.90}$$

Then the inequality

$$u(x) \le a(x) + q(x)\int_{Q(\alpha;x)} b(s)u(s)\,ds, \qquad x \in Q(\alpha;\beta), \tag{12.91}$$

implies

$$u(x) \le a(x) + q(x)\int_{Q(\alpha;x)} a(s)b(s)\exp\left(\int_{Q(s;x)} q(\tau)b(\tau)\,d\tau\right) ds, \qquad x \in Q(\alpha;\beta). \tag{12.92}$$

PROOF. Since $b \ge 0$, $q \ge 0$, as in theorem 12.19 with $G(x,s) = q(x)b(s)$, (12.91) implies

$$u(x) \le \sum_{j=0}^{m} v_j(x) + R_{m+1}(x), \qquad x \in Q(\alpha;\beta), \tag{12.93}$$

where $v_0(x) = a(x)$,

$$v_j(x) = q(x)\int_{Q(\alpha;x)} b(s)v_{j-1}(s)\,ds, \qquad x \in Q(\alpha;\beta), \qquad j = 1,2,\dots,$$

$$R_{m+1}(x) = \int_{Q(\alpha;x)} q(x)b(s_1)\left(\int_{Q(\alpha;s_1)} q(s_1)b(s_2)\dots\right.$$
$$\left.\dots\left(\int_{Q(\alpha;s_m)} q(s_m)b(s)u(s)\,ds\right)\dots\right) ds_1.$$

By lemma 12.4,

$$0 \le v_j(x) \le q(x)\int_{Q(\alpha;x)} a(s)b(s)\frac{1}{(j-1)!}\left(\int_{Q(s;x)} q(\tau)b(\tau)\,d\tau\right)^{j-1} ds, \tag{12.94}$$

$$0 \le R_{m+1}(x) \le q(x)\int_{Q(\alpha;x)} u(s)b(s)\frac{1}{m!}\left(\int_{Q(s;x)} q(\tau)b(\tau)\,d\tau\right)^{m} ds, \tag{12.95}$$

for $x \in Q(\alpha;\beta)$, $j = 1,2,\dots$, $m = 0,1,\dots$.

(12.90) and (12.95) imply that $R_{m+1}(x) \to 0$ as $m \to \infty$. Hence, by (12.93),

$$u(x) \le \sum_{j=0}^{\infty} v_j(x).$$

This and (12.94) imply (12.92). □

COROLLARY 12.8. *Let* $u, a, b, q\colon \mathbf{R}_+^2 \to \mathbf{R}$ *be nonnegative continuous functions, with*

$$u(x,y) \le a(x,y) + q(x,y)\int_x^\infty\int_y^\infty b(s,t)u(s,t)\,ds\,dt, \qquad x \ge 0, \quad y \ge 0,$$

where

$$\int_0^\infty\int_0^\infty u(s,t)b(s,t)\,ds\,dt < \infty, \qquad \int_0^\infty\int_0^\infty q(s,t)b(s,t)\,ds\,dt < \infty,$$
$$\int_0^\infty\int_0^\infty a(s,t)b(s,t)\,ds\,dt < \infty.$$

Then for $x \ge 0$, $y \ge 0$,

$$u(x,y) \le a(x,y) + q(x,y)\int_x^\infty\int_y^\infty a(s,t)b(s,t)\exp\left(\int_x^s\int_y^t q(\sigma,\tau)b(\sigma,\tau)\,d\sigma\,d\tau\right)ds\,dt. \tag{12.96}$$

In particular, if $a(x,y) \equiv a \ge 0$ *and* $q(x,y) \equiv 1$, *then*

$$u(x,y) \le a\left[1 + \int_x^\infty\int_y^\infty b(s,t)\exp\left(\int_x^s\int_y^t b(\sigma,\tau)\,d\sigma\,d\tau\right)ds\,dt\right], \tag{12.97}$$

and this estimate is better than the estimate

$$u(x,y) \le a\exp\left(\int_x^\infty\int_y^\infty b(\sigma,\tau)\,d\sigma\,d\tau\right). \qquad \square \tag{12.98}$$

COROLLARY 12.9. *Let* $u, a, b, q\colon \mathbf{R}_+^2 \to \mathbf{R}$ *be nonnegative continuous functions, with*

$$u(x,y) \le a(x,y) + q(x,y)\int_0^x\int_y^\infty b(s,t)u(s,t)\,ds\,dt, \qquad x \ge 0, \quad y \ge 0,$$

where

$$\int_0^\infty\int_0^\infty u(s,t)b(s,t)\,ds\,dt < \infty, \qquad \int_0^\infty\int_0^\infty q(s,t)b(s,t)\,ds\,dt < \infty,$$
$$\int_0^\infty\int_0^\infty a(s,t)b(s,t)\,ds\,dt < \infty.$$

Then for $x \ge 0$, $y \ge 0$,

$$u(x,y) \le a(x,y) + q(x,y)\int_0^x\int_y^\infty a(s,t)b(s,t)\exp\left(\int_s^x\int_y^t q(\sigma,\tau)b(\sigma,\tau)\,d\sigma\,d\tau\right)ds\,dt. \tag{12.99}$$

In particular, if $a(x,y) \equiv a \ge 0$ *and* $q(x,y) \equiv 1$, *then*

$$u(x,y) \le a\left[1 + \int_0^x\int_y^\infty b(s,t)\exp\left(\int_s^x\int_y^t b(\sigma,\tau)\,d\sigma\,d\tau\right)ds\,dt\right], \tag{12.100}$$

and this estimate is better than the estimate

$$u(x,y) \le a\exp\left(\int_0^x\int_y^\infty b(\sigma,\tau)\,d\sigma\,d\tau\right). \qquad \square \tag{12.101}$$

Indeed, these corollaries follow from theorem 12.20 in case $n = 2$ and

$$\begin{aligned}
Q[(s,t);(x,y)] &= \{(\sigma,\tau) \in \mathbf{R}_+^2 : x \le \sigma \le s, \quad y \le \tau \le t\},\\
\alpha = (\infty,\infty), &\qquad \beta = (0,0) \qquad \text{(for corollary 12.8)};\\
Q[(s,t);(x,y)] &= \{(\sigma,\tau) \in \mathbf{R}_+^2 : s \le \sigma \le x, \quad y \le \tau \le t\},\\
\alpha = (0,\infty), &\qquad \beta = (\infty,0) \qquad \text{(for corollary 12.9)}.
\end{aligned}$$

13. Integro-differential inequalities

We begin with the following comparison result for integro-differential inequalities.

THEOREM 13.1. (Lakshmikantham, Leela, and Rao, 1987) *Let $f\colon \mathbf{R}_+^2 \to \mathbf{R}$ and $H\colon \mathbf{R}_+^3 \to \mathbf{R}$ be continuous functions in their domains, with $H(t,s,u)$ nondecreasing with respect to u for any $(t,s) \in \mathbf{R}_+^2$, while for $t \ge t_0 \ge 0$,*

$$D^+u(t) \le f(t,u(t)) + \int_{t_0}^{t} H(t,s,u(s))\,ds, \tag{13.1}$$

where $u(t)$ is a nonnegative continuous function in $\mathbf{R}_+$. Suppose that $r(t)$ is the maximal solution of the equation

$$r'(t) = f(t,r(t)) + \int_{t_0}^{t} H(t,s,r(s))\,ds, \qquad r(t_0) = r_0 \ge 0, \tag{13.2}$$

in $[t_0,\infty)$. Then

$$u(t) \le r(t), \qquad t \ge t_0,$$

provided that $u(t_0) \le r_0$.

For a proof of the existence of the maximal solution of (13.2) see Hu, Zhuang, and Khavanin [99].

PROOF. We define functions

$$F(t,u) = f(t,m(t,u)), \qquad K(s,t,u) = H(t,s,m(s,u)),$$

where $m(t,u) = \max[u(t),u]$, and let $v(t)$ be any solution of

$$v'(t) = F(t,v(t)) + \int_{t_0}^{t} K(t,s,v(s))\,ds, \qquad v(t_0) \ge u(t_0).$$

We will show that $u(t) \le v(t)$ for $t \ge t_0$. If this were not true, then $u(t^*) > v(t^*)$ for some $t^* > t_0$. Then there is a $t_1 > t_0$ such that

$$v(t_1) \le u(t_1) \text{ and } v'(t_1) < D^+u(t_1).$$

It is clear from the definition of $m(t,u)$ that $u(t) \le m(t, v(t))$ and $u(t_1) = m(t_1, v(t_1))$. Using the monotonicity of H, we have

$$\begin{aligned} D^+u(t_1) &\le f(t_1, u(t_1)) + \int_{t_0}^{t_1} H(t_1, s, u(s))\,ds \le \\ &\le f(t_1, m(t_1, v(t_1))) + \int_{t_0}^{t_1} H(t_1, s, m(s, v(s)))\,ds = \\ &= F(t_1, v(t_1)) + \int_{t_0}^{t_1} K(t_1, s, v(s))\,ds = v'(t_1), \end{aligned}$$

which is a contradiction. Thus, $u(t) \le v(t)$ implies that $v(t)$ is also a solution of (13.2). Since $r(t)$ is the maximal solution of (13.2), it follows that $u(t) \le r(t)$, $t \ge t_0$. □

COROLLARY 13.1. (Lakshmikantham, Leela, and Rao, 1987) *Let (13.1) hold with $f = a(t)u$, $H = H(t,s)u$. Then*

$$m(t) \le R(t, t_0) r_0, \tag{13.3}$$

where $u(t_0) \le r_0$ and $R(t,s)$ is the solution of

$$\frac{\partial R}{\partial s}(t,s) + R(t,s)a(s) + \int_s^t R(t,\tau)H(\tau,s)\,d\tau = 0, \qquad R(t,t) = 1,$$

in the interval $t_0 \le s \le t$.

PROOF. It suffices to note that $r(t) = R(t, t_0) r_0$ is the solution of (13.2). □

Even when f and H are linear, in general it is difficult to find $R(t,s)$. Hence a comparison result which enables us to reduce integro-differential inequalities to differential inequalities will be of great advantage. Such a result is considered below, but first we need the following lemma.

LEMMA 13.1. *Let $g_0, g\colon \mathbf{R}_+^2 \to \mathbf{R}$ be continuous functions, and let*

$$g_0(t,v) \le g(t,v), \qquad (t,v) \in \mathbf{R}_+^2. \tag{13.4}$$

Then for every solution $w(t)$ of

$$w' = g(t,w), \qquad w(t_0) = w_0 \ge 0, \qquad t \in [t_0, T], \tag{13.5}$$

and the left maximal solution $\eta(t; T, v_T)$ of

$$v' = g_0(t,v), \qquad v(T) = v_T \ge 0, \qquad t \in [t_0, T], \tag{13.6}$$

there is the relation

$$w(t) \le \eta(t; T, v_T), \qquad t \in [t_0, T], \tag{13.7}$$

whenever $w(T) \le v_T$.

PROOF. It is well known that

$$\eta(t;T,v_T)=\lim_{\epsilon\to 0_+}\eta(t,\epsilon),$$

where $v(t,\epsilon)$ is some solution of

$$\eta'=g_0(t,v)-\epsilon, \qquad \eta(T)=v_T+\epsilon,$$

on the left of T, and $\epsilon\in(0,\epsilon_0]$, with $\epsilon_0>0$ sufficiently small. We note that (13.7) follows if we can prove that

$$w(t)<\eta(t,\epsilon), \qquad t\in[t_0,T], \qquad \epsilon\in(0,\epsilon_0].$$

Since $g_0-\epsilon<g_0\le g$ and $w(T)<v_T+\epsilon=\eta(T)$, it is readily seen that there is a $\delta>0$ such that

$$w(t)<\eta(t,\epsilon), \qquad t\in[T-\delta,T].$$

We claim that

$$w(t)<\eta(t,\epsilon), \qquad t\in[t_0,T], \qquad \epsilon\in(0,\epsilon_0].$$

Assuming this to be not true, for some $\epsilon^*\in(0,\epsilon_0]$ there is a $t^*\in[t_0,T)$ such that

$$\begin{aligned} w(t^*)&=\eta(t^*,\epsilon^*),\\ w(t)&<\eta(t,\epsilon^*), \qquad t\in(t^*,T]. \end{aligned}\tag{13.8}$$

Then

$$w'(t^*)=g(t^*,w(t^*))>g_0(t^*,\eta(t^*,\epsilon^*))-\epsilon^*=\eta'(t^*,\epsilon^*),$$

whence $w(t)>\eta(t,\epsilon^*)$ in a neighbourhood $(t^*,t^*+\sigma]$, contradicting (13.8). □

THEOREM 13.2. (Lakshmikantham, Leela, and Rao, 1987) *Let $f\colon \mathbf{R}_+^2\to\mathbf{R}$ and $H\colon\mathbf{R}_+^3\to\mathbf{R}$ be continuous functions in their domains, and let*

$$D_-u(t)\le f(t,u(t))+\int_{t_0}^t H(t,s,u(t))\,ds, \qquad t\in I_0, \tag{13.9}$$

where $u(t)$ is a nonnegative continuous function in $\mathbf{R}_+$, $I_0=\{t\ge t_0\colon u(s)\le \eta(s;t,u(t)),\ t_0\le s\le t\}$, and $\eta(t;T,v_T)$ is the left maximal solution of (13.6) *in $[t_0,T]$. Suppose that $g_0\colon\mathbf{R}_+^2\to\mathbf{R}$ is a continuous function satisfying*

$$g_0(t,v)\le F(t,v;t_0), \qquad (t,v)\in\mathbf{R}_+^2,$$

where

$$F(t,v;t_0)=f(t,v)+\int_{t_0}^t H(t,s,v)\,ds.$$

If $r(t)$ is the maximal solution of

$$r'=F(t,r;t_0), \qquad r(t_0)=r_0, \tag{13.10}$$

in $[t_0,\infty)$ and $u(t_0)\le r_0$, then

$$u(t)\le r(t), \qquad t\ge t_0. \tag{13.11}$$

PROOF. It is well known that on each compact set $[t_0, T] \subset [t_0, \infty)$,

$$r(t) = \lim_{\epsilon \to 0_+} r(t, \epsilon),$$

where $r(t, \epsilon)$ is some solution of

$$r' = F(t, r; t_0) + \epsilon, \qquad r(t_0) = r_0 + \epsilon,$$

and $\epsilon \in (0, \epsilon_0]$, with $\epsilon_0 > 0$ sufficienty small. For this it suffices to prove that

$$u(t) < r(t, \epsilon), \qquad t \in [t_0, T], \qquad \epsilon \in (0, \epsilon_0].$$

Assuming this to be not true, for some $\epsilon^* \in (0, \epsilon_0]$ there is a $t^* \in (t_0, T]$ such that

$$u(t) < r(t, \epsilon^*), \qquad t \in [t_0, t^*), \tag{13.12}$$

$$u(t^*) = r(t^*, \epsilon^*). \tag{13.13}$$

Then (13.13) and lemma 13.1 imply that

$$r(s, \epsilon^*) \le \eta(s; t^*, u(t^*)), \qquad s \in [t_0, t^*], \tag{13.14}$$

where $\eta(s; t^*, u(t^*))$ is the left maximal solution of

$$\frac{d\eta}{ds} = g_0(s, \eta), \qquad \eta(t^*) = u(t^*), \qquad s \in [t_0, t^*].$$

From (13.12)–(13.14) it follows that $u(s) \le \eta(s; t^*, u(t^*))$ for $s \in [t_0, t^*]$. Hence $t^* \in I_0$, and by (13.9),

$$D_- u(t^*) \le F(t^*, u(t^*); t_0) < F(t^*, r(t^*, \epsilon^*); t_0) + \epsilon^* = r'(t^*, \epsilon^*).$$

This and (13.13) imply that

$$u(t) > r(t, \epsilon^*)$$

in a small left neighbourhood $[t^* - \sigma, t^*]$, contradicting (13.12). □

REMARK 13.1. In theorem 13.2 we can choose $g_0 = f$ if $H \ge 0$. □

We will use the comparison theorem 13.2 to study the integro-differential inequality (13.1). The advantage of this approach will be clear from the following special cases.

THEOREM 13.3. (Lakshmikantham, Leela, and Rao, 1987) *Let $f\colon \mathbf{R}_+^2 \to \mathbf{R}$ and $H\colon \mathbf{R}_+^3 \to \mathbf{R}$ be continuous functions in their domains, with $H(t, s, u)$ nondecreasing with respect to u for each $(t, s) \in \mathbf{R}_+^2$, and suppose that for $t \ge t_0 \ge 0$,*

$$D_- u(t) \le f(t, u(t)) + \int_{t_0}^{t} H(t, s, u(s))\, ds, \tag{13.15}$$

where $u(t)$ is a nonnegative continuous function in $\mathbf{R}_+$. Suppose that $g_0\colon \mathbf{R}_+^2 \to \mathbf{R}$ is a continuous function satisfying

$$g_0(t, v) \le G(t, v; t_0), \qquad (t, v) \in \mathbf{R}_+^2,$$

where

$$G(t, v; t_0) = f(t, v) + \int_{t_0}^{t} K(t, s, v)\, ds,$$

$$K(t, s, u) = H(t, s, \eta(s; t, u)),$$

and $\eta(s;T,v_T)$ is the left maximal solution of (13.6) *in $t_0 \le s \le T$. If $r(t)$ is the maximal solution of*

$$\frac{dr}{dt} = G(t,r;t_0), \qquad r(t_0) = r_0, \tag{13.16}$$

in $[t_0,\infty)$ and $u(t_0) \le r_0$, then

$$u(t) \le r(t), \qquad t \ge t_0. \tag{13.17}$$

PROOF. We set $I_0 = \{t \ge t_0\colon\ u(s) \le \eta(s;t,u(t)),\ t_0 \le s \le t\}$. Since H is nondecreasing with respect to u, (13.15) implies the inequality

$$D_-u(t) \le f(t,u(t)) + \int_{t_0}^{t} H(t,s,\eta(s;t,u(t)))\,ds, \qquad t \in I_0.$$

Hence theorem 13.2 implies the required result. □

COROLLARY 13.2. (Lakshmikantham, Leela, and Rao, 1987) *For the special case of $f \ge 0$, $H \ge 0$, we have $u(t) \le r(t)$, $t \ge t_0$, where $r(t)$ is the maximal solution of*

$$r'(t) = f(t,r(t)) + \int_{t_0}^{t} H(t,s,r(t))\,ds, \qquad r(t_0) = r_0, \tag{13.18}$$

in $[t_0,\infty)$, provided that $u(t_0) \le r_0$. In particular, if $f = a(t)u$, $H = H(t,s)u$, with $a(t) \ge 0$, $H(s,t,) \ge 0$, then

$$u(t) \le u(t_0)\exp\left(\int_{t_0}^{t} A(s,t_0)\,ds\right), \qquad t \ge t_0, \tag{13.19}$$

where $A(t,t_0) = a(t) + \int_{t_0}^{t} H(t,s)\,ds$. □

COROLLARY 13.3. (Lakshmikantham, Leela, and Rao, 1987) *For the special case of $g_0 = f$, $H \ge 0$, we have $u(t) \le r(t)$, $t \ge t_0$, where $r(t)$ is the maximal solution of* (13.16) *in $[t_0,\infty)$, provided that $u(t_0) \le r_0$. In particular, if $g_0 = f = -\alpha u$, $\alpha > 0$, $H = H(t,s)u$, with $H(t,s) \ge 0$, then*

$$u(t) \le u(t_0)\exp\left(\int_{t_0}^{t} B(s,t_0)\,ds\right), \qquad t \ge t_0, \tag{13.20}$$

where $B(t,t_0) = -\alpha + \int_{t_0}^{t} H(t,s)e^{\alpha(t-s)}\,ds$. □

COROLLARY 13.4. (Lakshmikantham, Leela, and Rao, 1987) *For the special case of $g_0 = f = -c(u)$, $H \ge 0$, where $c(u)$ is a nondecreasing continuous function for $u \ge 0$ with $c(0) = 0$ and $c(u) > 0$ for $u > 0$, we have $u(t) \le r(t)$, $t \ge t_0$, where $r(t)$ is the maximal solution of*

$$r'(t) = -c(r(t)) + \int_{t_0}^{t} H(t,s,J^{-1}[J(r(t)) + t - s])\,ds, \qquad r(t_0) = r_0, \tag{13.21}$$

in $[t_0,\infty)$, provided that $u(t_0) \le r_0$. Here $J(u) = \int_{u_0}^{u} dx/c(x)$ $u > 0$, $u_0 > 0$. □

In the theorems below we consider integro-differential inequalities involving higher-order derivatives. In proving them we use the fact that if a function $u(t)$ has continuous k-th derivative $u^{(k)}(t) = f(t)$, $t \geq 0$, then

$$u^{(j)}(t) = \sum_{i=j}^{k-1} u^{(i)}(0)\frac{t^{i-j}}{(i-j)!} + \frac{1}{(k-j-1)!}\int_0^t (t-\tau)^{k-j-1} f(\tau)\, d\tau \qquad (13.22)$$

for $j = 0, \ldots, k-1$.

THEOREM 13.4. (Thandapani and Agarwal, 1980) *Let $a(t)$, $q(t)$, $b_j(t)$, $u^{(j)}(t)$, $j = 0, \ldots, k$, be nonnegative continuous functions for $t \geq 0$, and suppose*

$$u^{(k)}(t) \leq a(t) + q(t)\sum_{j=0}^{k}\int_0^t b_j(s)u^{(j)}(s)\, ds, \qquad t \geq 0, \qquad (13.23)$$

where $k \geq 0$ is an integer. Then

$$u^{(k)}(t) \leq a(t) + q(t)\int_0^t \phi_1(s)\exp\left(\int_s^t \phi_2(\tau)\, d\tau\right) ds, \qquad t \geq 0, \qquad (13.24)$$

where

$$\begin{aligned}\phi_1(t) =& a(t)b_k(t) + \sum_{j=0}^{k-1}\sum_{i=0}^{j} u^{(j)}(0)b_i(t)\frac{t^{j-i}}{(j-i)!} + \\ &+ \sum_{j=0}^{k-1}\frac{b_{k-j-1}(t)}{j!}\int_0^t (t-x)^j a(x)\, dx,\\ \phi_2(t) =& q(t)b_k(t) + \sum_{j=0}^{k-1}\frac{b_{k-j-1}(t)}{j!}\int_0^t (t-x)^j q(x)\, dx.\end{aligned}$$

PROOF. We set

$$R(t) = \sum_{j=0}^{k}\int_0^t b_j(s)u^{(j)}(s)\, ds,$$

$R(0) = 0$. Then (13.23) implies

$$u^{(k)}(t) \leq a(t) + q(t)R(t). \qquad (13.25)$$

Integrating (13.25) $k - j$ times, we arrive at

$$u^{(j)}(t) \leq \sum_{i=j}^{k-1} u^{(i)}(0)\frac{t^{i-j}}{(i-j)!} + \frac{1}{(k-j-1)!}\int_0^t (t-\tau)^{k-j-1}[a(\tau) + q(\tau)R(\tau)]\, d\tau$$

for $j = 0, \dots, k-1$. Thus, the definiton of $R(t)$ implies that

$$R'(t) = \sum_{j=0}^{k} b_j(t) u^{(j)}(t) \leq$$

$$\leq b_k(t) u^{(k)}(t) + \sum_{j=0}^{k-1} b_j(t) \times$$

$$\times \left\{ \sum_{i=j}^{k-1} u^{(i)}(0) \frac{t^{i-j}}{(i-j)!} + \frac{1}{(k-j-1)!} \int_0^t (t-\tau)^{k-j-1} [a(\tau) + q(\tau) R(\tau)]\, d\tau \right\}.$$

Using (13.25), the nondecrease of $R(t)$, and rearranging terms, we arrive at

$$R'(t) \leq \phi_1(t) + \phi_2(t) R(t), \qquad R(0) = 0,$$

so that lemma 1.1 implies $R(t) \leq \int_0^t \phi_1(s) \exp(\int_s^t \phi_2(\tau)\, d\tau)\, ds$. Together with (13.25) this gives (13.24). □

COROLLARY 13.5. (Thandapani and Agarwal, 1980) *Let in* (13.23), $u^{(j)}(0) = 0$, $j = 0, \dots, k-1$, *and* $a(t)$ *be nondecreasing. Then*

$$u^{(k)}(t) \leq a(t) \left[1 + q(t) \int_0^t \phi_3(s) \exp\left(\int_s^t \phi_4(\tau)\, d\tau \right) ds \right], \qquad t \geq 0, \quad (13.26)$$

where

$$\phi_3(t) = \sum_{j=0}^{k} b_{k-j}(t) \frac{t^j}{j!}, \qquad \phi_4(t) = \phi_2(t). \qquad \square$$

COROLLARY 13.6. (Thandapani and Agarwal, 1980) *If in* (13.23) $a(t)$ *is differentiable and* $a'(t) \geq 0$, *and also* $q(t) \equiv 1$, *then*

$$u^{(k)}(t) \leq \exp\left(\int_0^t \phi_3(s)\, ds \right) \left[a(0) + \int_0^t \phi_5(s) \exp\left(-\int_0^s \phi_3(\tau)\, d\tau \right) ds \right], \qquad t \geq 0, \quad (13.27)$$

where

$$\phi_5(t) = a'(t) + \sum_{j=0}^{k-1} \sum_{i=0}^{j} u^{(j)}(0) b_i(t) \frac{t^{j-i}}{(j-i)!}. \qquad \square$$

COROLLARY 13.7. (Thandapani and Agarwal, 1980) *If in* (13.23) $k = 1$ *and* $b_0(t) = b_1(t) = b(t)$, *then*

$$u'(t) \leq a(t) + q(t) \int_0^t \phi_6(s) \exp\left(\int_s^t \phi_7(\tau)\, d\tau \right) ds,$$

where

$$\phi_6(t) = b(t) \left[u(0) + a(t) + \int_0^t a(\tau)\, d\tau \right], \qquad \phi_7(t) = b(t) \left[q(t) + \int_0^t q(\tau)\, d\tau \right]. \qquad \square$$

The following theorem can be similarly proved.

THEOREM 13.5. (Megahed, Hamad, and Salem, 1987) *Let* $a(t)$, $q(t)$, $b_j(t)$, $u^{(j)}(t)$, $j = 0, \dots, k$, *be nonnegative continuous functions for* $t \geq 0$, *and suppose*

$$u^{(k)}(t) \leq a(t) + q(t) \sum_{j=0}^{k} \int_t^\infty b_j(s) u^{(j)}(s)\, ds,$$

where $k \geq 0$ *is an integer. Then*

$$u(t) \leq a(t) + q(t) \int_t^\infty \mu_1(s) \exp\left(\int_t^s \mu_2(\tau)\, d\tau \right) ds, \qquad t \geq 0, \tag{13.28}$$

where

$$\mu_1 = a(t) b_k(t) + \sum_{j=0}^{k-1} b_j(t) \left\{ \mu_0(t) + \frac{1}{(k-j-1)!} \int_0^t (t-\tau)^{k-j-1} a(\tau)\, d\tau \right\},$$

$$\mu_2 = q(t) b_k(t) + \sum_{j=0}^{k-1} b_j(t) \frac{1}{(k-j-1)!} \int_0^t (t-\tau)^{k-j-1} q(\tau)\, d\tau,$$

$$\mu_0 = \sum_{i=j}^{k-1} u^{(i)}(0) \frac{t^{i-j}}{(i-j)!}. \qquad \square$$

THEOREM 13.6. (Thandapani and Agarwal, 1987) *Let* $a(t)$, $b_j(t)$, $u^{(j)}(t)$, $j = 0, \dots, k$, *be nonnegative continuous functions for* $t \geq 0$, *with* $a(t)$ *nondecreasing, and suppose*

$$u^{(k)}(t) \leq a(t) + \sum_{j=0}^{k} \int_0^t b_j(s) u^{(k)}(s) u^{(j)}(s)\, ds, \qquad t \geq 0, \tag{13.29}$$

where $k \geq 1$ *is an integer. Then*

$$u(t) \leq \frac{a(t) \exp\left(\int_0^t \psi_1(s)\, ds \right)}{1 - \int_0^t \psi_2(s) \exp\left(\int_0^s \psi_1(\tau)\, d\tau \right) ds}, \tag{13.30}$$

while

$$1 - \int_0^t \psi_2(s) \exp\left(\int_0^s \psi_1(\tau)\, d\tau \right) ds > 0,$$

where

$$\psi_1(t) = \sum_{j=0}^{k-1} \sum_{i=0}^{j} u^{(j)}(0) b_i(t) \frac{t^{j-i}}{(j-i)!},$$

$$\psi_2(t) = a(t) \sum_{j=0}^{k} b_{k-j}(t) \frac{t^j}{j!}.$$

PROOF. We first assume that $a(t)$ is positive. Since it is also nondecreasing, (13.29) implies

$$\frac{u^{(k)}(t)}{a(t)} \leq 1 + \sum_{j=0}^{k} \int_0^t b_j(s) \frac{u^{(k)}(s)}{a(s)} u^{(j)}(s)\, ds. \tag{13.31}$$

We denote by $R(t)$ the righthand side of (13.31), and thus find, by differentiation,

$$R'(t) \le \sum_{j=0}^{k} b_j(t)R(t)u^{(j)}(t) \le$$
$$\le b_k(t)a(t)R^2(t) +$$
$$+\sum_{j=0}^{k-1} b_j(t)R(t)\left[\sum_{i=j}^{k-1} u^{(j)}(0)\frac{t^{i-j}}{(i-j)!} + \frac{1}{(k-j-1)!}\int_0^t (t-\tau)^{k-j-1}a(\tau)R(\tau)\,d\tau\right].$$

Using the fact that $R(t)$ and $a(t)$ are nondecreasing, we obtain

$$R'(t) \le \psi_1(t)R(t) + \psi_2(t)R^2(t).$$

Lemma 4.1 and (13.31) now imply (13.30).

If $a(t)$ is nonnegative, we carry out the above scheme with $a(t)+\epsilon$ instead of $a(t)$ ($\epsilon > 0$), and subsequently pass to the limit to obtain (13.30). □

The following theorem can be similarly proved.

THEOREM 13.7. (Thandapani and Agarwal, 1987) *Let $b_j(t)$, $u^{(j)}(t)$, $j = 0, \dots, k$, be nonnegative continuous functions for $t \ge 0$, let $a(t)$ be a positive nondecreasing function, and suppose*

$$u^{(k)}(t) \le a(t) + \sum_{j=0}^{k}\int_0^t b_j(s)u^{(l)}(s)u^{(j)}(s)\,ds,$$

where $k-1 \ge l \ge 0$. Then

$$u^{(k)}(t) \le \frac{a(t)\exp\left(\int_0^t \psi_3(s)\,ds\right)}{1-\int_0^t \psi_4(s)\exp\left(\int_0^s \psi_3(\tau)\,d\tau\right)\,ds}, \tag{13.32}$$

while

$$1-\int_0^t \psi_4(s)\exp\left(\int_0^s \psi_3(\tau)\,d\tau\right)\,ds > 0,$$

where

$$\psi_3(t) = \sum_{j=0}^{k-1}\sum_{i=0}^{j} u^{(j)}(0)b_i(t)\frac{t^{j-i}}{(j-i)!}\left(\sum_{i=l}^{k-1}\frac{u^{(i)}(0)}{a(t)}\frac{t^{i-l}}{(i-l)!} + \frac{t^{k-l}}{(k-l)!}\right),$$

$$\psi_2(t) = \frac{t^{k-l}}{(k-l)!}\sum_{j=0}^{k} b_{k-j}(t)\frac{t^j}{j!}. \qquad \square$$

REMARK 13.2. Theorems 13.4–13.6 remain valid if the derivatives $u^{(j)}(t)$, $j = 0, \dots, k$, are replaced by $|x^{(j)}(t)|$, where $x\colon \mathbf{R} \to \mathbf{R}^n$ is a function with continuous derivatives up to order k inclusive. This conclusion is based on the inequality

$$|x^{(j)}(t)| \le \sum_{i=j}^{k-1} |x^{(i)}(0)|\frac{t^{i-j}}{(i-j)!} + \frac{1}{(k-j-1)!}\int_0^t (t-\tau)^{k-j-1}|x^{(k)}(\tau)|\,d\tau. \qquad \square$$

14. Integral inequalities with delay

This section is concerned with a type of integral inequalities conventionally called 'inequalities with delay'.

Such inequalities arise in the investigation of equations with delay, describing evolution processes whose state at a given time depends on the history of the process till this time [31]. The 'delay' in the inequalities below is mainly expressed by:

I) The presence of integrals in which, besides the function to be estimated ($u(t)$) one or more functions of the form $u(h)$ occur, where $h = h(t) \leq t$ is the 'delaying' argument.

II) The dependence of the values for $t \geq \alpha$ of a function to be estimated on its values for $t < \alpha$.

Usually, we will 'solve' such inequalities by reduction to integral or differential inequalities without delay and already investigated. E.g., using Bihari's inequality (or the equivalent differential inequality (5.10)) the following theorem can be proved.

THEOREM 14.1. (Bobrowski, Popenda, and Werbowski, 1978) *Let $u(t)$, $a(t)$, $b(t)$, $c(t)$, $q(t)$, $h(t)$, be nonnegative continuous functions in $\mathbf{R}_+$, with $0 \leq h(t) \leq t$, and suppose*

$$u(t) \leq a(t) + q(t) \int_0^t [b(s)u^\alpha(s) + c(s)u^\beta(h(s))]\, ds, \qquad t \in \mathbf{R}_+, \tag{14.1}$$

where $\alpha > 0$, $\beta > 0$ are constants. Then

$$u(t) \leq a(t) + q(t)F^{-1}\left(\int_0^t [b(s)r^\alpha(s) + c(s)r^\beta(h(s))]\, ds\right) \tag{14.2}$$

for $t \in J = \{t \in \mathbf{R}_+ \colon \int_0^t [b(s)r^\alpha(s) + c(s)r^\beta(h(s))]\, ds \in \operatorname{Dom} F^{-1}\}$, where F^{-1} is the inverse function of

$$F(u) = \int_0^u \frac{dx}{(1+x)^\gamma}, \qquad u \geq 0,$$

and $r(t) = \max\{a(t), q(t)\}$, $\gamma = \max\{\alpha, \beta\}$. □

We do not give a proof of theorem 14.1, since it is a particular case of theorem 14.2. Before stating theorem 14.2 we make some remarks.

Let $N = \{1, \ldots, n\}$, $M = \{1, \ldots, m\}$, and let S be an arbitrary, fixed, set of pairs of integers such that $S \subset N \times M$.

We introduce the following conditions:

H14.1) $g_k, u_k \colon \mathbf{R}_+ \to \mathbf{R}_+$, $k \in N$, are continuous functions;

H14.2) $d_{ij} \colon \mathbf{R}_+ \to \mathbf{R}_+$ $((i,j) \in S)$ are continuous functions satisfying

$$0 \leq d_{ij}(t) \leq t \quad \text{for } t \geq 0 \qquad \text{and} \qquad \lim_{t\to\infty} d_{ij}(t) = \infty;$$

H14.3) $B \colon \mathbf{R}_+ \times \mathbf{R}^\mu \to \mathbf{R}$ is a continuous function such that

$$0 \leq B(t, y_1, \ldots, y_\mu) \leq B(t, \overline{y}_1, \ldots, \overline{y}_\mu)$$

for $0 \leq y_k \leq \overline{y}_k < \infty$ and $0 \leq t < \infty$, and

$$B(t, \lambda y_1, \ldots, \lambda y_\mu) \leq A(\lambda) B(t, y_1, \ldots, y_\mu)$$

for $t \geq 0$, $\lambda \geq 1$, and $0 \leq y_k < \infty$, $k = 1, \ldots, \mu$, where $A(\lambda)$ is a continuous function for $\lambda \geq 1$ which is positive for $\lambda > 1$, and $\mu = \sum_{(i,j)\in S} \mu_{ij}$ is an integer, where

$$\mu_{ij} = \begin{cases} 1 & \text{for } (i,j) \in S, \\ 0 & \text{otherwise;} \end{cases}$$

H14.4) $\phi\colon \mathbf{R}_+ \to \mathbf{R}_+$ is a nondecreasing continuous function;
H14.5) ψ is a continuous function satisfying the conditions

$$\begin{aligned} 0 \leq \psi(t,y) &\leq \psi(t,\overline{y}) && \text{for } 0 \leq t < \infty,\ 0 \leq y \leq \overline{y}, \\ \psi(t,\nu y) &\leq c(\nu)\psi(t,y) && \text{for } 0 \leq t < \infty,\ \nu \geq 0,\ y \geq 0, \end{aligned}$$

where $c(\nu)$ is a nonnegative continuous function in $\mathbf{R}_+$.

Below we set

$$[u_i(d_{ij})]_S = [\mu_{11}u_1(d_{11}), \ldots, \mu_{1m}u_1(d_{1m}), \ldots, \ldots, \mu_{n1}u_n(d_{n1}), \ldots, \mu_{nm}u_n(d_{nm})].$$

E.g., for $n = 2$, $m = 3$, and $S = \{(1,1),(1,2),(2,3)\}$ we have $[u_i(d_{ij})]_S = [u_1(d_{11}), u_1(d_{12}), u_2(d_{23})]$.

THEOREM 14.2. (Bobrowski, Popenda, and Werbowski, 1978) *Let the conditions* H14.1–4 *hold, and suppose*

$$u_k(t) \leq f_k(t) + g_k(t)\phi\left(\int_0^t B(s, [u_i(d_{ij}(s))]_S)\, ds\right), \qquad k \in N, \tag{14.3}$$

for $t \geq 0$. *Then*

$$u_k(t) \leq f_k(t) + g_k(t)\phi\left(F^{-1}\left(\int_0^t B(s, [r_i(d_{ij}(s))]_S)\, ds\right)\right), \qquad k \in N, \tag{14.4}$$

for $t \in J = \{t \in \mathbf{R}_+\colon \int_0^t B(s, [r_i(d_{ij}(s))]_S)\, ds \in \operatorname{Dom} F^{-1}\}$, *where* F^{-1} *is the inverse function of*

$$F(u) = \int_0^u \frac{dx}{A(1+\phi(x))}, \qquad u \geq 0,$$

and $r_k(t) = \max\{f_k(t), g_k(t)\}$, $k \in N$.

PROOF. For $t \geq 0$ we put

$$v(t) = \int_0^t B(s, [u_i(d_{ij}(s))]_S)\, ds \geq 0.$$

Then (14.3) takes the form

$$u_k(t) \leq f_k(t) + g_k(t)\phi(v(t)), \quad k \in N, \qquad t \geq 0. \tag{14.5}$$

In particular, for $(i,j) \in S$ and $t \geq 0$,

$$\begin{aligned} u_i(d_{ij}(t)) &\leq f_i(d_{ij}(t)) + g_i(d_{ij}(t))\phi(v(d_{ij}(t))) \leq \\ &\leq r_i(d_{ij}(t))[1 + \phi(v(t))], \end{aligned} \tag{14.6}$$

since $v(t)$ is nondecreasing and $d_{ij}(t) \le t$. Consequently, by H14.3 and (14.6) we find that

$$v'(t) = B(t, [u_i(d_{ij}(t))]_S) \le B(t, [(1 + \phi(v(t)))r_i(d_{ij}(t))]_S) \le \\ \le A(1 + \phi(v(t)))B(t, [r_i(d_{ij}(t))]_S),$$

and so, by lemma 5.1,

$$v(t) \le F^{-1}\left(\int_0^t B(s, [r_i(d_{ij}(s))]_S)\, ds\right), \qquad t \in J.$$

This and (14.5) imply (14.4). □

The next three theorems can be proved in a similar manner.

THEOREM 14.3. (Bobrowski, Popenda, and Werbowski, 1978) *Suppose that the conditions* H14.1–4 *hold, and that*

$$u_k(t) \le f_k(t) + \int_0^t \sum_{(i,j)\in S} h_{kj}(s)u_k(d_{ij}(s))\, ds + \\ + g_k(t)\phi\left(\int_0^t B(s, [u_i(d_{ij}(s))]_S)\, ds\right), \qquad k \in N, \qquad t \in \mathbf{R}_+, \tag{14.7}$$

where $h_{kj}(t)$ are nonnegative continuous functions in $\mathbf{R}_+$. Then

$$u_k(t) \le H_{f_k}(t) + H_{g_k}(t)\phi\left[F^{-1}\left(\int_0^t B(s, [r_i(d_{ij}(s))]_S)\, ds\right)\right], \qquad k \in N, \tag{14.8}$$

for $t \in J$, where J and F^{-1} are as in theorem 14.2,

$$r_k(t) = \max\{H_{f_k}(t), H_{g_k}(t)\},$$

$$H_f(t) = f(t) + \int_0^t \sum_{k\in N}\sum_{(i,j)\in S} h_{kj}(s)f(d_{ij}(s))\exp\left(\int_s^t \sum_{k\in N}\sum_{(i,j)\in S} h_{kj}(\tau)\, d\tau\right) ds. \qquad \square$$

THEOREM 14.4. (Bobrowski, Popenda, and Werbowski, 1978) *Suppose that the conditions* H14.1–3, H14.5 *hold, and that*

$$u_k(t) \le f_k(t) + g_k(t)\psi\left(t, \int_0^t B(s, [u_i(d_{ij}(s))]_S)\, ds\right), \qquad k \in N, \qquad t \in \mathbf{R}_+. \tag{14.9}$$

Then

$$u_k(t) \le f_k(t) + g_k(t)\psi\left(t, G^{-1}\left(\int_0^t B(s, [r_i(d_{ij}(s))]_S)\, ds\right)\right), \qquad k \in N, \qquad t \in \mathbf{R}_+, \tag{14.10}$$

where $r_k(t) = \max\{f_k(t), g_k(t)\psi(t,1)\}$ and G^{-1} is the inverse function of $G(u) = \int_0^u dx/A(1 + c(x))$, $u \ge 0$. □

THEOREM 14.5. (Bobrowski, Popenda, and Werbowski, 1978) *Suppose that the conditions* H14.1–3, H14.5 *hold, and that*

$$u_k(t) \leq f_k(t) + \int_0^t \sum_{(i,j)\in S} h_{kj}(s) u_k(d_{ij}(s))\, ds + \\ + g_k(t)\psi\left(t, \int_0^t B(s, [u_i(d_{ij}(s))]_S)\, ds\right), \qquad k \in N, \qquad t \in \mathbf{R}_+, \tag{14.11}$$

where $h_{kj}(t)$ *are nonnegative continuous functions in* $\mathbf{R}_+$. *Then*

$$u_k(t) \leq H_{f_k}(t) + H_{g_k}(t)\psi\left(t, G^{-1}\left(\int_0^t B(s, [r_i(d_{ij}(s))]_S)\, ds\right)\right), \qquad k \in N, \tag{14.12}$$

for $t \in J$, *where* J, H_{f_k} *and* H_{g_k} *are as in theorem* 14.3, G *is as in theorem* 14.4, *and* $r_k(t) = \max\{H_{f_k}(t), H_{g_k}(t)\psi(t,1)]\}$. □

In the proofs of these theorems, the following condition is used in an essential manner:

$$0 \leq d_{ij}(s) \leq s.$$

While the condition $d_{ij}(s) \leq s$ for the deviating argument d_{ij} is natural for delay equations, the condition $d_{ij}(s) \geq 0$ is too restrictive. For instance, theorems 14.1–14.5 cannot be used in the investigation of equations with fixed delay,

$$x'(t) = F(t, x(t), x(t-r)) \qquad (r > 0).$$

Before considering inequalities in which the condition $d_{ij} \geq 0$ is omitted, we introduce some notation.

Let $J = [\alpha, \infty)$.

For a function $h(t)$ we introduce the following conditions:

H14.6) $h\colon J \to \mathbf{R}$ is continuous and

$$h(t) \leq t, \qquad t \in J, \qquad \lim_{t\to\infty} h(t) = \infty.$$

H14.7) $h(t)$ is nondecreasing in J and has an inverse $h^{-1}\colon [h(\alpha), \infty) \to J$.

The following theorem holds:

THEOREM 14.6. (Marušiak, 1973) *Let* $u(t)$, $b(t)$, $c(t)$ *be nonnegative continuous functions in* $J = [\alpha, \infty)$, *and suppose*

$$u(t) \leq a + \int_\alpha^t [b(s)u(s) + c(s)u(h(s))]\, ds \qquad t \in J, \tag{14.13}$$

$$u(t) = \phi(t), \qquad t \in E = \{t \in J :\ h(t) < \alpha\} \cup \{\alpha\}, \tag{14.14}$$

where $a \geq 0$ *is a constant,* $\phi(t)$ *is a nonnegative continuous function in* E, *and the function* $h(t)$ *satisfies the condition* H14.6. *Then*

$$u(t) \leq \left[a + \int_E c(s)\phi(h(s))\, ds\right] \exp\left(\int_\alpha^t [b(s) + c(s)]\, ds\right), \qquad t \in J. \tag{14.15}$$

PROOF. We set $v(t) = \int_\alpha^t [b(s)u(s) + c(s)u(h(s))]\, ds + a$. Then

$$u(t) \le v(t), \tag{14.16}$$

and since $v(t)$ is nondecreasing in J we have

$$\begin{aligned} v'(t) =& b(t)u(t) + c(t)u(h(t)) \le \\ \le& b(t)v(t) + c(t) \begin{cases} \phi(h(t)) & \text{if } t \in E, \\ v(t) & \text{otherwise.} \end{cases} \end{aligned}$$

Integration of this inequality from α to t leads to

$$v(t) \le a + \int_E c(s)\phi(h(s))\, ds + \int_\alpha^t [b(s) + c(s)]v(s)\, ds,$$

so that (14.15) follows by theorem 1.1 and (14.16). □

The following theorem allows us to prove assertions similar to theorem 14.6.

Let C_0 denote the set of nonnegative continuous functions $u\colon J \to \mathbf{R}_+$.

Let $f\colon J \times C_0 \to \mathbf{R}$ be a functional satisfying the conditions

H14.8) for any $u \in C_0$ and $t \in J$ the value $f(t, u)$ does not depend on the values $u(s)$ for $s > t$;

H14.9) for any fixed $u \in C_0$ the function $\psi(t) = f(t, u)$ is continuous and has a finite, nonnegative, upper right Dini derivative $D^+\psi(t)$;

H14.10) $D^+ f(t, \lambda u) \le \lambda D^+ f(t, u)$ for $\lambda \ge 0$, $t \in J$, $u \in C_0$;

H14.11) $D^+ f(t, u) \le D^+ f(t, v)$ for $t \in J$ and $u, v \in C_0$ such that

$$u(s) \le v(s), \qquad s \in [\alpha, t];$$

H14.12) $D^+ f(t, u + v) \le D^+ f(t, u) + D^+ f(t, v)$ for $t \in J$, $u, v \in C_0$.

THEOREM 14.7. *Let f be a functional satisfying the conditions* H14.8–12, *let $u, a, q \in C_0$, and suppose*

$$u(t) \le a(t) + q(t) f(t, u), \qquad t \in J = [\alpha, \infty). \tag{14.17}$$

Then

$$\begin{aligned} u(t) \le& a(t) + q(t) \Big\{ f(\alpha, u) \exp\Big(\int_\alpha^t D^+ f(\tau, q)\, d\tau \Big) + \\ &+ \int_\alpha^t D^+ f(s, a) \exp\Big(\int_s^t D^+ f(\tau, q)\, ds \Big) \Big\}, \qquad t \in J. \end{aligned} \tag{14.18}$$

PROOF. Setting $v(t) = f(t, u)$, (14.17) takes the form

$$u(t) \le a(t) + q(t)v(t). \tag{14.19}$$

From H14.9 we conclude that $v(t)$ is nondecreasing in J. Then

$$\begin{aligned} D^+v(t) &= D^+f(t,u) \le && \text{(by H14.11)}\\ &\le D^+f(t,a+qv) \le && \text{(by H14.12)}\\ &\le D^+f(t,a)+D^+f(t,qv) \le && \text{(by H14.8, H14.11, and the nondecrease of } v)\\ &\le D^+f(t,a)+D^+f(t,qv(t)) \le && \text{(by H14.10)}\\ &\le D^+f(t,a)+D^+f(t,q)v(t), \quad t\in J, \end{aligned}$$

so that remark 6.1 and (14.19) imply (14.18). □

COROLLARY 14.1. *Suppose the conditions of theorem 14.7 hold.*

I. *If* $a(t) \equiv \phi(t)q(t)$ *and* $\phi \in C_0$, *then*

$$u(t) \le q(t)\left[\phi(t) + f(\alpha,u)E(t,\alpha) + \int_\alpha^t D^+f(s,q\phi)E(t,s)\,ds\right], \qquad t\in J, \tag{14.20}$$

where $E(t,s) = \exp(\int_s^t D^+f(\tau,q)\,d\tau)$, $\alpha \le s \le t$.

II. *If* $a(t) \equiv \phi(t)q(t)$ *and* $\phi \in C_0$ *is nondecreasing in* J, *then*

$$u(t) \le q(t)[\phi(t) + f(\alpha,u)]\exp\left(\int_\alpha^t D^+f(\tau,q)\,d\tau\right) \qquad t\in J. \tag{14.21}$$

III. *If* $q(t)\equiv 1$ *and* $a(t)$ *is nondecreasing in* J, *then*

$$u(t) \le [a(t) + f(\alpha,u)]\exp\left(\int_\alpha^t D^+f(\tau,1)\,d\tau\right) \qquad t\in J. \quad \square \tag{14.22}$$

REMARK 14.1. Since condition H14.9 guarantees the validity of the inequality

$$\int_s^t D^+f(\tau,q)\,d\tau \le f(t,q) - f(s,q)$$

for $q \in C_0$ and $\alpha \le s \le t$, in (14.18), (14.20)–(14.22) the function $E(t,s) = \exp(\int_s^t D^+f(\tau,q)\,d\tau)$ can be replaced by the function $G(t,s) = \exp[f(t,q) - f(s,q)]$, provided that $f(\alpha,u) \ge 0$. □

REMARK 14.2. The conditions H14.10–12 hold if the following conditions are met:

14.10*) $f(t,\lambda u) = \lambda f(t,u)$ for $\lambda \ge 0$ and $u \in C_0$;

14.11*) $f(t,u) - f(s,u) \le f(t,v) - f(s,v)$ for $\alpha \le s \le t$ and $u,v \in C_0$ such that $u(s) \le v(s)$ for $\alpha \le s \le t$;

14.12*) $f(t,u+v) - f(s,u+v) \le f(t,u) - f(s,u) + f(t,v) - f(s,v)$ for $\alpha \le s \le t$ and $u,v \in C_0$.

Thus, theorem 14.7 still holds if the conditions H14.8–12 are replaced by H14.8–9, H14.10*–12*. □

Theorem 1.5 is a direct consequence of theorem 14.7. Other corollaries of this theorem are:

THEOREM 14.8. *Let u, ϕ_0, q, b_i, $i = 0, \ldots, n$, be nonnegative continuous functions in $J = [\alpha, \infty)$, and suppose*

$$u(t) \le q(t) \left[\phi_0(t) + \sum_{i=0}^{n} \int_\alpha^t b_i(s) u(h_i(s))\, ds \right], \qquad t \in J, \tag{14.23}$$

$$u(t) = \psi(t) \qquad \text{for } t \in E = \cup_{i=0}^{n} E_i, \tag{14.24}$$

where $\psi(t)$ is a nonnegative continuous function in E, h_i, $i = 0, \ldots, n$, are functions satisfying the condition H14.6, *and*

$$E_i = \{t \in J : \ h_i(t) < \alpha\} \cup \{\alpha\}, \qquad i = 0, \ldots, n.$$

Then

$$u(t) \le q(t) \left[\phi(t) + \int_\alpha^t \sum_{i=0}^{n} \chi[t, I_t \setminus E_i] b_i(s) q(h_i(s)) \phi(h_i(s)) \times \right.$$

$$\left. \times \exp\left(\sum_{i=0}^{n} \int_{[s,t]\setminus E_i} b_i(s) q(h_i(\tau))\, d\tau \right) ds \right] \tag{14.25}$$

for $t \in J$, where

$$\chi[t, A] = \begin{cases} 1 & \text{if } t \in A, \\ 0 & \text{otherwise,} \end{cases}$$

$I_t = [\alpha, t]$, and

$$\phi(t) = \phi_0(t) + \sum_{i=0}^{n} \int_{I_t \cap E_i} b_i(s) \psi(h_i(s))\, ds. \tag{14.26}$$

PROOF. (14.23) and (14.24) imply that

$$u(t) \le q(t)[\phi(t) + f(t, u)] \qquad t \in J,$$

where $\phi(t)$ is defined by (14.26), and

$$f(t, u) = \sum_{i=0}^{n} \int_{I_t \setminus E_i} b_i(s) u(h_i(s))\, ds.$$

Now (14.25) follows from corollary 14.1I and remark 14.1. □

COROLLARY 14.2. *If the conditions of theorem 14.8 hold, and $\phi_0(t)$ is nondecreasing in J, then*

$$u(t) \le$$

$$\le q(t) \left[\phi_0(t) + \sum_{i=0}^{n} \int_{I_t \cap E_i} b_i(s) \psi(h_i(s))\, ds \right] \exp\left[\sum_{i=0}^{n} \int_{I_t \setminus E_i} b_i(s) q(h_i(s))\, ds \right]. \tag{14.27}$$

PROOF. Indeed, in this case ϕ is nondecreasing in J, and (14.27) follows from corollary 14.1II and remark 14.1. □

COROLLARY 14.3. *Let u, ϕ_0, b, c be nonnegative continuous functions in $J = [\alpha, \infty)$, with ϕ_0 nondeceasing in J, and suppose*

$$u(t) \le \phi_0(t) + \int_\alpha^t [b(s)u(s) + c(s)u(h(s))]\, ds, \qquad t \in J,$$
$$u(t) = \psi(t) \qquad \text{for } t \in E = \{t \in J\colon h(t) < \alpha\} \cup \{\alpha\},$$

where $\psi(t)$ is a nonnegative continuous function in E, and $h(t)$ is a function satisfying the condition H14.6. *Then, for $t \in J$,*

$$u(t) \le \left[\phi_0(t) + \int_{I_t \cap E} c(s)\psi(h(s))\, ds\right] \exp\left[\int_\alpha^t b(s)\, ds + \int_{I_t \setminus E_i} c(s)\, ds\right]. \tag{14.28}$$

PROOF. Indeed, (14.28) is a particular case of (14.27) with $n = 1$, $q(t) \equiv 1$, $b_0(s) = b(s)$, $b_1(s) = c(s)$, $h_0(t) = t$, $h_1(t) = h(t)$, $E_0 = \emptyset$, $E_1 = E$.

Moreover, (14.28) implies the, worse but simpler, estimate

$$u(t) \le \left[\phi_0(t) + \int_E c(s)\psi(h(s))\, ds\right] \exp\left(\int_\alpha^t [b(s) + c(s)]\, ds\right), \qquad t \in J. \tag{14.29}$$

In particular, if $\phi_0(t) \equiv a$, then (14.29) coincides with (14.15) in theorem 14.6. □

COROLLARY 14.4. *Under the conditions of corollary* 14.3, *let $h(t) = t - r$ $(r > 0)$. Then the inequality*

$$u(t) \le \phi_0(t) + \int_\alpha^t [b(s)u(s) + c(s)u(s - r)]\, ds, \qquad t \in J,$$

implies

$$u(t) \le \begin{cases} \left[\phi_0(t) + \int_\alpha^t c(s)\psi(s - r)\, ds\right] \times \\ \quad \times \exp\left(\int_\alpha^t b(s)\, ds\right), & \alpha \le t \le \alpha + r, \\ \left[\phi_0(t) + \int_\alpha^{r+\alpha} c(s)\psi(s - r)\, ds\right] \times \\ \quad \times \exp\left(\int_\alpha^t b(s)\, ds + \int_{r+\alpha}^t c(s)\, ds\right), & t > \alpha + r. \end{cases} \tag{14.30}$$

□

THEOREM 14.9. *Let u, ϕ_0, q, b be nonnegative continuous functions in $J = [\alpha, \infty)$, with ϕ_0 nondecreasing in J, and suppose*

$$u(t) \le q(t) \left[\phi_0(t) + \int_\alpha^t \left[b(s)u(s) + c(s) \max_{\tau \in \Delta_s} u(\tau)\right] ds\right], \qquad t \in J, \tag{14.31}$$
$$u(t) = \psi(t) \qquad \text{for } t \in \Delta = [\alpha - r, \alpha], \tag{14.32}$$

where $r > 0$, $\psi(t)$ is a nonnegative continuous function in Δ, and the closed interval $\Delta_t \subset [\alpha - r, t]$ for $t \in J$. Then, for $t \in J$,

$$\begin{aligned} u(t) \le{}& q(t) \left[\phi_0(t) + \int_\alpha^t c(s) \max_{\tau \in \Delta_s \cap \Delta} \psi(\tau)\, ds\right] \times \\ & \times \exp\left\{\int_\alpha^t \left[b(s)q(s) + c(s) \max_{\tau \in \Delta_s \setminus \Delta} q(\tau)\right] ds\right\}. \end{aligned} \tag{14.33}$$

PROOF. (14.31) and (14.32) imply that

$$u(t) \leq q(t)[\phi(t) + f(t, u)], \qquad t \in J,$$

where

$$\phi(t) = \phi_0(t) + \int_\alpha^t c(s) \max_{\tau \in \Delta_s \cap \Delta} \psi(\tau)\, ds,$$

$$f(t, u) = \int_\alpha^t \left[b(s)u(s) + c(s) \max_{\tau \in \Delta_s \setminus \Delta} u(\tau) \right] ds.$$

Now (14.33) follows from corollary 14.1II and remark 14.1. □

In the following two theorems, the integral inequalities are given in a form suitable for immediate application to delay equations.

Let $\mathbf{R}^n$ be the n-dimensional vector space with norm $|\cdot|$, $C([a, b], \mathbf{R}^n)$ $(-\infty < a < b)$ the Banach space of continuous functions $y\colon [a, b] \to \mathbf{R}^n$ with the topology of uniform convergence, and $L_1([a, \infty), \mathbf{R}^n)$ the Banach space of Lebesgue integrable functions $z\colon [a, \infty) \to \mathbf{R}^n$ with norm $\|z\|_{L_1} = \int_a^\infty |z(t)|\, dt$.

For given $r_0 \geq r > 0$, $t \geq 0$, and $x \in C([t - r_0, t], \mathbf{R}^n)$ we put

$$\|x_t\|_r = \max_{t-r \leq s \leq t} |x(s)|.$$

THEOREM 14.10. (Arino and Györi, 1984) *Let $J = [\alpha, \infty)$, and suppose that the following conditions hold:*

1) *$h\colon J \to \mathbf{R}$ is a function satisfying the conditions* H14.6–7*;*
2) *$r\colon J \to \mathbf{R}_+$ is a function such that $h(\alpha) \leq \alpha - r(\alpha) \leq t - r(t)$, $t \in J$;*
3) *$\rho, p\colon J \to \mathbf{R}_+$ are locally Lebesgue integrable functions;*
4) *$b\colon [h(\alpha), \infty) \times [h(\alpha), \infty) \to \mathbf{R}_+$ is a locally bounded, locally Lebesgue integrable function, and there is a locally bounded, locally Lebesgue integrable function $a\colon [h(\alpha), \infty) \to [1, \infty)$ such that*

$$\int_s^{h^{-1}(s)} b(t, s)a(t)\, dt \leq a(s) - 1, \qquad s \geq h(\alpha), \tag{14.34}$$

where h^{-1} is the function inverse to h;

5) *$x\colon [h(\alpha), \infty) \to \mathbf{R}^n$ is an absolutely continuous function, for which*

$$|\dot{x}(t)| \leq \int_{h(t)}^t b(t, s)|\dot{x}(s)|\, ds + p(t)\|x_t\|_{r(t)} + \rho(t), \qquad t \geq \alpha. \tag{14.35}$$

Then

i)

$$|x(t)| \leq \left[\|x_\alpha\|_{r(\alpha)} + \int_{h(\alpha)}^\alpha [a(s) - 1]|\dot{x}(s)|\, ds + \right.$$

$$\left. + \int_\alpha^t a(\tau)\rho(\tau)\, d\tau \right] e^{\int_\alpha^t a(\tau)p(\tau)\, d\tau}, \qquad t \geq \alpha. \tag{14.36}$$

ii) *If, in addition to the above, we also suppose that*

$$\int_\alpha^\infty a(t)[p(t)+\rho(t)]\,dt<\infty, \tag{14.37}$$

then

$$\int_\alpha^\infty |\dot{x}(t)|\,dt<\infty. \tag{14.38}$$

PROOF. i) Let c: $[h(\alpha),\infty)\to\mathbf{R}$ be an arbitrary, but fixed, locally Lebesgue integrable function. Then the function $a(t)\int_{h(t)}^t b(t,s)c(s)\,ds$ is defined and locally Lebesgue integrable in $[\alpha,\infty)$. Changing the order of integration we find

$$\left|\int_\alpha^t\left[a(\tau)\int_{h(\tau)}^\tau b(\tau,s)c(s)\,ds\right]d\tau\right|\le\int_{h(\alpha)}^t\left(\int_s^{h^{-1}(s)} b(\tau,s)a(\tau)\,d\tau\right)|c(s)|\,ds,$$

so that (14.34) implies

$$\left|\int_\alpha^t\left[a(\tau)\int_{h(\tau)}^\tau b(\tau,s)c(s)\,ds\right]d\tau\right|\le\int_{h(\alpha)}^t[a(s)-1]|c(s)|\,ds,\qquad t\ge\alpha. \tag{14.39}$$

Applying (14.39) to $c(t)=|\dot{x}(t)|$, (14.35) gives

$$\int_\alpha^t a(\tau)|\dot{x}(\tau)|\,d\tau\le\int_{h(\alpha)}^t[a(\tau)-1]|\dot{x}(\tau)|\,d\tau+$$
$$+\int_\alpha^t a(\tau)\rho(\tau)\|x_\tau\|_{r(\tau)}\,d\tau+\int_\alpha^t a(\tau)\rho(\tau)\,d\tau,$$

and thus

$$\int_\alpha^t|\dot{x}(\tau)|\,d\tau\le\int_{h(\alpha)}^t[a(\tau)-1]|\dot{x}(\tau)|\,d\tau+\int_\alpha^t a(\tau)\rho(\tau)\,d\tau+$$
$$+\int_\alpha^t a(\tau)\rho(\tau)\|x_\tau\|_{r(\tau)}\,d\tau,\qquad t\ge\alpha. \tag{14.40}$$

Taking into account condition 2) above, we find that

$$\|x_t\|_{r(t)}=\max_{t-r(t)\le s\le t}|x(s)|\le\|x_\alpha\|_{r(\alpha)}+\int_\alpha^t|\dot{x}(s)|\,ds\equiv y(t),\qquad t\ge\alpha,$$

and so (14.40) implies

$$y(t)\le A(t)+\int_\alpha^t a(\tau)p(\tau)u(\tau)\,d\tau,\qquad t\ge\alpha, \tag{14.41}$$

where

$$A(t)=\|x_\alpha\|_{r(\alpha)}+\int_{h(\alpha)}^\alpha[a(\tau)-1]|\dot{x}(\tau)|\,d\tau+\int_\alpha^t a(\tau)\rho(\tau)\,d\tau.$$

By corollary 1.2 and (14.41) we obtain

$$|x(t)|\le y(t)\le A(t)e^{\int_\alpha^t a(\tau)p(\tau)\,d\tau},\qquad t\ge\alpha, \tag{14.42}$$

which proves (14.36).

ii) Under the condition (14.37), inequality (14.38) follows from (14.40) and (14.42). □

COROLLARY 14.5. (Atkinson and Haddock, 1983) *Let* $b, p\colon [\alpha - r_0, \infty) \to \mathbf{R}_+$ *(*$r_0 > 0$*) be continuous, and suppose there is a continuous function* $a\colon [\alpha - r_0, \infty) \to [1, \infty)$ *such that*

$$\int_s^{r_0+s} b(t)a(t)\,dt \le a(s) - 1, \qquad s \ge \alpha - r_0,$$

and

$$\int_\alpha^\infty a(t)p(t)\,dt < \infty.$$

If $x\colon [\alpha - r_0, \infty) \to \mathbf{R}^n$ *is an absolutely continuous function such that*

$$|\dot{x}(t)| \le b(t) \int_{t-r_0}^t |\dot{x}(s)|\,ds + p(t)\|x_t\|_{r_0}, \qquad t \ge \alpha, \tag{14.43}$$

then

$$|x(t)| \le \left[\max_{\alpha - r_0 \le s \le \alpha} |x(s)| + \int_{\alpha - r_0}^\alpha [a(s) - 1]|\dot{x}(s)|\,ds\right] e^{\int_\alpha^t a(\tau)p(\tau)\,d\tau}, \qquad t \ge \alpha, \tag{14.44}$$

and

$$\int_\alpha^\infty |\dot{x}(t)|\,dt < \infty. \tag{14.45}$$

PROOF. Indeed, (14.44) and (14.45) follow from theorem 14.10 with $b(t,s) = b(s)$, $h(t) = t - r_0$, $r(t) = r_0$, $\rho(t) \equiv 0$. □

If, in corollary 14.5, $b(t) \equiv L$, $a(t) = 1/(1 - Lr_0)$, then the following corollary is obtained.

COROLLARY 14.6. *If* $x\colon [\alpha - r_0, \infty) \to \mathbf{R}^n$ *is an absolutely continuous function such that*

$$|\dot{x}(t)| \le L \int_{t-r_0}^t |\dot{x}(s)|\,ds + p(t)\|x_t\|_{r_0}, \qquad t \ge \alpha, \tag{14.46}$$

where $r_0 > 0$ *and* $0 < L < 1/r_0$ *are constants, and* $p\colon [\alpha - r_0, \infty) \to \mathbf{R}_+$ *is a continuous function, then*

$$|x(t)| \le \left[\max_{\alpha - r_0 \le s \le \alpha} |x(s)| + \frac{Lr_0}{1 - Lr_0} \int_{\alpha - r_0}^\alpha |\dot{x}(s)|\,ds\right] e^{\frac{1}{1 - Lr_0}\int_\alpha^t p(\tau)\,d\tau}, \qquad t \ge \alpha. \quad \square \tag{14.47}$$

COROLLARY 14.7. *If* $x\colon [\alpha - r_0, \infty) \to \mathbf{R}^n$ *is an absolutely continuous function such that*

$$|\dot{x}(t)| \le L \int_{t-r_0}^t |\dot{x}(s)|\,ds + p(t)|x(t)|, \qquad t \ge \alpha, \tag{14.48}$$

where $r_0 > 0$ *and* $0 < L < 1/r_0$ *are constants, and* $p\colon [\alpha - r_0, \infty) \to \mathbf{R}_+$ *is a continuous function, then*

$$|x(t)| \le \left[|x(\alpha)| + \frac{Lr_0}{1 - Lr_0} \int_{\alpha - r_0}^\alpha |\dot{x}(s)|\,ds\right] e^{\frac{1}{1 - Lr_0}\int_\alpha^t p(\tau)\,d\tau}, \qquad t \ge \alpha. \tag{14.49}$$

PROOF. Indeed, (14.49) follows from theorem 14.10 with $b(s,t) \equiv L$, $h(t) = t - r_0$, $r(t) \equiv 0$, $\rho(t) \equiv 0$, $a(t) = 1/(1 - Lr_0)$. □

REMARK 14.3. If the assumptions 1–4) in theorem 14.10 hold, and the absolutely continuous function $x\colon [\alpha - r_0, \infty) \to \mathbf{R}^n$ satisfies (14.35), then the limit

$$\lim_{t\to\infty} x(t) = x(\infty) \in \mathbf{R}^n$$

exists and is finite, by (14.37). □

However, theorem 14.10 gives no indication of the rate of convergence. To give an estimate of this, we prove the following assertion.

THEOREM 14.11. (Arino and Györi, 1984) *Let the conditions* 1–3) *in theorem* 14.10 *hold, and let* $b\colon [h(\alpha), \infty) \times [h(\alpha), \infty) \to \mathbf{R}_+$ *be a locally bounded, locally Lebesgue integrable function such that*

$$\int_s^{h^{-1}(s)} b(t,s)\,dt \le \mu, \qquad s \ge h(\alpha), \tag{14.50}$$

where $\mu \in (0,1)$ *is a constant.*

If $x\colon [h(\alpha), \infty) \to \mathbf{R}^n$ *is an absolutely continuous function satisfying* (14.35), *then*

i)

$$|x(t)| \le \left[\|x_\alpha\|_{r(\alpha)} + \frac{\mu}{1-\mu} \int_{h(\alpha)}^{\alpha} |\dot{x}(s)|\,ds + \right.$$
$$\left. + \frac{\mu}{1-\mu} \int_\alpha^t \rho(\tau)\,d\tau \right] e^{\frac{1}{1-\mu}\int_\alpha^t p(\tau)\,d\tau}, \qquad t \ge \alpha. \tag{14.51}$$

ii) *If, in addition to the above, we also suppose that*

$$\int_\alpha^\infty [p(t) + \rho(t)]\,dt < \infty, \tag{14.52}$$

then for any $\beta \in L_1([h(\alpha), \alpha], \mathbf{R}_+)$ *there is a* $\phi \in L_1([h(\alpha), \infty], \mathbf{R}_+)$ *such that*

$$\phi(t) \ge \begin{cases} \beta(t), & h(\alpha) \le t \le \alpha, \\ \int_{h(t)}^t b(t,s)\phi(s)\,ds + p(t) + \rho(t), & t \ge \alpha, \end{cases} \tag{14.53}$$

and such that

$$|\dot{x}(t)| \le \delta\phi(t), \qquad t \ge h(\alpha), \tag{14.54}$$

provided that

$$|\dot{x}(t)| \le \beta(t), \qquad h(\alpha) \le t \le \alpha, \tag{14.55}$$

where

$$\delta = \max\left\{1, \left[\|x_\alpha\|_{r(\alpha)} + \frac{\mu}{1-\mu}\int_{h(\alpha)}^{\alpha} \beta(s)\,ds + \frac{1}{1-\mu}\int_\alpha^\infty \rho(s)\,ds\right] e^{\frac{1}{1-\mu}\int_\alpha^\infty p(s)\,ds}\right\}. \tag{14.56}$$

PROOF. i) (14.51) follows from theorem 14.10 with $a(t) = 1/(1-\mu)$.

ii) The existence of a solution of inequality (14.53) is shown if we prove that the equation

$$c(t) = K[c](t), \qquad t \geq h(\alpha), \tag{14.57}$$

has a solution $c_0 \in L_1([h(\alpha), \infty), \mathbf{R}_+)$, where

$$K[c](t) = \begin{cases} \beta(t), & h(\alpha) \leq t \leq \alpha, \\ \int_{h(t)}^t b(t,s)c(s)\,ds + p(t) + \rho(t), & t \geq \alpha, \end{cases}$$

for each $c \in L_1([h(\alpha), \infty), \mathbf{R}_+) = L_1$. The conditions imposed on b, p, ρ imply that $K[c](t)$ is defined for all $(t,c) \in [h(\alpha), \infty) \times L_1$, and is locally integrable in $[h(\alpha), \infty)$. Moreover, (14.39) with $a(t) = 1/(1-\mu)$ implies that for $t \geq \alpha$,

$$\left| \int_\alpha^t \left(\int_{h(\tau)}^\tau b(\tau,s)c(s)\,ds \right) d\tau \right| \leq \int_\alpha^t \left(\int_{h(\tau)}^\tau b(\tau,s)|c(s)|\,ds \right) d\tau \leq \mu \int_{h(\alpha)}^t |c(s)|\,ds. \tag{14.58}$$

Thus, $K[c] \in L_1$, and for all $c_1, c_2 \in L_1$,

$$\int_{h(\alpha)}^\infty |K[c_1](t) - K[c_2](t)|\,dt \leq$$

$$\leq \int_\alpha^\infty \left[\int_{h(t)}^t b(t,s)|c_1(s) - c_2(s)|\,ds \right] dt \leq$$

$$\leq \mu \int_\alpha^\infty |c_1(t) - c_2(t)|\,dt \leq \mu \int_{h(\alpha)}^\infty |c_1(t) - c_2(t)|\,dt,$$

i.e. $K \colon L_1 \to L_1$ is a contraction.

Consequently, for any closed subset $B \subset L_1$ the inclusion $K(B) \subset B$ implies that (14.57) has only one solution $c_0 \in B$.

Let

$$B = \left\{ u \in L_1 : \; u(t) = \beta(t), \quad h(\alpha) \leq t \leq \alpha; \quad u(t) \geq 0, \quad t \geq \alpha; \quad \int_{h(\alpha)}^\infty u(t)\,dt < d_0 \right\}$$

be a closed set in L_1, where

$$d_0 = \frac{1}{1-\mu} \left[\int_{h(\alpha)}^\alpha \beta(t)\,dt + \int_\alpha^\infty [p(t) + \rho(t)]\,dt \right].$$

Since $K[c](t) \geq 0$ for all $c \in B$ and $t \geq h(\alpha)$, (14.58) implies

$$\int_{h(\alpha)}^\infty K[c](t)\,dt = \int_{h(\alpha)}^\alpha \beta(t)\,dt + \int_\alpha^\infty K[c](t)\,dt \leq$$

$$\leq \int_{h(\alpha)}^\alpha \beta(t)\,dt + \mu \int_{h(\alpha)}^\infty |c(t)|\,dt + \int_\alpha^\infty [p(t) + \rho(t)]\,dt \leq d_0,$$

i.e. $K(B) \subset B$. Hence (14.57) has only one solution $c_0 \in B \subset L_1$, which implies that (14.53) also has a solution in L_1.

We now compare $|\dot{x}(t)|$ and $\delta\phi(t)$, where $\phi \in L_1$ is an arbitrary, fixed, solution of (14.53), when (14.55) holds.

Let $K_1 \colon L_1 \to L_1$ be the operator defined by

$$K_1[c](t) = \begin{cases} \delta\beta(t), & h(\alpha) \le t \le \alpha, \\ \int_{h(t)}^{t} b(t,s)c(s)\,ds + \delta p(t) + \delta\rho(t), & t \ge \alpha. \end{cases} \tag{14.59}$$

Using (14.51) and (14.56) we find that $\|x_t\|_{r(t)} \le \delta$, $t \ge \alpha$. Then (14.35) and (14.55) imply that

$$|\dot{x}(t)| \le K_1[|\dot{x}|](t), \qquad t \ge h(\alpha). \tag{14.60}$$

Moreover, (14.53) and $\delta \ge 1$ imply that

$$K_1[\delta\phi](t) \le \delta\phi(t), \qquad t \ge h(\alpha). \tag{14.61}$$

Taking into account (14.58), it is easily seen that the operator K_1 is a contraction, and hence the equation

$$c(t) = K_1[c](t), \qquad t \ge h(\alpha), \tag{14.62}$$

has at most one solution in L_1.

Consider the set

$$B_1 = \{u \in L_1 : 0 \le u(t) \le \delta\phi(t), \quad t \ge h(\alpha)\};$$

it is closed in L_1. Since $b(t,s) \ge 0$ and (14.61) is valid, we have

$$0 \le K_1[c](t) \le K_1[\delta\phi](t) \le \delta\phi(t)$$

for each $c \in B_1$ and $t \ge h(\alpha)$.

The last inequality means that $K_1(B_1) \subset B_1$, and thus (14.62) has one solution $c_1 \in B_1$.

(14.55), (14.52), and theorem 14.10ii) imply

$$\|\dot{x}\|_{L_1} = \int_{h(\alpha)}^{\infty} |\dot{x}(t)|\,dt \le \int_{h(\alpha)}^{\alpha} \beta(t)\,dt + \int_{\alpha}^{\infty} |\dot{x}(t)|\,dt = M < \infty. \tag{14.63}$$

We define a constant

$$d = \max\left\{M, \frac{\delta}{1-\mu}\left(\int_{h(\alpha)}^{\alpha} \beta(t)\,dt + \int_{\alpha}^{\infty} [p(t) + \rho(t)]\,dt\right)\right\},$$

and a set

$$B_2 = \{u \in L_1 : u(t) = \delta\beta(t), \quad h(\alpha) \le t \le \alpha; \quad |\dot{x}(t)| \le u(t), \quad t \ge \alpha; \quad \|u\|_{L_1} \le d\}.$$

(14.63) implies that B_2 is nonempty, and B_2 is obviously closed in L_1. Taking into account (14.60) we may conclude that

$$0 \le |\dot{x}(t)| \le K_1[|\dot{x}|](t) \le K_1[c](t), \qquad c \in B_2, \qquad t \ge h(\alpha).$$

By (14.58) and (14.59) we obtain for $c \in B_2$,

$$\int_{h(\alpha)}^{\infty} K_1[c](t)\,dt \le \delta \int_{h(\alpha)}^{\alpha} \beta(t)\,dt + \mu \int_{h(\alpha)}^{\infty} c(t)\,dt + \delta \int_{\alpha}^{\infty} [p(t) + \rho(t)]\,dt \le d.$$

Hence $K(B_2) \subset B_2$, and (14.62) has only one solution $c_2 \in B_2$. But it has at most one solution in L_1, hence $c_1(t) = c_2(t)$, $t \geq h(\alpha)$, belongs to $B_1 \cap B_2$. The definitions of B_1 and B_2 then imply that

$$|\dot{x}(t)| \leq c_2(t) = c_1(t) \leq \delta\phi(t). \qquad \square$$

COROLLARY 14.8. *Let $r_0 > 0$, $L \geq 0$, $m \geq 0$, $\mu \in (0,1)$, and $\gamma > 0$ be constants such that*

$$Lr_0 \leq \mu, \qquad L(e^{\gamma r_0} - 1) < \gamma,$$

and suppose that $x\colon [\alpha - r_0, \infty) \to \mathbf{R}^n$ is an absolutely continuous function satisfying

$$|\dot{x}(t)| \leq L \int_{t-r_0}^{t} |\dot{x}(s)|\, ds + me^{-\gamma(t-\alpha)} \|x_t\|_{r_0}, \qquad t \geq \alpha. \tag{14.64}$$

Then

$$|x(t)| \leq \left[\|x_\alpha\|_{r_0} + \frac{\mu}{1-\mu} \int_{\alpha-r_0}^{\alpha} |\dot{x}(s)|\, ds \right] e^{m/(\gamma(1-\mu))}, \qquad t \geq \alpha, \tag{14.65}$$

and

$$|\dot{x}(t)| \leq \delta m_0 e^{-\gamma(t-\alpha)}, \qquad t \geq \alpha, \tag{14.66}$$

if only

$$|\dot{x}(t)| \leq m_0 e^{-\gamma(t-\alpha)}, \qquad \alpha - r_0 \leq t \leq \alpha, \tag{14.67}$$

where $m_0 \geq m\gamma/(\gamma - L(e^{\gamma_1\gamma_0} - 1))$ and

$$\delta = \max\left\{1, \left[\|x_\alpha\|_{r_0} + \frac{\mu m_0(e^{\gamma r_0} - 1)}{\gamma(1-\mu)}\right] e^{m/(\gamma(1-\mu))}\right\}. \qquad \square$$

15. Discrete inequalities

In this section we consider discrete inequalities. Many of these are discrete analogues of integral and integro-differential inequalities already known to us.

For discrete inequalities it is characteristic that the functions occurring in them are defined on countable sets, which can, without loss of generality, be assumed to be subsets of the set $\mathbf{Z}$ of integers.

Before stating the main results we introduce some notation. Let $\alpha, \beta \in \mathbf{Z}$, $\alpha \leq \beta$, $\mathbf{N}_\alpha = \{n \in \mathbf{Z}\colon\ n \geq \alpha\}$, $\mathbf{N}_{\alpha,\beta} = \{n \in \mathbf{Z}\colon\ \alpha \leq n \leq \beta\}$, and let $x_n = x(n)$ be a sequence of real numbers, defined for $n \in \mathbf{N}_\alpha$. Let $\sum_{j=\alpha}^{\beta} x_j$, and $\prod_{j=\alpha}^{\beta} x_j$, be the sum, respectively the product, of the numbers x_j, $j \in \mathbf{N}_{\alpha,\beta}$, and assume that $\sum_{j=\alpha}^{\alpha-1} x_j = 0$, $\prod_{j=\alpha}^{\alpha-1} x_j = 1$. The *first difference* of the sequence x_j is the sequence $\Delta x_n = x_{n+1} - x_n$, $n \in \mathbf{N}_\alpha$, and the *k-th difference* (*difference of order* k) $\Delta^k x_n$ is inductively defined by

$$\Delta^k x_n = \Delta(\Delta^{k-1} x_n) = \Delta^{k-1} x_{n+1} - \Delta^{k-1} x_n, \qquad n \in \mathbf{N}_\alpha.$$

All sequences considered below consist of real numbers.

For two sequences f_i, g_i, $i \in \mathbf{N}_\alpha$, the following '*summation by parts*' formula holds:

$$\sum_{i=\alpha}^{n-1} f_i \Delta g_i = f_n g_n - f_\alpha g_\alpha - \sum_{i=\alpha}^{n-1} g_{i+1} \Delta f_i, \qquad n \in \mathbf{N}_\alpha. \tag{15.1}$$

Moreover,

$$\sum_{i=\alpha}^{n-1} \Delta g_i = g_n - g_\alpha, \qquad n \in \mathbf{N}_\alpha, \tag{15.2}$$

and

$$1 + \sum_{i=\alpha}^{n-1} f_i \prod_{j=i+1}^{n-1} (1 + f_j) = \prod_{j=\alpha}^{n-1} (1 + f_j), \qquad n \in \mathbf{N}_\alpha, \tag{15.3}$$

provided $1 + f_j \neq 0$, $j \in \mathbf{N}_\alpha$.

First we consider the following general result.

LEMMA 15.1. (Pachpatte, 1971) *Let u_n, v_n be sequences defined for $n \in \mathbf{N}_\alpha$ satisfying the inequalities*

$$u_{n+1} \leq g(n, u_n), \qquad v_{n+1} \geq g(n, v_n), \tag{15.4}$$

for all $n \in \mathbf{N}_\alpha$, where g is a function defined in $\mathbf{N}_\alpha \times \mathbf{R}$ such that $g(n,x)$ is nondecreasing with respect to x for any fixed n. Then $u_\alpha \leq v_\alpha$ implies that

$$u_n \leq v_n, \qquad n \in \mathbf{N}_\alpha. \tag{15.5}$$

PROOF. Suppose (15.5) is not valid, i.e. there is a $k \geq \alpha$ such that $u_k \leq v_k$ and $u_{k+1} > v_{k+1}$. Using (15.4) and the nondecreasing nature of g, we arrive at a contradiction: $u_{k+1} \leq g(n, u_k) \leq g(n, v_k) \leq v_{k+1}$. □

One of the simplest discrete inequalities is considered in the following lemma.

LEMMA 15.2. *Let u_n, f_n, and $b_n \geq -1$ be sequences defined for $n \in \mathbf{N}_\alpha$ satisfying the inequality*

$$\Delta u_n \leq b_n u_n + f_n, \qquad n \in \mathbf{N}_\alpha,. \tag{15.6}$$

Then, for $n \in \mathbf{N}_\alpha$,

$$u_n \leq u_\alpha \prod_{s=\alpha}^{n-1} (1+b_s) + \sum_{s=\alpha}^{n-1} f_s \prod_{i=s+1}^{n-1} (1+b_i). \tag{15.7}$$

PROOF. We write (15.6) as

$$u_{s+1} - (1+b_s)u_s \leq f_s,$$

and multiply this termwise by $\prod_{i=s+1}^{n-1}(1+b_i)$. This gives

$$u_{s+1} \prod_{i=s+1}^{n-1} (1+b_i) - u_s \prod_{i=s}^{n-1} (1+b_i) \leq f_s \prod_{i=s+1}^{n-1} (1+b_i).$$

Summation from α to $n-1$ yields

$$u_n \prod_{i=n}^{n-1} (1+b_i) - u_\alpha \prod_{i=\alpha}^{n-1} (1+b_i) \leq \sum_{s=\alpha}^{n-1} f_s \prod_{i=s+1}^{n-1} (1+b_i),$$

which implies (15.7). □

COROLLARY 15.1. *If $\Delta u_n \leq f_n$, then $u_n \leq u_\alpha + \sum_{s=\alpha}^{n-1} f_s$.* □

COROLLARY 15.2. *If $\Delta u_n \leq b_n u_n$, then $u_n \leq u_\alpha \prod_{s=\alpha}^{n-1}(1+b_s)$.* □

We will look for analogies between the differential inequality

$$u'(t) \leq b(t)u(t) + f(t), \qquad t \geq \alpha, \tag{15.8}$$

and its solution

$$u(t) \leq u(\alpha)e^{\int_\alpha^t b(\tau)\,d\tau} + \int_\alpha^t f(s)e^{\int_s^t b(\tau)\,d\tau}\,ds, \qquad t \geq \alpha, \tag{15.9}$$

on the one hand, and the difference inequality (15.6) and its solution (15.7) on the other hand.

First we note that the derivative and the integral in (15.8) and (15.9) are replaced in (15.6) and (15.7), respectively, by a first difference and a sum, while instead of the expression $\exp(\int_s^t b(\tau)\,d\tau)$ in (15.9) the expression $\prod_{i=s+1}^{n-1}(1+b_i)$ occurs in (15.7).

A second common feature of the two inequalities is that the functions appearing on the righthand sides of (15.9) and (15.7) are solutions of the respective comparison equations

$$u'(t) = b(t)u(t) + f(t), \qquad t \geq \alpha,$$

and

$$\Delta u_n = b_n u_n + f_n, \qquad n \in \mathbf{N}_\alpha.$$

This last feature is preserved for a wider class of differential and difference inequalities, which can be seen from the following comparison lemma for difference inequalities, which is an analogue of theorem 6.3.

LEMMA 15.3. (Pachpatte, 1971) *Let u_n, v_n, w_n be sequences defined for $n \in \mathbf{N}_\alpha$ satisfying the inequalities*

$$f_1(n, u_n) \le \Delta u_n \le f_2(n, u_n), \qquad n \in \mathbf{N}_\alpha,$$
$$\Delta v_n = f_1(n, v_n), \qquad n \in \mathbf{N}_\alpha,$$
$$\Delta w_n = f_2(n, w_n), \qquad n \in \mathbf{N}_\alpha,$$

where f_1, f_2 are functions defined in $\mathbf{N}_\alpha \times \mathbf{R}$ which are nondecreasing with respect to x for any fixed n. Then $v_0 \le u_0 \le w_0$ implies that

$$v_n \le u_n \le w_n, \qquad n \in \mathbf{N}_\alpha.$$

PROOF. The inequality $v_n \le u_n$ ($u_n \le w_n$) follows immediately from lemma 15.1 with $g(n,x) = x + f_1(n,x)$ ($g(n,x) = x + f_2(n,x)$). □

A corollary of this lemma is the readily proved

LEMMA 15.4. (Pachpatte, 1971) *Let u_n, a_n, $q_n \ge 0$ be sequences defined for $n \in \mathbf{N}_\alpha$ satisfying the inequality*

$$u_n \le a_n + q_n \sum_{s=\alpha}^{n-1} W(s, u_s), \qquad n \in \mathbf{N}_\alpha,$$

where W is a function defined in $\mathbf{N}_\alpha \times \mathbf{R}$ such that $W(n,x)$ is nondecreasing with respect to x for any fixed n. Then

$$u_n \le a_n + q_n r_n,$$

where r_n is the solution of

$$\Delta r_n = W(n, a_n + q_n r_n), \qquad r_\alpha = 0. \qquad \square$$

Explicit representation of the solution r_n of a comparison difference equation of the form $\Delta r_n = f(n, r_n)$ is not always possible. Therefore, in applications of lemmas 15.3 and 15.4 this solution is often replaced by an upper (lower) bound for it, as in the following lemmas.

LEMMA 15.5. *Suppose the inequality*

$$\Delta u_n \le b_n u_n + k_n u_n^p, \qquad n \in \mathbf{N}_\alpha, \tag{15.10}$$

holds for sequences $u_n \ge 0$, $b_n \ge 0$, $k_n \ge 0$ defined for $n \in \mathbf{N}_\alpha$, and $p \ge 0$, $p \ne 1$. Then

$$u_n \le \frac{\left(u_\alpha^q + q \sum_{s=\alpha}^{n-1} k_s e_s^q\right)^{1/q}}{e_{n-1}}, \qquad n \in \mathbf{N}_\alpha, \tag{15.11}$$

provided that $u_\alpha + q \sum_{s=\alpha}^{n-1} k_s e_s^q > 0$, where $q = 1 - p$ and $e_n = \prod_{i=\alpha}^{n} (1 + b_i)^{-1}$.

PROOF. By lemma 15.3, $u_s \leq w_s$, where

$$\Delta w_s = b_s w_s + k_s w_s^p, \qquad w_\alpha = u_\alpha. \tag{15.12}$$

Multiplying (15.12) by the factor e_s and summing, we obtain

$$e_s w_{s+1} - e_{s-1} w_s = k_s e_s w_s^p.$$

The mean value theorem implies that

$$(e_s w_{s+1})^q - (e_{s-1} w_s)^q = q x^{q-1} (e_s w_{s+1} - e_{s-1} w_s),$$

where $q = 1 - p$ and x lies between $e_{s-1} w_s$ and $e_s w_{s+1}$. Since the sequence w_s is nondecreasing, the sequence e_s is nonincreasing, and $q - 1 \leq 0$, so that $x > e_s w_s$ and

$$x^{q-1} \leq (e_s w_s)^{q-1}.$$

Consequently,

$$(e_s w_{s+1})^q - (e_{s-1} w_s)^q \underset{\geq}{\leq} q k_s e_s^q, \qquad s \in \mathbf{N}_\alpha, \tag{15.13}$$

where $\leq$ ($\geq$) holds if $q > 0$ (i.e. $0 \leq p < 1$) (respectively, $q < 0$ ($p > 1$)).

Summation of (15.13) from α to $n - 1$ gives

$$(e_{n-1} w_n)^q \underset{\geq}{\leq} (e_{\alpha-1} w_\alpha)^q + q \sum_{s=\alpha}^{n-1} k_s e_s^q,$$

and since $e_{\alpha-1} = 1$, $w_\alpha = u_\alpha$, taking a q-th root we obtain

$$e_{n-1} w_n \leq \left[u_\alpha^q + q \sum_{s=\alpha}^{n-1} k_s e_s^q \right]^{1/q}. \tag{15.14}$$

Since $u_n \leq w_n$ and $e_{n-1} > 0$, (15.14) implies (15.11). □

LEMMA 15.6. *Let u_n, $k_n \geq 0$, $\sigma_n \geq 0$ be sequences defined for $n \in \mathbf{N}_{\alpha,\beta}$ satisfying the inequality*

$$\Delta u_n \leq k_n g(u_n) + \sigma_n, \qquad n \in \mathbf{N}_{\alpha,\beta}, \tag{15.15}$$

where $g(x)$ is a positive, nondecreasing, continuous function in an interval I containing u_α. Let

$$\gamma_1 = \sup \left\{ \nu \in \mathbf{N}_{\alpha,\beta} : \, u_\alpha + \sum_{s=\alpha}^{\nu-1} \sigma_s \in I \right\},$$

$$\gamma_2 = \sup \left\{ \nu \in \mathbf{N}_{\alpha,\beta} : \, G\left(u_\alpha + \sum_{s=\alpha}^{t-1} \sigma_s \right) + \sum_{s=\alpha}^{t-1} k_s \in G(I), \quad t \in \mathbf{N}_{\alpha,\nu} \right\},$$

$$\gamma = \min(\gamma_1, \gamma_2),$$

where $G(y) = \int_{y_0}^{y} dx/g(x)$, $y \in I$ ($y_0 \in I$). Then for $t \in \mathbf{N}_{\alpha,\gamma}$,

$$u_n \leq G^{-1} \left[G\left(u_\alpha + \sum_{s=\alpha}^{n-1} \sigma_s \right) + \sum_{s=\alpha}^{n-1} k_s \right]. \tag{15.16}$$

PROOF. We put

$$w_n = G^{-1}\left[G\left(u_\alpha + \sum_{s=\alpha}^{n-1}\sigma_s\right) + \sum_{s=\alpha}^{n-1}k_s\right].$$

By lemma 15.3, inequality (15.6) holds if for $n \in \mathbf{N}_{\alpha,\gamma}$,

$$\Delta w_n \geq k_n g(w_n) + \sigma_n, \qquad w_\alpha \geq u_\alpha. \tag{15.17}$$

The definition of w_n implies that $w_\alpha = u_\alpha$. Moreover, it is readily verified that for w_n and $r_n = u_\alpha + \sum_{s=\alpha}^{n-1}\sigma_s$ we have $\Delta r_n = \sigma_n$,

$$r_n \leq r_{n+1}, \qquad w_n \leq w_{n+1}, \qquad r_n \leq w_n, \tag{15.18}$$

$$G(w_{n+1}) - G(w_n) = G(r_{n+1}) - G(r_n) + k_n, \tag{15.19}$$

for $n \in \mathbf{N}_{\alpha,\gamma-1}$.

Let $\Delta w_n > 0$ and $\Delta r_n = \sigma_n > 0$. Then (15.19) and the mean value theorem imply that

$$\frac{\Delta w_n}{g(x)} = k_n + \frac{\sigma_n}{g(y)}, \tag{15.20}$$

where $x \in (w_n, w_{n+1})$, $y \in (r_n, r_{n+1})$. Since the function $1/g(z)$ is positive, nonincreasing, and continuous in I, (15.18) implies that the mean values $1/g(x)$ and $1/g(y)$ of the integrals $\int_{w_n}^{w_{n+1}} dz/g(z)$ and $\int_{r_n}^{r_{n+1}} dz/g(z)$ satisfy the inequalities

$$\frac{1}{g(y)} = \frac{1}{\Delta r_n}\int_{r_n}^{r_{n+1}}\frac{dz}{g(z)} \geq \frac{1}{\Delta w_n}\int_{w_n}^{w_{n+1}}\frac{dz}{g(z)} = \frac{1}{g(x)},$$

i.e. $g(x) \geq g(y)$. Moreover, $g(x) \geq g(w_n)$, and then (15.20) implies (15.17):

$$\Delta w_n = k_n g(x) + \sigma_n \frac{g(x)}{g(y)} \geq k_n g(w_n) + \sigma_n.$$

If $w_n = 0$ ($k_n = 0$, $\sigma_n = 0$) or $\Delta r_n = 0$ ($\sigma_n = 0$), then (15.17) follows directly from (15.19). □

Now we will consider the discrete inequalities known in the mathematics literature as 'discrete inequalities of Gronwall type'. The inequality in the following theorem is a discrete analogue of the integral inequality (1.20) in theorem 1.5.

THEOREM 15.1. (Pachpatte, 1973) *Let u_n, a_n, $b_n \geq 0$, $q_n \geq 0$ be sequences defined for $n \in \mathbf{N}_\alpha$ satisfying the inequality*

$$u_n \leq a_n + q_n \sum_{s=\alpha}^{n-1} b_s u_s, \qquad n \in \mathbf{N}_\alpha. \tag{15.21}$$

Then

$$u_n \leq a_n + q_n \sum_{s=\alpha}^{n-1} b_s a_s \prod_{i=s+1}^{n-1}(1 + b_i q_i), \qquad n \in \mathbf{N}_\alpha. \tag{15.22}$$

PROOF. Setting

$$w_n = \sum_{s=\alpha}^{n-1} b_s u_s, \qquad w_\alpha = 0,$$

we successively obtain

$$u_n \leq a_n + q_n w_n, \tag{15.23}$$
$$\Delta w_n = b_n u_n,$$
$$\Delta w_n \leq b_n a_n + b_n q_n w_n. \tag{15.24}$$

We apply lemma 15.2 to the difference inequality (15.24), and obtain

$$w_n \leq \sum_{s=\alpha}^{n-1} b_s a_s \prod_{i=s+1}^{n-1} (1 + b_i q_i).$$

Then (15.23) and (15.24) imply (15.22). □

COROLLARY 15.3. (Sugiyama, 1969) *If*

$$u_n \leq a_n + \sum_{s=\alpha}^{n-1} b_s u_s, \qquad n \in \mathbf{N}_\alpha, \tag{15.25}$$

where $b_s \geq 0$, $s \in \mathbf{N}_\alpha$, *then*

$$u_n \leq a_n + \sum_{s=\alpha}^{n-1} b_s a_s \prod_{i=s+1}^{n-1} (1 + b_i), \qquad n \in \mathbf{N}_\alpha. \tag{15.26}$$

COROLLARY 15.4. (Lees, 1959) *If* $b_n \geq 0$ *and* a_n *is nondecreasing for* $n \in \mathbf{N}_\alpha$, *then* (15.25) *implies*

$$u_n \leq a_n \prod_{s=\alpha}^{n-1} (1 + b_s) \leq a_n \exp\left(\sum_{s=\alpha}^{n-1} b_s\right), \qquad n \in \mathbf{N}_\alpha. \tag{15.27}$$

COROLLARY 15.5. (Beesack, 1975) *If* $b_n \geq 0$ *and* $a_n \leq A$ *for* $n \in \mathbf{N}_\alpha$, *then* (15.25) *implies*

$$u_n \leq A \prod_{s=\alpha}^{n-1} (1 + b_s) \leq A \exp\left(\sum_{s=\alpha}^{n-1} b_s\right), \qquad n \in \mathbf{N}_\alpha. \tag{15.28}$$

COROLLARY 15.6. *Under the conditions of theorem* 15.1, *inequality* (15.21) *implies*

$$\sum_{s=\alpha}^{n-1} b_s u_s \leq \sum_{s=\alpha}^{n-1} b_s a_s \prod_{i=s+1}^{n-1} (1 + b_s q_s), \qquad n \in \mathbf{N}_\alpha. \tag{15.29}$$

PROOF. (15.26) is obtained from (15.22) with $q_i = 1$, and (15.27) and (15.28) additionally use (15.3). (15.29) was obtained in the course of proving theorem 15.1. □

From lemma 15.2 we can obtain the following theorem, which is also a corollary of theorem 15.1.

THEOREM 15.2. *Let u_n, p_n, and $b_n \geq 0$ be sequences defined for $n \in \mathbf{N}_\alpha$ such that*

$$u_n \leq u_\alpha + \sum_{s=\alpha}^{n-1}(b_s u_s + p_s), \qquad n \in \mathbf{N}_\alpha. \tag{15.30}$$

Then

$$u_n \leq u_\alpha \prod_{s=\alpha}^{n-1}(1+b_s) + \sum_{s=\alpha}^{n-1} p_s \prod_{i=s+1}^{n-1}(1+b_i), \qquad n \in \mathbf{N}_\alpha. \qquad \square \tag{15.31}$$

The following theorem can be proved by simple induction; it is a discrete analogue of theorem 2.1 of Chu and Metcalf.

THEOREM 15.3. *Let u_n and a_n be sequences defined for $n \in \mathbf{N}_\alpha$ such that*

$$u_n \leq a_n + \sum_{s=\alpha}^{n-1} k_{ns} u_s, \qquad n \in \mathbf{N}_\alpha, \tag{15.32}$$

where $k_{ns} \geq 0$, $n \in \mathbf{N}_\alpha$, $s \in \mathbf{N}_\alpha$. Then

i) *$u_n \leq w_n$, $n \in \mathbf{N}_\alpha$, where w_n is the solution of the equation*

$$w_n = a_n + \sum_{s=\alpha}^{n-1} k_{ns} w_s, \qquad n \in \mathbf{N}_\alpha; \tag{15.33}$$

ii) *The solution w_n has the form*

$$w_n = \sum_{m=0}^{n-\alpha} y_m(n), \qquad n \in \mathbf{N}_\alpha, \tag{15.34}$$

where $y_0(n) = a_n$,

$$y_{m+1}(n) = \sum_{s=m+\alpha}^{n-1} k_{ns} y_m(s), \qquad n \in \mathbf{N}_\alpha, \qquad m \in \mathbf{N}_0. \qquad \square$$

In some cases we can estimate the solution w_n of (15.33) as follows.
Multiplying the equalities

$$w_i = a_i + \sum_{j=\alpha}^{i-1} k_{ij} w_j, \qquad i \in \mathbf{N}_\alpha,$$

by k_{ni}, summation over i from α to $n-1$ gives

$$\sum_{i=\alpha}^{n-1} k_{ni} w_i = \sum_{i=\alpha}^{n-1} k_{ni} a_i + \sum_{i=\alpha}^{n-1}\sum_{j=\alpha}^{i-1} k_{ni} k_{ij} w_j, \qquad n \in \mathbf{N}_\alpha.$$

From (15.33),

$$w_n = a_n + \sum_{i=\alpha}^{n-1} k_{ni} a_i + \sum_{i=\alpha}^{n-1}\sum_{j=\alpha}^{i-1} k_{ni} k_{ij} w_j,$$

and by changing the order of summation we find

$$w_n = a_n + \sum_{i=\alpha}^{n-1} k_{ni} a_i + \sum_{j=\alpha}^{n-2} \left(\sum_{i=j+1}^{n-1} k_{ni} k_{ij} \right) w_j. \tag{15.35}$$

Let $A_n = a_n + \sum_{i=\alpha}^{n-1} k_{ni} a_i$, and assume that

$$\sum_{i=j+1}^{n-1} k_{ni} k_{ij} \leq B_j, \qquad j \in \mathbf{N}_{\alpha, n-1},$$

independently of $n \in \mathbf{N}_\alpha$. Since $B_{n-1} = 0$, (15.35) implies the inequality

$$w_n \leq A_n + \sum_{j=\alpha}^{n-1} B_j w_j,$$

to which we can apply theorem 15.1.

By the above reasoning we can prove the following assertion.

Theorem 15.4. (McKee, 1982) *Let u_i be a sequence defined for $i \in \mathbf{N}_0$ such that*

$$u_i \leq \delta + h^{1/2} M \sum_{j=0}^{i-1} \frac{u_j}{(i-j)^{1/2}}, \qquad i = 1, \ldots, n, \tag{15.36}$$

where $u_0 \leq \delta$, $M > 0$ is a constant which is independent of $h > 0$, $\delta > 0$, and $T = nh$. Then

$$\max_{1 \leq i \leq n} u_i \leq \delta(1 + h^{1/2} M + hM^2\pi + 2MT^{1/2}) e^{M^2 \pi T}. \tag{15.37}$$

Proof. The proof can be carried out by repeating the transformations above with

$$a_i = \delta + \delta M h^{1/2} i^{-1/2}, \qquad k_{ij} = M h^{1/2} (i-j)^{-1/2},$$

and using the estimates

$$\sum_{i=1}^{n-1} \frac{1}{(n-i)^{1/2}} \leq \int_0^{n-1} \frac{dx}{x^{1/2}} < 2\sqrt{n},$$

$$\sum_{i=j+1}^{n-1} \frac{1}{(n-i)^{1/2}(i-j)^{1/2}} \leq \int_j^n \frac{dx}{(n-x)^{1/2}(x-j)^{1/2}} = \pi. \qquad \square$$

The discrete inequalities in the following two theorems are analogues of theorems 4.2 and 4.9 of Willett and Wong.

Theorem 15.5. (Willett and Wong, 1965) *Let $u_n \geq 0$, $a_n \geq 0$, $b_n \geq 0$, $k_n \geq 0$ be sequences defined for $n \in \mathbf{N}_\alpha$ such that*

$$u_n \leq a_n + b_n \left(\sum_{s=\alpha}^{n-1} k_s u_s^p \right)^{1/p}, \qquad n \in \mathbf{N}_\alpha, \tag{15.38}$$

where $p \geq 1$ is a constant. Then

$$u_n \leq a_n + b_n \frac{\left(\sum_{s=\alpha}^{n-1} k_s a_s^p e_s\right)^{1/p}}{1-(1-e_{n-1})^{1/p}}, \qquad n \in \mathbf{N}_\alpha, \tag{15.39}$$

where $e_n = \prod_{i=\alpha}^{n}(1+k_i b_i^p)^{-1}$, $n \in \mathbf{N}_\alpha$.

PROOF. For the sequences e_s and $y_n = \sum_{s=\alpha}^{n-1} k_s u_s^p$, $y_\alpha = 0$, we have

$$e_s - e_{s-1} = -e_s k_s b_s^p, \tag{15.40}$$

$$\Delta y_s = k_s u_s^p, \tag{15.41}$$

$$u_s \leq a_s + b_s y_s^{1/p}. \tag{15.42}$$

Taking into account (15.40)-(15.42), we obtain

$$e_s y_{s+1} - e_{s-1} y_s \leq \{e_s^{1/p} k_s^{1/p} a_s + e_s^{1/p} k_s^{1/p} b_s y_s^{1/p}\}^p - e_s k_s b_s^p y_s. \tag{15.43}$$

Summation of (15.43) from α to $n-1$ gives

$$e_{n-1} y_n + \sum_{s=\alpha}^{n-1} e_s k_s b_s^p y_s \leq \sum_{s=\alpha}^{n-1} \{e_s^{1/p} k_s^{1/p} a_s + e_s^{1/p} k_s^{1/p} b_s y_s^{1/p}\}^p.$$

Taking the p-th root and applying Minkowski's inequality, we arrive at

$$(e_{n-1} y_n + A)^{1/p} - A^{1/p} \leq \left(\sum_{s=\alpha}^{n-1} e_s k_s a_s^p\right)^{1/p}, \tag{15.44}$$

where $A = \sum_{s=\alpha}^{n-1} e_s k_s b_s^p y_s$.

Note that (15.44) still holds if A is made larger. (15.40) implies that

$$A = \sum_{s=\alpha}^{n-1} e_s k_s b_s^p y_s = \sum_{s=\alpha}^{n-1} (e_{s-1} - e_s) y_s \leq \sum_{s=\alpha}^{n-1} (e_{s-1} - e_s) y_n =$$
$$= y_n (e_{\alpha-1} - e_{n-1}) = y_n (1 - e_{n-1}).$$

Consequently, (15.44) implies

$$y_n^{1/p} - y_n^{1/p}(1-e_{n-1})^{1/p} \leq \left(\sum_{s=\alpha}^{n-1} e_s k_s a_s^p\right)^{1/p}.$$

Then

$$y_n^{1/p} \leq \frac{\left(\sum_{s=\alpha}^{n-1} e_s k_s a_s^p\right)^{1/p}}{1-(1-e_{n-1})^{1/p}},$$

and (15.42) implies (15.39). □

THEOREM 15.6. (Willett and Wong, 1965) *Let $u_n \geq 0$, $b_n \geq 0$, $k_n \geq 0$ be sequences defined for $n \in \mathbf{N}_\alpha$ such that*

$$u_n \leq a + \sum_{s=\alpha}^{n-1} b_n u_n + \sum_{s=\alpha}^{n-1} k_s u_s^p, \qquad n \in \mathbf{N}_\alpha, \tag{15.45}$$

where $a > 0$ and $p \geq 0$, $p \neq 1$, are constants. Then

$$u_n \leq e_{n-1}^{-1} \left\{ a^{1-p} + (1-p) \sum_{s=\alpha}^{n-1} k_s e_s^{1-p} \right\}^{1/(1-p)}, \qquad n \in \mathbf{N}_{\alpha,\beta}, \tag{15.46}$$

where $e_n = \prod_{i=\alpha}^{n} (1+b_i)^{-1}$, $n \in \mathbf{N}_\alpha$, and

$$\beta = \sup \left\{ n \in \mathbf{N}_\alpha : \; a^{1-p} + (1-p) \sum_{s=\alpha}^{n-1} k_s e_s^{1-p} > 0 \right\}.$$

PROOF. For the sequence $y_n = a + \sum_{s=\alpha}^{n-1} b_s u_s + \sum_{s=\alpha}^{n-1} k_s u_s^p$, (15.45) implies

$$u_n \leq y_n, \tag{15.47}$$

$$\Delta y_n = b_n u_n + k_n u_n^p \leq b_n y_n + k_n y_n^{1/p}, \qquad y_\alpha = a. \tag{15.48}$$

By lemma 15.5 it follows from (15.48) that

$$y_n \leq \frac{1}{e_{n-1}} \left\{ a^{1-p} + (1-p) \sum_{s=\alpha}^{n-1} k_s e_s^{1-p} \right\}^{1/(1-p)},$$

which together with (15.47) implies (15.46). □

In the theorems below we consider various discrete inequalities of Bihari type. Their proofs are based on lemma 15.6.

THEOREM 15.7. *Let $u_n \geq 0$, $a_n > 0$, $\Delta a_n \geq 0$, and $k_n \geq 0$ be sequences defined for $n \in \mathbf{N}_\alpha$ such that*

$$u_n \leq a_n + \sum_{s=\alpha}^{n-1} k_s g(u_s), \qquad n \in \mathbf{N}_\alpha, \tag{15.49}$$

where $g(u)$ is a positive, nondecreasing, continuous function in $(0, \infty)$. Then

$$u_n \leq G^{-1} \left[G(a_n) + \sum_{s=\alpha}^{n-1} k_s \right], \qquad n \in \mathbf{N}_{\alpha,\beta}, \tag{15.50}$$

where $G(y) = \int_{y_0}^{y} dx/g(x)$, $y > 0$ ($y_0 > 0$), and

$$\beta = \sup \left\{ n \in \mathbf{N}_\alpha : \; G(a_n) + \sum_{s=\alpha}^{n-1} k_s \in \operatorname{Dom} G^{-1} \right\}.$$

PROOF. For the sequence $y_n = a_n + \sum_{s=\alpha}^{n-1} k_s g(u_s)$ from (15.49) we have

$$u_n \leq y_n, \tag{15.51}$$

$$\Delta y_n \leq \Delta a_n + k_n g(y_n), \qquad y_\alpha = a_\alpha. \tag{15.52}$$

By lemma 15.6 it follows from (15.52) that for $n \in \mathbf{N}_{\alpha,\beta}$,

$$y_n \leq G^{-1} \left[G\left(a_\alpha + \sum_{s=\alpha}^{n-1} \Delta a_s \right) + \sum_{s=\alpha}^{n-1} k_s \right],$$

which together with (15.51) implies (15.50). □

COROLLARY 15.7. (Beesack, 1975) *If* $u_n \le a+\sum_{s=\alpha}^{n-1} k_s g(u_s)$, $n \in \mathbf{N}_\alpha$, *where* $a > 0$ *is a constant and* $k_s \ge 0$, $s \in \mathbf{N}_\alpha$, *then*

$$u_n \le G^{-1}\left[G(a) + \sum_{s=\alpha}^{n-1} k_s\right], \qquad n \in \mathbf{N}_{\alpha,\beta}. \qquad \square \tag{15.53}$$

COROLLARY 15.8. (Hull and Luxemburg, 1960) *If* $u_n \le a+k\sum_{s=\alpha}^{n-1} g(u_s)$, $n \in \mathbf{N}_\alpha$, *where* $a > 0$ *and* $k > 0$ *are constants, then*

$$u_n \le G^{-1}[G(a) + k(n-\alpha)], \qquad n \in \mathbf{N}_{\alpha,\beta}. \qquad \square \tag{15.54}$$

COROLLARY 15.9. *Let* $u_n \ge 0$, $k_n \ge 0$, *and* $c_n \ge 0$ *be sequences defined for* $n \in \mathbf{N}_\alpha$ *such that*

$$u_n \le a_n + c_n \sum_{s=\alpha}^{n-1} k_s g(u_s), \qquad n \in \mathbf{N}_\alpha, \tag{15.55}$$

where g *is a positive, nondecreasing, continuous function in* $(0,\infty)$. *Now,*

I) *if* $a_\alpha > 0$, $0 < c_n \le 1$, $\Delta(a_n/c_n) \ge 0$, *then* (15.55) *implies*

$$u_n \le c_n G^{-1}\left[G\left(\frac{a_n}{c_n}\right) + \sum_{s=\alpha}^{n-1} k_s\right], \qquad n \in \mathbf{N}_{\alpha,\beta}; \tag{15.56}$$

II) *if* $a_\alpha > 0$, $\Delta a_n \ge 0$, $c_n > 0$, $\Delta c_n \le 0$, *then* (15.55) *implies*

$$u_n \le G^{-1}\left[G(a_n) + \sum_{s=\alpha}^{n-1} k_s c_s\right], \qquad n \in \mathbf{N}_{\alpha,\beta}; \tag{15.57}$$

III) *if* $a_\alpha > 0$, $c_n > 0$, $\Delta(a_n/c_n) \ge 0$, *and* g *is submultiplicative, then* (15.55) *implies*

$$u_n \le c_n G^{-1}\left[G\left(\frac{a_n}{c_n}\right) + \sum_{s=\alpha}^{n-1} k_s g(c_s)\right], \qquad n \in \mathbf{N}_{\alpha,\beta}, \tag{15.58}$$

where $G(x) = \int_{x_0}^{x} dz/g(z)$, $x > 0$ $(x_0 > 0)$, *and* $\beta \ge \alpha$ *is so chosen that the righthand sides of* (15.56)–(15.58) *are defined for* $n \in \mathbf{N}_{\alpha,\beta}$.

PROOF. (15.55) implies the inequalities

$$\frac{u_n}{c_n} \le \frac{a_n}{c_n} + \sum_{s=\alpha}^{n-1} k_s g\left(\frac{u_s}{c_s}\right) \text{(in case I)},$$

$$u_n \le a_n + \sum_{s=\alpha}^{n-1} k_s c_s g(u_s) \text{(in case II)},$$

$$\frac{u_n}{c_n} \le \frac{a_n}{c_n} + \sum_{s=\alpha}^{n-1} k_s g(c_s) g\left(\frac{u_s}{c_s}\right) \text{(in case III)}.$$

Applying theorem 15.7 to these inequalities, we obtain respectively (15.56)–(15.58). $\square$

THEOREM 15.8. *Let $u_n \geq 0$, $k_n \geq 0$, and $a_n \geq 0$ be sequences defined for $n \in \mathbf{N}_\alpha$ such that*

$$u_n \leq a_n + f\left(\sum_{s=\alpha}^{n-1} k_s g(u_s)\right), \qquad n \in \mathbf{N}_\alpha, \tag{15.59}$$

where $k_\alpha g(a_\alpha) > 0$, $\Delta(k_n g(a_n)) \geq 0$, and f, g are positive, nondecreasing, continuous functions in $(0, \infty)$, with g subadditive. Then

$$u_n \leq a_n + f\left\{F^{-1}\left[F(k_n g(a_n)) + \sum_{s=\alpha}^{n-1} k_s\right]\right\}, \qquad n \in \mathbf{N}_{\alpha,\beta}, \tag{15.60}$$

where $F(x) = \int_{x_0}^{x} dz/g(f(z))$, $x > 0$ ($x_0 > 0$), and $\beta = \sup\{n \in \mathbf{N}_\alpha\colon F(k_n g(a_n)) + \sum_{s=\alpha}^{n-1} k_s \in \operatorname{Dom} F^{-1}\}$. □

THEOREM 15.9. *Let $u_n \geq 0$, $k_n \geq 0$, $a_n \geq 0$, $c_n \geq 0$ be sequences defined for $n \in \mathbf{N}_\alpha$ such that*

$$u_n \leq a_n + c_n f\left(\sum_{s=\alpha}^{n-1} k_s g(u_s)\right), \qquad n \in \mathbf{N}_\alpha, \tag{15.61}$$

where $k_\alpha g(a_\alpha) > 0$, $\Delta(k_n g(a_n)) \geq 0$, and f, g are positive, nondecreasing, continuous functions in $(0, \infty)$, with g submultiplicative and subadditive. Then

$$u_n \leq a_n + c_n f\left\{F^{-1}\left[F(k_n g(a_n)) + \sum_{s=\alpha}^{n-1} k_s g(c_s)\right]\right\}, \qquad n \in \mathbf{N}_{\alpha,\beta}, \tag{15.62}$$

where $F(x) = \int_{x_0}^{x} dz/g(f(z))$, $x > 0$ ($x_0 > 0$), and $\beta = \sup\{n \in \mathbf{N}_\alpha\colon F(k_n g(a_n)) + \sum_{s=\alpha}^{n-1} k_s g(c_s) \in \operatorname{Dom} F^{-1}\}$. □

The proofs of these theorems are similar to the proof of theorem 15.7, so we omit them.

In obtaining new results for discrete inequalities, the following lemma, which is a discrete analogue of lemma 9.1, proves useful.

LEMMA 15.7. *Let $u_n \geq 0$ and $a_n \geq 0$ be sequences defined for $n \in \mathbf{N}_\alpha$ such that*

$$u_n \leq a_n + \sum_{s=\alpha}^{n-1} k(n,s) g(u_s), \qquad n \in \mathbf{N}_\alpha, \tag{15.63}$$

where $k(n,s) \geq 0$ is a function defined for $n \in \mathbf{N}_\alpha$, $s \in \mathbf{N}_{\alpha,n-1}$, which is nondecreasing with respect to n for s fixed, while g is a nondecreasing continuous function in $\mathbf{R}_+$, with $g(u) > 0$ for $u > 0$. Now,

i) *if $G(x) = \int_{x_0}^{x} dz/g(z)$, $x > 0$ ($x_0 > 0$), then*

$$u_n \leq G^{-1}\left[G(\hat{a}_n) + \sum_{s=\alpha}^{n-1} k(n,s)\right], \qquad n \in \mathbf{N}_{\alpha,\beta_1}, \tag{15.64}$$

where $\hat{a}_n = \max\{a_i\colon i \in \mathbf{N}_{\alpha,n}\}$, $n \in \mathbf{N}_\alpha$;

ii) *if* $a_n > 0$, $n \in \mathbf{N}_\alpha$, $g \in H_{r,w}$, *and* $W(x) = \int_{x_0}^{x} dz/w(z)$, $x > 0$ $(x_0 > 0)$, *then*

$$u_n \leq \hat{a}_n W^{-1}\left[W(1) + \frac{r(\hat{a}_n)}{\hat{a}_n}\sum_{s=\alpha}^{n-1} k(n,s)\right], \qquad n \in \mathbf{N}_{\alpha,\beta_2}. \tag{15.65}$$

Moreover, in this case

$$u_n \leq \hat{a}_n W^{-1}\left[W(1) + \sum_{s=\alpha}^{n-1} k(n,s)\frac{r(\hat{a}_s)}{\hat{a}_s}\right], \qquad n \in \mathbf{N}_{\alpha,\beta_3}. \tag{15.66}$$

In (15.64)–(15.66) *the numbers* β_i, $i = 1,2,3$, *are chosen so that the quantity between square brackets is in the range of* G *(or* W*) for* $n \in \mathbf{N}_{\alpha,\beta_i}$. □

THEOREM 15.10. *Let* $u_n \geq 0$, $a_n \geq 0$, $\Delta a_n \geq 0$, $b_n \geq 0$, *and* $k_n \geq 0$ *be sequences defined for* $n \in \mathbf{N}_\alpha$ *such that*

$$u_n \leq a_n + \sum_{s=\alpha}^{n-1} b_s u_s + \sum_{s=\alpha}^{n-1} k_s g(u_s), \qquad n \in \mathbf{N}_\alpha, \tag{15.67}$$

where g *is a nondecreasing continuous function in* $\mathbf{R}_+$ *with* $g(u) > 0$ *for* $u > 0$. *Then*

$$u_n \leq G^{-1}\left[G(A_n) + \sum_{s=\alpha}^{n-1} k_s \prod_{i=s+1}^{n-1}(1+b_i)\right], \qquad n \in \mathbf{N}_{\alpha,\beta}, \tag{15.68}$$

where $A_n = a_n + \sum_{s=\alpha}^{n-1} a_s b_s \prod_{j=s+1}^{n-1}(1+b_j)$, $n \in \mathbf{N}_\alpha$, $G(x) = \int_{x_0}^{x} dz/g(z)$, $x > 0$ $(x_0 > 0)$,

$$\beta = \sup\left\{n \in \mathbf{N}_\alpha :\ G(A_n) + \sum_{s=\alpha}^{n-1} k_s \prod_{i=s+1}^{n-1}(1+b_i) \in \operatorname{Dom} G^{-1}\right\}.$$

PROOF. We set $c_n = a_n + \sum_{s=\alpha}^{n-1} k_s g(u_s)$. Then (15.67) implies that

$$u_n \leq c_n + \sum_{s=\alpha}^{n-1} b_s u_s,$$

so that, by corollary 15.3,

$$u_n \leq c_n + \sum_{s=\alpha}^{n-1} c_s b_s \prod_{i=s+1}^{n-1}(1+b_i). \tag{15.69}$$

Changing the order of summation in (15.69), we obtain

$$u_n \leq A_n + \sum_{i=\alpha}^{n-1} k_i \prod_{s=i+1}^{n-1}(1+b_s) g(u_i). \tag{15.70}$$

By theorem 15.7, (15.70) implies (15.68). □

REMARK 15.1. Theorem 15.10 still holds if A_n is replaced by $B_n = a_n \prod_{s=\alpha}^{n-1}(1+b_s)$. □

THEOREM 15.11. *Let u_n, a_n, $b_n \geq 0$ be sequences defined for $n \in \mathbf{N}_\alpha$ such that*

$$u_n \leq a_n + \sum_{s=\alpha}^{n-1} \left[b_s u_s + \sum_{j=\alpha}^{s-1} c_{sj} u_j \right], \qquad n \in \mathbf{N}_\alpha, \tag{15.71}$$

where $c_{sj} \geq 0$, $s \in \mathbf{N}_\alpha$, $j \in \mathbf{N}_\alpha$. Then

$$u_n \leq a_n + \sum_{s=\alpha}^{n-1} A_s \prod_{j=s+1}^{n-1} (1 + B_j), \quad n \in \mathbf{N}_\alpha, \tag{15.72}$$

where $A_s = a_s b_s + \sum_{j=\alpha}^{s-1} c_{sj} a_j$, $B_s = b_s + \sum_{j=\alpha}^{s-1} c_{sj}$.

PROOF. We put

$$y_n = \sum_{s=\alpha}^{n-1} \left[b_s u_s + \sum_{j=\alpha}^{s-1} c_{sj} u_j \right].$$

Then $y_\alpha = 0$, and (15.71) successively implies

$$u_n \leq a_n + y_n, \tag{15.73}$$

$$\Delta y_n = b_n u_n + \sum_{j=\alpha}^{n-1} c_{nj} u_j \leq A_n + B_n y_n. \tag{15.74}$$

By lemma 15.2, (15.74) implies

$$y_n \leq \sum_{s=\alpha}^{n-1} A_s \prod_{j=s+1}^{n-1} (1 + B_j),$$

which together with (15.73) implies (15.72). □

COROLLARY 15.10. *If, in theorem 15.11, $a_n \equiv a$, then (15.71) implies*

$$u_n \leq a \prod_{s=\alpha}^{n-1} \left(1 + b_s + \sum_{j=\alpha}^{n-1} c_{sj} \right), \qquad n \in \mathbf{N}_\alpha. \qquad \square \tag{15.75}$$

THEOREM 15.12. (Pachpatte, 1974) *Let $u_n \geq 0$, $a_n \geq 0$, $p_n \geq 0$, $f_n \geq 0$, $g_n \geq 0$ be sequences defined for $n \in \mathbf{N}_\alpha$ such that*

$$u_n \leq a_n + p_n \left[\sum_{s=\alpha}^{n-1} f_s u_s + \sum_{s=\alpha}^{n-1} f_s p_s \sum_{i=\alpha}^{s-1} g_i u_i \right], \qquad n \in \mathbf{N}_\alpha. \tag{15.76}$$

Then, for $n \in \mathbf{N}_\alpha$,

$$u_n \leq a_n + p_n \left[\sum_{s=\alpha}^{n-1} f_s \left(a_s + p_s \sum_{i=\alpha}^{s-1} a_i (f_i + g_i) \prod_{j=i+1}^{s-1} (1 + p_j (f_j + g_j)) \right) \right]. \tag{15.77}$$

PROOF. Let m_n denote the expression between square brackets in (15.76). Then we successively obtain

$$u_n \le a_n + p_n m_n, \tag{15.78}$$

$$\Delta m_n = f_n u_n + f_n p_n \sum_{i=\alpha}^{n-1} g_i u_i, \qquad m_\alpha = 0,$$

$$\Delta m_n \le f_n a_n + f_n p_n \left[m_n + \sum_{i=\alpha}^{n-1} g_i (a_i + p_i m_i) \right]. \tag{15.79}$$

Setting $v_n = m_n + \sum_{i=\alpha}^{n-1} g_i(a_i + p_i m_i)$, $v_\alpha = 0$, then (15.79) implies

$$\Delta v_n \le a_n (f_n + g_n) + p_n (f_n + g_n) v_n.$$

From lemma 15.2 we obtain

$$v_n \le \sum_{s=\alpha}^{n-1} a_s (f_s + g_s) \prod_{i=s+1}^{n-1} (1 + p_i (f_i + g_i)),$$

and from (15.79),

$$\Delta m_n \le f_n \left[a_n + p_n \sum_{s=\alpha}^{n-1} a_s (f_s + g_s) \prod_{i=s+1}^{n-1} (1 + p_i (f_i + g_i)) \right].$$

Corollary 15.1 and lemma 15.2 now imply

$$m_n \le \sum_{s=\alpha}^{n-1} f_s \left[a_s + p_s \sum_{i=\alpha}^{s-1} a_i (f_i + g_i) \prod_{j=i+1}^{s-1} (1 + p_j (f_j + g_j)) \right],$$

which together with (15.78) implies (15.77). □

COROLLARY 15.11. *Let $u_n \ge 0$, $a_n \ge 0$, $\Delta a_n \ge 0$, $f_n \ge 0$, $g_n \ge 0$ be sequences defined for $n \in \mathbf{N}_\alpha$ such that*

$$u_n \le a_n + \sum_{s=\alpha}^{n-1} f_s u_s + \sum_{s=\alpha}^{n-1} f_s \sum_{i=\alpha}^{s-1} g_i u_i, \qquad n \in \mathbf{N}_\alpha. \tag{15.80}$$

Then, for $n \in \mathbf{N}_\alpha$,

$$u_n \le a_n \left[1 + \sum_{s=\alpha}^{n-1} f_s \prod_{i=\alpha}^{s-1} (1 + f_i + g_i) \right]. \tag{15.81}$$

PROOF. Indeed, (15.81) follows from (15.77) by setting $p_n \equiv 1$, replacing a_s by a_n, which is larger, and applying (15.3). □

THEOREM 15.13. (Agarwal and Thandapani, 1982) *Let $u(t)$, $a(t)$, $b(t)$, $f_i(t)$, $i = 1, \ldots, r$, and g_j, $j = 1, \ldots, r-1$, be nonnegative functions, defined for $t \in \mathbf{N}_\alpha$, such that*

$$u(t) \le a(t) + b(t)\left[\sum_{t_1=\alpha}^{t-1} f_1(t_1)u(t_1) + \sum_{t_1=\alpha}^{t-1} g_1(t_1)\sum_{t_2=\alpha}^{t_1-1} f_2(t_2)u(t_2) + \right.$$
$$\left. + \cdots + \sum_{t_1=\alpha}^{t-1} g_1(t_1)\sum_{t_2=\alpha}^{t_1-1} g_2(t_2)\cdots\sum_{t_{r-1}=\alpha}^{t_{r-2}-1} g_{r-1}(t_{r-1})\sum_{t_r=\alpha}^{t_{r-1}-1} f_r(t_r)u(t_r)\right], \tag{15.82}$$

for $t \in \mathbf{N}_\alpha$. Then

$$u(t) \le a(t) + b(t)Q_k(t), \qquad k = 1, \ldots, r, \qquad t \in \mathbf{N}_\alpha, \tag{15.83}$$

where

$$Q_k(t) = \sum_{s=\alpha}^{t-1}\left[a(s)\sum_{i=1}^{k} f_i(s) + g_k(s)Q_{k+1}(s)\right]\prod_{\tau=s+1}^{t-1}(1 + M_k(\tau) - g_k(\tau)),$$
$$k = 1, \ldots, r,$$
$$M_k(t) = \max\left[b(t)\sum_{i=1}^{k} f_i(t), g_1(t), \ldots, g_{k-1}(t)\right], \qquad k = 2, \ldots, r,$$
$$M_1(t) = b(t)f_1(t), \qquad Q_{r+1}(t) \equiv 0, \qquad g_r(t) \equiv 0. \qquad \square$$

Theorem 15.13 is a discrete analogue of theorem 9.10. The proof is similar, so we omit it.

THEOREM 15.14. *Let $u_n \ge 0$, $a_n \ge 1$, $\Delta a_n \ge 0$, be sequences defined for $n \in \mathbf{N}_\alpha$ such that*

$$u_n \le a_n + \sum_{r=1}^{m}\sum_{s=\alpha}^{n-1} h_{rs}W_r(u_s), \qquad n \in \mathbf{N}_\alpha, \tag{15.84}$$

where $h_{rs} \ge 0$, $n \in \mathbf{N}_\alpha$, $r \in \mathbf{N}_{1,m}$, and $W_r \in \mathcal{F}$, $r \in \mathbf{N}_{1,m}$. Then

$$u_n \le a_n \prod_{r=1}^{m} E_r(n), \qquad n \in \mathbf{N}_{\alpha,\beta}, \tag{15.85}$$

where $E_1(n) = F_1^{-1}[F_1(1) + \sum_{s=\alpha}^{n-1} h_{1s}]$,

$$E_r(n) = F_r^{-1}\left[F_r(1) + \sum_{s=\alpha}^{n-1} h_{rs}\prod_{i=1}^{r-1} E_i(s)\right], \qquad r = 2, \ldots, m,$$

F_r^{-1} is the inverse function of $F_r(x) = \int_{x_0}^{x} dz/W_r(z)$, $x > 0$ ($x_0 > 0$), and

$$\beta = \sup\left\{n \in \mathbf{N}_\alpha : \ F_r(1) + \sum_{s=\alpha}^{n-1} h_{rs}\prod_{i=1}^{r-1} E_i(s) \in \operatorname{Dom} F_r^{-1}, \quad r \in \mathbf{N}_{1,m}\right\}.$$

PROOF. We prove this theorem by induction with respect to m. Let $m = 1$. Then

$$u_n \le a_n + \sum_{s=\alpha}^{n-1} h_{1s} W_1(u_s), \qquad n \in \mathbf{N}_\alpha,$$

and since $W_1 \in \mathcal{F}$, $a_n \ge 1$, and a_n is nondecreasing, we obtain

$$\frac{u_n}{a_n} \le 1 + \sum_{s=\alpha}^{n-1} h_{1s} W_1\left(\frac{u_s}{a_s}\right).$$

Consequently, by theorem 15.7,

$$\frac{u_n}{a_n} \le F_1^{-1}\left[F_1(1) + \sum_{s=\alpha}^{n-1} h_{1s}\right] = E_1(n),$$

and thus (15.85) is proved for $m = 1$.

Suppose that the assertion of theorem 15.14 holds for $m = k$. Then, for $m = k + 1$,

$$u_n \le a_n + \sum_{s=\alpha}^{n-1} h_{k+1,s} W_{k+1}(u_s) + \sum_{r=1}^{k} \sum_{s=\alpha}^{n-1} h_{rs} W_r(u_s),$$

and by the induction assumption,

$$u_n \le \left[a_n + \sum_{s=\alpha}^{n-1} h_{k+1,s} W_{k+1}(u_s)\right] \prod_{r=1}^{k} E_r(n).$$

since $\Delta_n = \prod_{r=1}^{k} E_r(n) \ge 1$ and Δ_n is a nondecreasing sequence, the above inequality implies

$$\frac{u_n}{\prod_{r=1}^{k} E_r(n)} \le a_n + \sum_{s=\alpha}^{n-1} h_{k+1,s} \prod_{r=1}^{k} E_r(s) \cdot W_{k+1}\left(\frac{u_s}{\prod_{r=1}^{k} E_r(s)}\right).$$

By theorem 15.7 we conclude that

$$\frac{u_n}{\prod_{r=1}^{k} E_r(n)} \le a_n E_{r+1}(n). \qquad \square$$

THEOREM 15.15. (Agarwal and Thandapani, 1981) *Suppose the inequality*

$$u(t) \le p(t) + q(t) \sum_{r=1}^{n} E_r(t, u) + \sum_{r=1}^{m} \sum_{s=0}^{t-1} h_r(s) W_r(u(s)) \tag{15.86}$$

holds for $t \in \mathbf{N}_0$, where

$$E_r(t, u) = \sum_{t_1=0}^{t-1} f_1(t_1) \sum_{t_2=0}^{t_1-1} f_2(t_2) \cdots \sum_{t_r=0}^{t_{r-1}-1} f_r(t_r) u(t_r)$$

for $r = 1, \ldots, n$, $W_r \in \mathcal{F}$, and u, p, q, h_r, f_r are real-valued, nonnegative functions in $\mathbf{N}_0$, with $p(t) \ge 1$ and p nondecreasing for $t \in \mathbf{N}_0$. Then

$$u(t) \le p(t)\phi(t) \prod_{r=1}^{m} E_r(t), \qquad t \in \mathbf{N}_{0,\beta}, \tag{15.87}$$

where

$$\phi(t) = 1 + q(t)\sum_{s=0}^{t-1}\sum_{r=1}^{n}\Delta E_r(s,1)\prod_{\tau=s+1}^{t-1}\left[1+\sum_{r=1}^{n}\Delta E_r(\tau,q)\right],$$

$$E_1(t) = F_1^{-1}\left[F_1(1)+\sum_{s=0}^{t-1}h_1(s)\phi(s)\right],$$

$$E_k(t) = F_k^{-1}\left[F_k(1)+\sum_{s=0}^{t-1}h_k(s)\phi(s)\prod_{r=1}^{k-1}E_r(s)\right], \qquad k=2,\dots,m,$$

F_k^{-1} *is the inverse function of* $F_k(t) = \int_{x_0}^{x}\frac{dz}{W_k(z)}, \qquad x>0 \quad (x_0>0),$

$$\beta = \sup\left\{t\in \mathbf{N}_0:\ F_k(1)+\sum_{s=0}^{t-1}h_k(s)\phi(s)\prod_{r=1}^{k-1}E_r(s)\in \operatorname{Dom}F_k^{-1}, \quad k\in\mathbf{N}_{1,m}\right\}.$$

PROOF. We can write (15.86) as

$$u(t) \le p^*(t) + q(t)R(t), \tag{15.88}$$

where

$$p^*(t) = p(t) + \sum_{r=1}^{m}\sum_{s=0}^{t-1}h_r(s)W_r(u(s)),$$

$$R(t) = \sum_{r=1}^{n}E_r(t,u), \qquad R(0)=0.$$

Since $R(t)$ is nondecreasing in $\mathbf{N}_0$, we find

$$\Delta R(t) = \sum_{r=1}^{n}\Delta E_r(t,u) \le \sum_{r=1}^{n}\Delta E_r(t,p^*+qR) \le$$
$$\le \sum_{r=1}^{n}\Delta E_r(t,p^*) + \sum_{r=1}^{n}\Delta E_r(t,q)\cdot R(t).$$

It follows from lemma 15.2 that

$$R(t) \le \sum_{s=0}^{t-1}\left[\sum_{r=1}^{n}\Delta E_r(s,p^*)\right]\prod_{\tau=s+1}^{t-1}\left[1+\sum_{r=1}^{n}\Delta E_r(\tau,q)\right]. \tag{15.89}$$

Since $p^*(t)$ is nondecreasing, (15.88) and (15.89) imply

$$u(t) \le p^*(t)\phi(t). \tag{15.90}$$

Setting

$$v(t) = \frac{u(t)}{p(t)\phi(t)}, \tag{15.91}$$

(15.90) and the fact that $W_r \in \mathcal{F}$ imply that

$$v(t) \le 1 + \sum_{r=1}^{m}\sum_{s=0}^{t-1}h_r(s)\phi(s)W_r(v(s)). \tag{15.92}$$

Theorem 15.14 and (15.92) imply the estimate

$$v(t) \leq \prod_{r=1}^{m} E_r(t),$$

which together with (15.91) implies (15.87). □

In the following theorems we consider certain discrete difference inequalities.

THEOREM 15.16. (Pachpatte, 1977) *Suppose the inequality*

$$\Delta u_n \leq f_n \left[u_n + \sum_{s=\alpha}^{n-1} g_s \Delta u_s \right], \qquad n \in \mathbf{N}_\alpha, \tag{15.93}$$

holds, where $u_n \geq 0$, $\Delta u_n \geq 0$, $f_n \geq 0$, $g_n \geq 0$ are sequences defined for $n \in \mathbf{N}_\alpha$. Then

$$u_n \leq u_\alpha \left[1 + \sum_{s=\alpha}^{n-1} f_s \prod_{i=\alpha}^{s-1} (1 + f_i + f_i g_i) \right], \qquad n \in \mathbf{N}_\alpha. \tag{15.94}$$

PROOF. We set $m_n = u_n + \sum_{s=\alpha}^{n-1} g_s \Delta u_s$, $m_\alpha = u_\alpha$. Then we successively obtain

$$\Delta u_n \leq f_n m_n, \tag{15.95}$$

$$\Delta m_n = \Delta u_n + g_n \Delta u_n,$$

$$\Delta m_n \leq (f_n + f_n g_n) m_n. \tag{15.96}$$

(15.96) and corollary 15.2 (following from lemma 15.2) imply the estimate

$$m_n \leq u_\alpha \prod_{s=\alpha}^{n-1} (1 + f_s + f_s g_s),$$

which together with (15.95) yields

$$\Delta u_n \leq u_\alpha f_n \prod_{s=\alpha}^{n-1} (1 + f_s + f_s g_s).$$

Applying corollary 15.1, we obtain (15.94). □

THEOREM 15.17. *Let $u_n \geq 0$, $\Delta u_n \geq 0$, $a_n \geq 0$, $b_n \geq 0$, and $c_n \geq 0$ be sequences defined for $n \in \mathbf{N}_\alpha$ such that*

$$\Delta u_n \leq a_n + c_n f \left(u_n + \sum_{s=\alpha}^{n-1} b_s \Delta u_s \right), \qquad n \in \mathbf{N}_\alpha, \tag{15.97}$$

where $f(u)$ is a positive, nondecreasing continuous function in $(0, \infty)$, and $u_\alpha > 0$. Then

$$u_n \leq u_\alpha + \sum_{s=\alpha}^{n-1} \{ a_s + c_s f[F^{-1}(R_s)] \}, \qquad n \in \mathbf{N}_{\alpha,\beta}, \tag{15.98}$$

where

$$R_n = F\left(u_\alpha + \sum_{s=\alpha}^{n-1} a_s(1+b_s)\right) + \sum_{s=\alpha}^{n-1} c_s(1+b_s),$$
$$F(x) = \int_{x_0}^{x} \frac{dz}{f(z)}, \qquad x > 0 \qquad (x_0 > 0),$$
$$\beta = \sup\{n \in \mathbf{N}_\alpha : \ R_n \in \operatorname{Dom} F^{-1}\}.$$

PROOF. We set

$$w_n = u_n + \sum_{s=\alpha}^{n-1} b_s \Delta u_s, \qquad w_\alpha = u_\alpha.$$

Then w_n is a positive sequence. We successively obtain

$$\Delta u_n \leq a_n + c_n f(w_n), \tag{15.99}$$
$$\Delta w_n = \Delta u_n + b_n \Delta u_n,$$
$$\Delta w_n \leq a_n(1+b_n) + c_n(1+b_n) f(w_n). \tag{15.100}$$

By lemmas 15.3 and 15.6, (15.100) implies that

$$w_n \leq F^{-1}\left[F\left(u_\alpha + \sum_{s=\alpha}^{n-1} a_s(1+b_s)\right) + \sum_{s=\alpha}^{n-1} c_s(1+b_s)\right] \equiv F^{-1}(R_n).$$

Thus, (15.99) implies

$$\Delta u_n \leq a_n + c_n f[F^{-1}(R_n)],$$

so that (15.98) follows from corollary 15.1. □

In the following two theorems we consider inequalities involving higher-order differences.

THEOREM 15.18. (Agarwal and Thandapani, 1980) *Let $u(t) \geq 0$, $p(t) \geq 0$, $q(t) \geq 0$, $\Delta^j u(t) \geq 0$, $h_j(t) \geq 0$, $j = 0, \ldots, k$, be sequences defined for $t \in \mathbf{N}_0$ such that*

$$\Delta^k u(t) \leq p(t) + q(t) \sum_{j=0}^{k} \sum_{s=0}^{t-1} h_j(s) \Delta^j u(s), \qquad n \in \mathbf{N}_\alpha. \tag{15.101}$$

Then

$$\Delta^k u(t) \leq p(t) + q(t) \sum_{s=0}^{t-1} \phi_1(s) \prod_{\tau=s+1}^{t-1} [1 + \phi_2(\tau)], \tag{15.102}$$

where

$$\phi_1(t) = p(t)h_k(t) + \sum_{j=0}^{k-1}\sum_{i=0}^{j-1} \Delta^j u(0) h_i(t)\binom{t}{j-i} + \\ + \sum_{j=0}^{k-1} h_{k-j-1}(t) \sum_{s=0}^{t-j-1}\binom{t-s-1}{j} p(s),$$

$$\phi_2(t) = q(t)h_k(t) + \sum_{j=0}^{k-1} h_{k-j-1}(t) \sum_{s=0}^{t-j-1}\binom{t-s-1}{j} q(s).$$

PROOF. Let $0 \le j \le k-1$. Then it is readily verified that

$$\Delta^j u(t) = \sum_{i=j}^{k-1}\binom{t}{i-j}\Delta^i u(0) + \sum_{s=0}^{t-k+j}\binom{t-s-1}{k-j-1}\Delta^k u(s). \tag{15.103}$$

Set $R(t) = \sum_{j=0}^{k}\sum_{s=0}^{t-1} h_j(s)\Delta^j u(s)$; then (15.101) can be written as

$$\Delta^k u(t) \le p(t) + q(t)R(t), \tag{15.104}$$

with $R(0) = 0$. This and (15.103) implies that

$$\Delta R(t) = \sum_{j=0}^{k} h_j(t)\Delta u(t) = \\ = h_k(t)\Delta^k u(t) + \sum_{j=0}^{k-1} h_j(t)\left[\sum_{i=j}^{k-1}\binom{t}{i-j}\Delta^i u(0) + \sum_{s=0}^{t-k+j}\binom{t-s-1}{k-j-1}\Delta^k u(s)\right] \le \\ \le \sum_{j=0}^{k-1}\sum_{i=0}^{j}\Delta^j u(0)h_i(t)\binom{t}{j-i} + h_k(t)[p(t)+q(t)R(t)] + \\ + \sum_{j=0}^{k-1} h_{k-j-1}(t)\sum_{s=0}^{t-j-1}\binom{t-s-1}{j}[p(s)+q(s)R(s)].$$

Since $R(t)$ is nondecreasing, we find

$$\Delta R(t) \le \phi_1(t) + \phi_2(t)R(t).$$

Lemma 15.2 now implies the estimate

$$R(t) \le \sum_{s=0}^{t-1}\phi_1(s)\prod_{\tau=s+1}^{t-1}[1+\phi_2(\tau)],$$

which implies (15.102) because of (15.104). □

COROLLARY 15.12. (Agarwal and Thandapani, 1980) *If, in theorem* 15.18, $\Delta^j u(0) = 0$, $j = 0, \dots, k-1$, *and* $p(t)$ *is nondecreasing in* $\mathbf{N}_0$, *then* (15.101) *implies that*

$$\Delta^k u(t) \le p(t)\left[1 + q(t)\sum_{s=0}^{t-1}\phi_3(s)\sum_{\tau=s+1}^{t-1}[1+\phi_4(\tau)]\right], \tag{15.105}$$

where $\phi_3(t) = \sum_{j=0}^{k} h_{k-j}(t)\binom{t}{j}$, $\phi_4(t) = \phi_2(t)$. $\square$

THEOREM 15.19. (Agarwal and Thandapani, 1980) *Let* $\Delta^j u(t)$ *and* $h_j(t)$, $j = 0, \ldots, k$, *be nonnegative sequences defined for* $n \in \mathbf{N}_0$ *such that*

$$\Delta^k u(t) \le p(t) + \sum_{j=0}^{k} \sum_{s=0}^{t-1} h_j(s) \Delta^k u(s) \Delta^j u(s), \tag{15.106}$$

where $p(t)$ *is a positive nondecreasing sequence. Then*

$$\Delta^k u(t) \le \frac{p(t)e(t)}{1 - \sum_{s=0}^{t-1} \psi_1(s)e(s+1)}, \qquad t \in \mathbf{N}_{0,\beta}, \tag{15.107}$$

where

$$e(t) = \prod_{s=0}^{t-1} (1 + \psi_2(s)),$$

$$\psi_1(t) = p(t) \sum_{j=0}^{k} h_{k-j}(t) \binom{t}{j},$$

$$\psi_2(t) = \sum_{j=0}^{k-1} \sum_{i=0}^{j} \Delta^j u(0) h_i(t) \binom{t}{j-i},$$

$$\beta = \sup\left\{ t \in \mathbf{N}_0 : 1 - \sum_{s=0}^{t-1} \psi_1(s)e(s+1) > 0 \right\}. \qquad \square$$

Using induction with respect to n we can prove the following comparison result.

THEOREM 15.20. *Let* u_n *and* a_n *be sequences defined for* $n \in \mathbf{N}_\alpha$ *such that*

$$u_n \le a_n + \sum_{s=\alpha}^{n-1} Q(n, s, u_s), \qquad n \in \mathbf{N}_\alpha,$$

where $Q(n, s, x)$ *is a function defined for* $n \in \mathbf{N}_\alpha$, $s \in \mathbf{N}_\alpha$, $x \in \mathbf{R}$, *which is nondecreasing with respect to* $x \in \mathbf{R}$. *Then*

$$u_n \le w_n, \qquad n \in \mathbf{N}_\alpha,$$

where w_n *is the solution of the equation*

$$w_n = a_n + \sum_{s=\alpha}^{n-1} Q(n, s, w_s), \qquad n \in \mathbf{N}_\alpha. \qquad \square \tag{15.108}$$

The estimation of the solution of (15.108) is sometimes carries out by comparing w_n with the solution $w(t)$ of a suitably chosen integral inequality. This is justified by the following theorem.

THEOREM 15.21. (Tsalyuk, 1988) *Suppose the following conditions hold:*

1) $K(t, s, x)$ *is a nonnegative function for* $t \in \mathbf{R}$, $s \in [0, t]$, $x \in \mathbf{R}$, *which is nondecreasing with respect to* x *and satisfies the Carathéodory conditions* H18.1–5;
2) $a(t)$ *is a right-continuous function in* $\mathbf{R}_+$;

3) u_n, a_n *are sequences defined in* $\mathbf{N}_0$ *such that*

$$u_n \le a_n + \sum_{s=\alpha}^{n-1} Q(n,s,u_s), \qquad n \in \mathbf{N}_0,$$

where $Q(n,s,x)$ *is a function defined for* $n \in \mathbf{N}_\alpha$, $s \in \mathbf{N}_{0,n}$, $x \in \mathbf{R}$, *which is nondecreasing with respect to* $x \in \mathbf{R}$;

4) $Q(n,k,x) \le K(t,s,x)$, $a_n \le a(t)$, $t \in [n, n+1)$, $s \in [k, k+1)$, $x \in \mathbf{R}$.

Then

$$u_n \le w(n), \qquad n \in \mathbf{N}_0,$$

where $w(t)$ *is the maximal solution of the equation*

$$w(t) = a(t) + \int_0^t K(t,s,w(s))\,ds, \qquad t \ge 0. \qquad \square$$

An example of the application of theorem 15.21 to the investigation of discrete inequalities involving sums is provided by the following theorem.

THEOREM 15.22. (Dixon and McKee, 1983) *Let* $u_i \ge 0$ *be a sequence defined for* $i \in \mathbf{N}_{0,n}$ *such that*

$$u_i \le \phi_i, \qquad i = 0, \dots, m,$$

$$u_i \le \phi_i + Mh^{1-(\alpha-m)} \sum_{i_m=0}^{i-1} \sum_{i_{m-1}=0}^{i_m-1} \cdots \sum_{j=0}^{i_1-1} \frac{u_j}{(i_1-j)^\alpha}, \qquad i = m, m+1, \dots, n, \tag{15.109}$$

where $\alpha < 1$, $m \ge 1$, $M > 0$ *is independent of* $h > 0$, *and* $\phi_i \ge 0$ *is a nondecreasing sequence for* $i \in \mathbf{N}_{0,n}$. *Then*

$$u_i \le \phi_i E_{1-(\alpha-m)}[M\Gamma(1-\alpha)(ih)^{1-(\alpha-m)}], \qquad i \in \mathbf{N}_{0,n}, \tag{15.110}$$

where $E_{1-\beta}(z)$ *is the Mittag-Leffler function:*

$$E_{1-\beta}(z) = \sum_{n=0}^{\infty} \frac{z^n}{\Gamma(n(1-\beta)+1)},$$

and $\Gamma(a)$ *is the gamma-function, which is for* $\operatorname{Re} a > 0$ *defined by*

$$\Gamma(a) = \int_0^\infty w^{a-1} e^{-w}\,dw. \qquad \square$$

To conclude this section we give some discrete inequalities of Gronwall type on distributive lattices.

Let E be a distributive lattice (for definitions and basic properties see §19), and let $\wedge_{j=1}^n x_j$ be the greatest lower bound and $\vee_{j=1}^n x_j$ the least upper bound of a system $\{x_j\}$, $j \in \mathbf{N}_{1,n}$, of elements of E. For two arbitrary elements $x, y \in E$ these bounds are denoted by $x \wedge y$ and $x \vee y$, respectively (by the definition of a lattice, these bounds exist). We use the symbol '$\le$' for the order relation. For any $x \in E$,

$$x \vee \bigvee_{j=1}^{0} = x \wedge \bigwedge_{j=1}^{0} = x.$$

We recall the following facts:

i) $x \le y$ implies $x \wedge y = x$ and $x \vee y = y$;

ii) $x \le x \vee y$ and $x \wedge y \le x$ for $x, y \in E$;

iii) $x_j \le y_j$, $j = 1, 2$, implies

$$x_1 \wedge x_2 \le x_j \le y_j \le y_1 \vee y_2, \qquad x_1 \wedge x_2 \le y_1 \wedge y_2, \qquad x_1 \vee x_2 \le y_1 \vee y_2;$$

iv) $\vee_{j=1}^{k+1} x_j = (\vee_{j=1}^{k} x_j) \vee x_{k+1}$, $\wedge_{j=1}^{k+1} x_j = (\wedge_{j=1}^{k} x_j) \wedge x_{k+1}$ for any elements $x_j \in E$, $j = 1, \ldots, k+1$.

THEOREM 15.23. (Popenda, 1985) *Let $x, a, b\colon \mathbf{N}_1 \to E$ and*

$$x_{n+1} \le a_n \vee \bigvee_{j=1}^{n} (b_j \wedge x_j), \qquad n \in \mathbf{N}_1. \tag{15.111}$$

Then

$$x_{n+1} \le a_n \vee (x_1 \wedge b_1) \vee \bigvee_{j=1}^{n-1} (b_{j+1} \wedge a_j), \qquad n \in \mathbf{N}_1. \tag{15.112}$$

PROOF. For $n \in \mathbf{N}_1$ we set

$$p_n = \bigvee_{j=1}^{n} (b_j \wedge x_j), \qquad q_n = (x_1 \wedge b_1) \vee \bigvee_{j=1}^{n-1} (b_{j+1} \wedge a_j).$$

Note that if p_n is comparable with q_n and

$$p_n \le q_n, \tag{15.113}$$

then, by iii), $a_n \vee p_n \le a_n \vee q_n$. Hence, by (15.111) and the transitivity of the order relation, x_{n+1} is comparable with $a_n \vee q_n$ and (15.112) holds for this n, i.e.

$$x_{n+1} \le a_n \vee q_n.$$

It thus suffices to prove (15.113) for all $n \in \mathbf{N}_1$. This part of the proof is by induction. For $n = 1$ we have $p_1 = b_1 \wedge x_1 = q_1$. Now we assume that (15.113) holds for some $n = k$, and we consider p_{k+1}. First, by iv) we have

$$p_{k+1} = p_k \vee (b_{k+1} \wedge x_{k+1}), \qquad q_{k+1} = q_k \vee (b_{k+1} \wedge a_k). \tag{15.114}$$

Taking into account the first part of th proof, we have $x_{k+1} \le a_k \vee q_k$, whence (by iii))

$$b_{k+1} \wedge x_{k+1} \le b_{k+1} \wedge (a_k \vee q_k).$$

Applying to this inequality the finite distributive laws, i) and iii) imply

$$b_{k+1} \wedge x_{k+1} \le (b_{k+1} \wedge a_k) \vee (b_{k+1} \wedge q_k) \le (b_{k+1} \wedge a_k) \vee q_k,$$

which together with (15.114) leads to

$$x_{k+1} \wedge b_{k+1} \le q_{k+1}.$$

Together with the induction hypothesis this gives $p_{k+1} = p_k \vee (b_{k+1} \wedge x_{k+1}) \le q_k \vee q_{k+1}$. However, $q_k \le q_{k+1}$, so that $p_{k+1} \le q_{k+1}$. □

THEOREM 15.24. (Popenda, 1985) *Let $x, a, b\colon \mathbf{N}_1 \to E$. Then for $t \in \mathbf{N}_1$,*

1) $x_{n+1} \leq a_n \wedge \bigvee_{j=1}^{n}(b_j \wedge x_j)$ *implies* $x_{n+1} \leq a_n \wedge x_1 \wedge b_1$;
2) $x_{n+1} \leq a_n \wedge \bigwedge_{j=1}^{n}(b_j \wedge x_j)$ *implies* $x_{n+1} \leq x_1 \wedge \bigwedge_{j=1}^{n}(a_j \wedge b_j)$;
3) $x_{n+1} \leq a_n \vee \bigwedge_{j=1}^{n}(b_j \wedge x_j)$ *implies* $x_{n+1} \leq a_n \vee (x_1 \wedge \bigwedge_{j=1}^{n} b_j)$;
4) $x_{n+1} \leq a_n \wedge \bigvee_{j=1}^{n}(b_j \vee x_j)$ *implies* $x_{n+1} \leq a_n \wedge (x_1 \vee \bigvee_{j=1}^{n} b_j)$;
5) $x_{n+1} \leq a_n \wedge \bigwedge_{j=1}^{n}(b_j \vee x_j)$ *implies* $x_{n+1} \leq a_n \wedge (x_1 \vee b_1) \wedge \bigwedge_{j=1}^{n}(a_j \vee b_{j+1})$;
6) $x_{n+1} \leq a_n \vee \bigvee_{j=1}^{n}(b_j \vee x_j)$ *implies* $x_{n+1} \leq x_1 \vee \bigvee_{j=1}^{n}(b_j \vee a_j)$;
7) $x_{n+1} \leq a_n \vee \bigwedge_{j=1}^{n}(b_j \vee x_j)$ *implies* $x_{n+1} \leq a_n \vee x_1 \vee b_1$. □

16. Impulse integral inequalities

Such inequalities arise in the study of impulse differential equations. Before starting their investigation, we give a brief description of the following, sufficiently wide, class of impulse differential equations:

$$x' = f(t,x) \text{if } \phi(t,x) \neq 0,$$
$$\Delta x = I(t,x) \text{if } \phi(t,x) = 0, \tag{16.1}$$

where $f: \mathbf{R}_+ \times \Omega \to \mathbf{R}^n$, $\phi: \mathbf{R}_+ \times \Omega \to \mathbf{R}$, $I: \mathbf{R}_+ \times \Omega \to \mathbf{R}^n$, Ω is a domain in $\mathbf{R}^n$, $t \in \mathbf{R}_+$, $x \in \Omega$.

Equations of the type (16.1) describe processes and phenomena that at certain moments in their development (called *instants of impulse effect*) are subject to short-time effects and change their state jumpwise. The instants of impulse effect occur when the mapping point $(t, x(t))$ meets the hypersurface σ defined by $\phi(t,x) = 0$, i.e. when $\phi(t, x(t)) = 0$. At the instant of impulse effect $t = t_i$ the mapping point jumps instantaneously from position $(t_i, x(t_i))$ to position $(t_i, x(t_i^+))$, where $x(t_i^+) = x(t_i) + \Delta x(t_i) = x(t_i) + I(t_i, x(t_i))$. We assume that in an interval $(t_i, t_{i+1}]$ between two successive instants of impulse effect the solution $x(t)$ of (16.1) coincides with the solution of the initial value problem

$$y' = f(t,y), \qquad y(t_i) = x(t_i^+) \qquad (t_i \leq t \leq t_{i+1}),$$

and, moreover, $x(t)$ is left-continuous at each instant of impulse effect, i.e. $x(t_i^-) = x(t_i)$.

Impulse differential equations for which the impulses occur at fixed times, given by an increasing sequence $\{t_i\}$, $t_i < t_{i+1}$, are the most simple ones to study. In this case the impulse differential equation (16.1) can be written as

$$x' = f(t,x), \qquad t \neq t_k,$$
$$\Delta x = I_k(x), \qquad t = t_k, \tag{16.2}$$

or as

$$x' = f(t,x), \qquad t \neq t_k,$$
$$x(t_k^+) = \psi_k(x(t_k)),$$

where $\psi_k(x) = x + I_k(x)$, $k = 1, 2, \ldots$.

The following equations are more general than (16.2):

$$\begin{aligned} x' &= f(t,x), \qquad t \neq \tau_k(x), \\ \Delta x &= I_k(x), \qquad t = \tau_k(x), \end{aligned} \tag{16.3}$$

where $\tau_k \colon \Omega \to \mathbf{R}_+$ and $0 < \tau_1(x) < \tau_2(x) < \dots$ for $x \in \Omega$.

The instants t_i of impulse effect for the solution $x(t)$ of (16.3) occur when the mapping point $(t, x(t))$ meets some hypersurface σ_k with equation $t = \tau_k(x)$, i.e. when $t_i = \tau_k(x(t_i))$ for some k.

In the qualitative and quantitative investigation of solution s of impulse differential equations, the technique of integral inequalities is often used. The functions estimated in these inequalities, as well as the solutions of (16.1), have first-kind discontinuities at the instants t_k of impulse effect; moreover, the magnitude of the discontinuity at $t = t_k$ depends on the value at t_k of the function to be estimated. This characteristic feature distinguishes a certain class of integral inequalities, which we call the *impulse integral inequalities*.

A simple, but widely used impulse integral inequality is:

$$u(t) \le a + \int_\alpha^t b(s)u(s)\,ds + \sum_{\alpha < t_k < t} \beta_k u(t_k).$$

It was used by A.M. Samoilenko and N.A. Perestyuk (in 1977) in the investigation of stability of solutions of (16.2).

Before considering impulse integral inequalities, we introduce some notation.

Below we assume that $\{t_k\}_0^\infty$ is a fixed sequence such that $0 \le t_0 < t_1 < t_2 < \dots$ and $\lim_{k\to\infty} t_k = \infty$.

By $PC(\mathbf{R}_+, \mathbf{R}^n)$ we denote the class of piecewise continuous functions $u \colon \mathbf{R}_+ \to \mathbf{R}^n$ that have first kind discontinuities at the points $t = t_k$, $k = 1, 2, \dots$, only and that are left-continuous at $t = t_k$.

By $PC^1(\mathbf{R}_+, \mathbf{R}^n)$ we denote the class of piecewise differentiable functions $u \colon \mathbf{R}_+ \to \mathbf{R}^n$ such that $u' \in PC(\mathbf{R}_+, \mathbf{R}^n)$.

THEOREM 16.1. (Samoilenko and Perestyuk, 1977) *Suppose that for $t \ge t_0$ the following inequality holds:*

$$u(t) \le a + \int_{t_0}^t b(s)u(s)\,ds + \sum_{t_0 < t_k < t} \beta_k u(t_k),$$

where $u \in PC(\mathbf{R}_+, \mathbf{R})$, $b \in PC(\mathbf{R}_+, \mathbf{R}_+)$, and $\beta_k \ge 0$, $k \in \mathbf{N}_1$, and a are constants. Then for $t \ge t_0$,

$$u(t) \le a \prod_{t_0 < t_k < t} (1 + \beta_k) \exp\left(\int_{t_0}^t b(s)\,ds\right). \tag{16.4}$$

PROOF. The proof of this theorem can be given by induction with respect to $k \in \mathbf{N}_1$, applying for $t \in (t_k, t_{k+1}]$ theorem 1.1 to the Gronwall–Bellman inequality

$$u(t) \le a_k + \int_{t_k}^{t} b(s)u(s)\,ds, \qquad t_k \le t \le t_{k+1},$$

where

$$a_k = u(t_k^+) \le a + \int_{t_0}^{t_k} b(s)u(s)\,ds + \sum_{i=1}^{k} \beta_i u(t_i) \le a \prod_{i=1}^{k}(1+\beta_i)\exp\left(\int_{t_0}^{t_k} b(s)\,ds\right). \qquad \square$$

Impulse integral inequalities are often 'solved' by reduction to impulse differential inequalities and an application of the following comparison lemma.

LEMMA 16.1. *Suppose that*

1) $v \in PC^1(\mathbf{R}_+, \mathbf{R})$ *satisfies the inequality*

$$v'(t) \le g(t, v(t)), \qquad t \ne t_k, \qquad 0 \le t < T \le \infty,$$
$$v(t_k^+) \le \psi_k(v(t_k)), \qquad v(t_0) \le u_0, \tag{16.5}$$

where $\psi_k\colon \mathbf{R} \to \mathbf{R}$, $k \in \mathbf{N}_1$, *are nondecreasing functions,* $g\colon \mathbf{R}_+ \times \mathbf{R} \to \mathbf{R}$ *is a continuous function in the sets* $(t_{k-1}, t_k] \times \mathbf{R}$, $k \in \mathbf{N}_1$, *and for any* $x \in \mathbf{R}$ *and* $k \in \mathbf{N}_1$ *the limit of* $g(t,y)$ *as* $(t,y) \to (t_k, x)$, $t > t_k$, *exists.*

2) *The maximal solution* $u^+(t)$ *of the initial value problem*

$$u' = g(t,u), \qquad t \ne t_k, \qquad t \ge t_0,$$
$$u(t_k^+) = \psi_k(u(t_k)), \qquad u(t_0) = u_0, \tag{16.6}$$

is defined for $t \in [t_0, T)$.

Then

$$v(t) \le u^+(t), \qquad t \in [t_0, T). \tag{16.7}$$

PROOF. Let $t \in [t_0, t_1] \cap [t_0, T)$. Then, by theorem 6.3, $v(t) \le u^+(t)$. In particular, $v(t_1) \le u^+(t_1)$, and the monotonicity of $\psi_1(u)$ implies that

$$v(t_1^+) \le \psi_1(v(t_1)) \le \psi_1(u^+(t_1)) = u^+(t_1).$$

By repeating the arguments above we can prove that (16.7) holds on every subsequent interval $(t_k, t_{k+1}] \cap [t_0, T)$, $k = 1, 2, \dots$. $\square$

REMARK 16.1. Lemma 6.1 remains valid if $\le$ is replaced by $\ge$ in (16.5) and (16.7), while the maximal solution $u^+(t)$ of (16.6) is replaced by its minimal solution $u^-(t)$. $\square$

REMARK 16.2. Lemma 16.1 still holds if the derivative $v'(t)$ in (16.5) is replaced by some Dini derivative D^+v, D_+v, D^-v, D_-v of the function $v \in PC(\mathbf{R}_+, \mathbf{R})$. $\square$

COROLLARY 16.1. *Suppose a function $v \in PC^1(\mathbf{R}_+, \mathbf{R})$ satisfies the inequality*

$$\begin{aligned} v'(t) &\le b(t)v(t) + f(t), \qquad t \neq t_k, \qquad t \ge t_0, \\ v(t_k^+) &\le b_k v(t_k) + f_k, \qquad k \in \mathbf{N}_1, \end{aligned}$$

where $b \in PC(\mathbf{R}_+, \mathbf{R})$, $f \in PC(\mathbf{R}_+, \mathbf{R})$, and $b_k \ge 0$ and f_k, $k \in \mathbf{N}_1$, are constants. Then

$$\begin{aligned} v(t) \le & v(t_0) \prod_{t_0<t_k<t} b_k \exp\left(\int_{t_0}^t b(s)\,ds\right) + \int_{t_0}^t f(s) \prod_{s<t_k<t} b_k \exp\left(\int_s^t b(\tau)\,d\tau\right) ds + \\ & + \sum_{t_0<t_k<t} f_k \prod_{t_k<t_i<t} b_i \exp\left(\int_{t_k}^t b(\tau)\,d\tau\right). \end{aligned} \tag{16.8}$$

PROOF. For the proof of this corollary it suffices to put, in lemma 16.1, $g(t,v) = b(t)v + f(t)$, $\psi_k(v) = b_k v + f_k$, and to verify that the unique solution of the initial value problem

$$\begin{aligned} u' &= b(t)u + f(t), \qquad t \neq t_k, \qquad t \ge t_0, \\ u(t_k^+) &= b_k u(t_k) + f_k, \qquad u(t_0) = v(t_0), \end{aligned}$$

coincides with the righthand side of (16.8). □

THEOREM 16.2. *Suppose that for $t \ge t_0$ the following inequality holds:*

$$u(t) \le a(t) + \int_{t_0}^t b(s)u(s)\,ds + \sum_{t_0<t_k<t} \beta_k u(t_k), \tag{16.9}$$

where $u \in PC(\mathbf{R}_+, \mathbf{R})$, $a \in PC(\mathbf{R}_+, \mathbf{R})$, $b \in PC(\mathbf{R}_+, \mathbf{R}_+)$, and $\beta_k \ge 0$, $k \in \mathbf{N}_1$, are constants. Then, for $t \ge t_0$,

$$\begin{aligned} u(t) \le & a(t) + \int_{t_0}^t a(s)b(s) \prod_{s<t_k<t} (1+\beta_k) \exp\left(\int_s^t b(\tau)\,d\tau\right) ds + \\ & + \sum_{t_0<t_k<t} a(t_k)\beta_k \prod_{t_k<t_j<t} (1+\beta_j) \exp\left(\int_{t_k}^t b(\tau)\,d\tau\right). \end{aligned} \tag{16.10}$$

PROOF. Setting $v(t) = \int_{t_0}^t b(s)u(s)\,ds + \sum_{t_0<t_k<t} \beta_k u(t_k)$ we obtain

$$\begin{aligned} v'(t) &= b(t)u, \qquad t \neq t_k, \\ v(t_k^+) &= v(t_k) + \beta_k u(t_k), \qquad v(t_0) = 0. \end{aligned}$$

Since $u(t) \le a(t) + v(t)$, we have

$$\begin{aligned} v'(t) &\le b(t)v(t) + a(t)b(t), \qquad t \neq t_k, \\ v(t_k^+) &\le (1+\beta_k)v(t_k) + a(t_k)\beta_k, \qquad v(t_0) = 0. \end{aligned}$$

By corollary 16.1,

$$v(t) \leq \int_{t_0}^{t} a(s)b(s) \prod_{s<t_k<t} (1+\beta_k) \exp\left(\int_s^t b(\tau)\,d\tau\right) ds +$$
$$+ \sum_{t_0<t_k<t} a(t_k)\beta_k \prod_{t_k<t_j<t} (1+\beta_j) \exp\left(\int_{t_k}^t b(\tau)\,d\tau\right).$$

Together with $u(t) \leq a(t) + v(t)$, this implies (16.10). □

COROLLARY 16.2. *If, in theorem 16.2, the function $a(t)$ is nondecreasing in $\mathbf{R}_+$, then (16.9) implies*

$$u(t) \leq a(t) \prod_{t_0<t_k<t} (1+\beta_k) \exp\left(\int_{t_0}^t b(s)\,ds\right), \qquad t \geq t_0. \tag{16.11}$$

PROOF. Inequality(16.11) follow from (16.10), the monotonicity of $a(t)$, and the equality

$$1 + \int_{t_0}^{t} b(s) \prod_{s<t_k<t} (1+\beta_k) \exp\left(\int_s^t b(\tau)\,d\tau\right) ds +$$
$$+ \sum_{t_0<t_k<t} \beta_k \prod_{t_k<t_j<t} (1+\beta_j) \exp\left(\int_{t_k}^t b(\tau)\,d\tau\right) =$$
$$= \prod_{t_0<t_k<t} (1+\beta_k) \exp\left(\int_{t_0}^t b(\tau)\,d\tau\right). \qquad \square$$

THEOREM 16.3. *Suppose that for $t \geq t_0$ the following inequality holds:*

$$u(t) \leq a(t) + q(t) \left[\int_{t_0}^t b(s)u(s)\,ds + \sum_{t_0<t_k<t} \beta_k u(t_k)\right], \tag{16.12}$$

where $a, u \in PC(\mathbf{R}_+, \mathbf{R})$, $b, q \in PC(\mathbf{R}_+, \mathbf{R}_+)$, and $\beta_k \geq 0$, $k \in \mathbf{N}_1$, are constants. Then for $t \geq t_0$,

$$u(t) \leq a(t) + q(t) \left[\int_{t_0}^t a(s)b(s) \prod_{s<t_k<t} (1+\beta_k q(t_k)) \exp\left(\int_s^t b(\tau)q(\tau)\,d\tau\right) ds +\right.$$
$$\left. + \sum_{t_0<t_k<t} a(t_k)\beta_k \prod_{t_k<t_j<t} (1+\beta_j q(t_j)) \exp\left(\int_{t_k}^t b(\tau)q(\tau)\,d\tau\right)\right]. \tag{16.13}$$

PROOF. Setting $v(t) = \int_{t_0}^t b(s)u(s)\,ds + \sum_{t_0<t_k<t} \beta_k u(t_k)$ and taking into account (16.12), we obtain the impulse differential inequality

$$v'(t) \leq b(t)q(t)v(t) + a(t)b(t), \qquad t \neq t_k,$$
$$v(t_k^+) \leq (1+\beta_k q(t_k))v(t_k) + a(t_k)\beta_k, \qquad v(t_0) = 0.$$

Since $u(t) \leq a(t) + q(t)v(t)$, corollary 16.1 implies(16.13). □

THEOREM 16.4. *Suppose that for $t \geq t_0$ the following inequality holds:*

$$u(t) \leq a(t) + \int_{t_0}^{t} K(t,s)u(s)\,ds + \sum_{t_0<t_k<t} \beta_k(t)u(t_k), \tag{16.14}$$

where $\beta_k(t) \geq 0$, $k \in \mathbf{N}_1$, are nondecreasing functions in $\mathbf{R}_+$, $a \in PC(\mathbf{R}_+, \mathbf{R}_+)$ is a nondecreasing function in $\mathbf{R}_+$, $u \in PC(\mathbf{R}_+, \mathbf{R}_+)$, and $K \in C(\mathbf{R}_+^2, \mathbf{R}_+)$ is a nondecreasing function with respect to $t \in \mathbf{R}_+$ for s fixed. Then for $t \geq t_0$,

$$u(t) \leq a(t) \prod_{t_0<t_k<t} (1 + \beta_k(t)) \exp\left(\int_{t_0}^{t} K(t,s)\,ds\right). \tag{16.15}$$

PROOF. (16.14) and the monotonicity of β_k and K imply that

$$u(s) \leq a(s) + \int_{t_0}^{s} K(t,\tau)u(\tau)\,d\tau + \sum_{t_0<t_k<s} \beta_k(t)u(t_k)$$

for $t_0 \leq s \leq t$. By corollary 16.2,

$$u(s) \leq a(s) \prod_{t_0<t_k<s} (1 + \beta_k(t)) \exp\left(\int_{t_0}^{s} K(t,\tau)\,d\tau\right), \qquad t_0 \leq s \leq t,$$

and (16.15) follows for $s = t$. □

THEOREM 16.5. *Suppose that for $t \geq t_0$ the following inequality holds:*

$$\begin{aligned} u(t) \leq & a \prod_{t_0<t_k<t} p_k + \int_{t_0}^{t} \prod_{s<t_k<t} p_k[b(s)u(s) + c(s)]\,ds + \\ & + \sum_{t_0<t_k<t} \prod_{t_k<t_j<t} p_j[\beta_k u(t_k) + \gamma_k], \end{aligned} \tag{16.16}$$

where $u, c \in PC(\mathbf{R}_+, \mathbf{R})$, $b \in PC(\mathbf{R}_+, \mathbf{R}_+)$, and $\beta_k \geq 0$, $p_k \geq 0$, γ_k, a are constants. Then for $t \geq t_0$,

$$\begin{aligned} u(t) \leq & a \prod_{t_0<t_k<t} (p_k + \beta_k) \exp\left(\int_{t_0}^{t} b(s)\,ds\right) + \\ & + \int_{t_0}^{t} c(s) \prod_{s<t_k<t} (p_k + \beta_k) \exp\left(\int_{s}^{t} b(\tau)\,d\tau\right) ds + \\ & + \sum_{t_0<t_k<t} \gamma_k \prod_{t_k<t_j<t} (p_j + \beta_j) \exp\left(\int_{t_k}^{t} b(\tau)\,d\tau\right). \end{aligned} \tag{16.17}$$

PROOF. Let $v(t)$ denote the righthand side of inequality (16.16). We obtain for it the impulse differential inequality

$$\begin{aligned} v'(t) &\leq b(t)v(t) + c(t), \qquad t \neq t_k, \qquad t \geq t_0, \\ v(t_k^+) &\leq (p_k + \beta_k)v(t_k) + \gamma_k, \qquad v(t_0) = a, \end{aligned}$$

and corollary 16.1 implies (16.17). □

REMARK 16.3. We can similarly prove that under the conditions of theorem 16.5 the inequality

$$
\begin{aligned}
u(t) \le & a \prod_{t_0<t_k<t} p_k + \int_{t_0}^{t} \prod_{s<t_k<t} p_k[b(s)u(s) + c(s)]\,ds + \\
& + \sum_{t_0<t_k<t} \prod_{t_k \le t_j < t} p_j[\beta_k u(t_k) + \gamma_k],
\end{aligned}
$$

implies

$$
\begin{aligned}
u(t) \le & a \prod_{t_0<t_k<t} p_k(1+\beta_k) \exp\left(\int_{t_0}^{t} b(s)\,ds\right) + \\
& + \int_{t_0}^{t} c(s) \prod_{s<t_k<t} p_k(1+\beta_k) \exp\left(\int_{s}^{t} b(\tau)\,d\tau\right) ds + \\
& + \sum_{t_0<t_k<t} \gamma_k \prod_{t_k<t_j<t} p_j(1+\beta_j) \exp\left(\int_{t_k}^{t} b(\tau)\,d\tau\right). \quad \square \qquad (16.18)
\end{aligned}
$$

THEOREM 16.6. *Suppose that:*

1) $u, k \in PC(\mathbf{R}_+, \mathbf{R}_+)$;
2) $a > 0$, $p_j \ge 0$, $\beta_j \ge 0$, $j \in \mathbf{N}_1$;
3) $f, g \in C(\mathbf{R}_+, \mathbf{R}_+)$ *are nondecreasing functions in* $\mathbf{R}_+$ *which are positive in* $(0, \infty)$;
4) *for any* $n \in \mathbf{N}_1$ *there is an* $M_n \ge 0$ *such that*

$$G(r_n(x)) - G(r_n(y)) \le M_n[G(x) - G(y)]$$

for $x \ge y > 0$, *where* $r_n(x) = p_n x + \beta_n f(x)$, $G(u) = \int_{u_0}^{u} dx/g(x)$, $u > 0$ $(u_0 > 0)$;
5) *for* $t \ge t_0$ *the following inequality holds:*

$$
\begin{aligned}
u(t) \le & a \prod_{t_0<t_i<t} p_i + \int_{t_0}^{t} \left(\prod_{s<t_j<t} p_j\right) k(s) g(u(s))\,ds + \\
& + \sum_{t_0<t_i<t} \left(\prod_{t_i<t_j<t} p_j\right) \beta_i f(u(t_i)). \qquad (16.19)
\end{aligned}
$$

Then

$$u(t) \le G^{-1}\left[G(R(t)) + \int_{t_0}^{t} \left(\prod_{s<t_j<t} M_j\right) k(s)\,ds\right], \qquad t \in [t_0, T), \qquad (16.20)$$

where $R(t) \equiv R_0 = a$ *for* $t \in [t_0, t_1]$, $R(t) \equiv R_k = r_k(R_{k-1})$ *for* $t \in (t_k, t_{k+1}]$, $k \in \mathbf{N}_1$, *and*

$$T = \sup\left\{t \ge t_0 : G(R(t)) + \int_{t_0}^{t} \left(\prod_{s<t_j<t} M_j\right) k(s)\,ds \in \operatorname{Dom} G^{-1}\right\}.$$

PROOF. Let $v(t)$ denote the righthand side of (16.19). Since g and f are nondecreasing, while k and β_k are nonnegative, we find

$$v'(t) \leq k(t)g(v(t)), \qquad t \neq t_k, \qquad t \geq t_0, \tag{16.21}$$

$$v(t_n^+) \leq p_n v(t_n) + \beta_n f(v(t_n)) = r_n(v(t_n)), \qquad v(t_0) = a. \tag{16.22}$$

To prove (16.20) it suffices to show that

$$v(t) \leq G^{-1}\left[G(R(t)) + \int_{t_0}^{t} \left(\prod_{s<t_j<t} M_j\right) k(s)\,ds\right] \tag{16.23}$$

for $t \in [t_0, T)$.

For $t \in [t_0, t_1] \cap [t_0, T)$, (16.21) and lemma 5.1 imply that

$$v(t) \leq G^{-1}\left[G(a) + \int_{t_0}^{t} k(s)\,ds\right],$$

i.e. (16.23).

Suppose that (16.23) holds for $t \in [t_0, t_n] \cap [t_0, T)$. Then, in particular, if $t_n < T$,

$$G(v(t_n)) \leq G(R_{n-1}) + \int_{t_0}^{t_n} \left(\prod_{s<t_j<t_n} M_j\right) k(s)\,ds. \tag{16.24}$$

Let $t \in (t_n, t_{n+1}] \cap [t_0, T)$. Then lemma 5.1 and (16.21) imply that

$$G(v(t)) \leq G(v(t_n^+)) + \int_{t_n}^{t} k(s)\,ds,$$

so that, by (16.22),

$$G(v(t)) \leq G(r_n(v(t_n))) + \int_{t_n}^{t} k(s)\,ds. \tag{16.25}$$

If $v(t_n) \leq R_{n-1}$, then (16.25) implies (16.20). Otherwise, taking into account theorem 16.6,4) and (16.24), we obtain

$$\begin{aligned} G(r_n(v(t_n))) - G(R_n) &= G(r_n(v(t_n))) - G(r_n(R_{n-1})) \leq \\ &\leq M_n[G(v(t_n)) - G(R_{n-1})] \leq \\ &\leq M_n \int_{t_0}^{t_n} \left(\prod_{s<t_j<t_n} M_j\right) k(s)\,ds. \end{aligned}$$

Consequently, by (16.25),

$$\begin{aligned} G(v(t)) &\leq G(r_n(v(t_n))) - G(R_n) + G(R_n) + \int_{t_n}^{t} k(s)\,ds \leq \\ &\leq G(R_n) + M_n \int_{t_0}^{t_n} \left(\prod_{s<t_j<t_n} M_j\right) k(s)\,ds + \int_{t_n}^{t} k(s)\,ds = \\ &= G(R(t)) + \int_{t_0}^{t} \left(\prod_{s<t_j<t} M_j\right) k(s)\,ds. \qquad \square \end{aligned}$$

COROLLARY 16.3. *Suppose that for $t \geq t_0$ the following inequality holds:*

$$u(t) \leq a \prod_{t_0<t_i<t} p_i + \int_{t_0}^{t} \left(\prod_{s<t_j<t} p_j \right) k(s) u^m(s)\, ds + \\ + \sum_{t_0<t_i<t} \left(\prod_{t_i<t_j<t} p_j \right) \beta_i u(t_i), \tag{16.26}$$

where $u, k \in PC(\mathbf{R}_+, \mathbf{R}_+)$, and $a > 0$, $\beta_k \geq 0$, $p_k > 0$, $m > 0$, $m \neq 1$ are constants. Then

$$u(t) \leq a \prod_{t_0<t_i<t} (p_i + \beta_i) \left[1 - (m-1) \int_{t_0}^{t} \left(a \prod_{t_0<t_j \leq s} (p_j + \beta_j) \right)^{m-1} k(s)\, ds \right]^{1/(1-m)} \tag{16.27}$$

for those $t \geq t_0$ for which

$$(m-1) \int_{t_0}^{t} \left(a \prod_{t_0<t_j \leq s} (p_j + \beta_j) \right)^{m-1} k(s)\, ds < 1.$$

PROOF. Indeed, in this case we set $g(s) = s^m$, $f(s) = s$, $r_n(s) = (p_n + \beta_n)s$, and theorem 16.6, 4) holds with $M_n = (p_n+\beta_n)^{1-m}$, while in (16.20), $R(t) = a \prod_{t_0<t_j<t}(p_j + \beta_j)$ and $G(u) = u^{1-m}/(1-m) + \sigma$. □

The impulse integral inequalities considered above are usually obtained when estimating the difference $u(t) = |x_1(t) - x_2(t)|$ of two solutions $x_1(t)$, $x_2(t)$ of an equation with impulses at fixed times (16.2). If $x_1(t)$, $x_2(t)$ are solutions of equation (16.3), then, as a rule, their impulses occur at distinct times t_k^1 and t_k^2. Recall that t_k^1 and t_k^2 are the times at which the integral curves $(t, x_1(t))$ and $(t, x_2(t))$ meet the hypersurfaces σ_k with equations $t = \tau_k(x)$. If we assume that the functions $\tau_k(x)$ satisfy the conditions

$$\sup_{x\in\Omega} \tau_i(x) < \inf_{x\in\Omega} \tau_{i+1}(x), \qquad \lim_{k\to\infty} \tau_k(x) = \infty \qquad (x \in \Omega, \quad i \in \mathbf{N}_1),$$

then the times t_k^1, t_k^2 satisfy the conditions

$$t_k^i < t_{k+1}^j, \qquad \lim_{k\to\infty} t_k^i = \infty \qquad (i = 1, 2, \quad j = 1, 2, \quad k = 1, 2, \dots).$$

We introduce the notation $\underline{t}_k = \min(t_k^1, t_k^2)$, $\overline{t}_k = \max(t_k^1, t_k^2)$. Then

$$0 < \underline{t}_1 \leq \overline{t}_1 < \underline{t}_2 \leq \overline{t}_2 < \cdots < \underline{t}_k \leq \overline{t}_k < \dots, \qquad \lim_{k\to\infty} \underline{t}_k = \infty. \tag{16.28}$$

The following situation is characteristic particularly for problems related to stability of solutions of the equation (16.3): The lengths of the intervals $[\underline{t}_k, \overline{t}_k]$, $k \in \mathbf{N}_1$, are 'small' in comparison to the lengths of the intervals $[\overline{t}_i, \underline{t}_{i+1}]$, $i \in \mathbf{N}_1$, while the quantity $u(t) = |x_1(t) - x_2(t)|$ is 'small' in the intervals $[\overline{t}_i, \underline{t}_{i+1}]$ and 'large' in the intervals $[\underline{t}_k, \overline{t}_k]$. Therefore, when solving such problems we may confine ourselves to the evaluation of $u(t)$ on the 'large' intervals $[\overline{t}_i, \underline{t}_{i+1}]$. The following two theorems are concerned with impulse integral inequalities in this way.

Let $\{\underline{t}_k\}$ and $\{\bar{t}_k\}$ be sequences satisfying conditions (16.28), let $\{t_k\} = \{\underline{t}_k\} \cup \{\bar{t}_k\}$, and $0 \le t_0 < \underline{t}_1$.

Let $E[s,t) = [s,t) \setminus \cup_{k=1}^{\infty}(\underline{t}_k, \bar{t}_k]$.

THEOREM 16.7. *Suppose that for $t \in E[t_0,\infty)$ the following inequality holds:*

$$
\begin{aligned}
u(t) \le & a \prod_{t_0<\bar{t}_k<t} p_k + \int_{E[t_0,t)} \prod_{s<\bar{t}_k<t} p_k[b(s)u(s) + c(s)]\, ds + \\
& + \sum_{t_0<\bar{t}_k<t} \prod_{\bar{t}_k<\bar{t}_j<t} p_j[\beta_k u(\underline{t}_k) + \gamma_k],
\end{aligned}
\tag{16.29}
$$

where $u, c \in PC(\mathbf{R}_+, \mathbf{R})$, $b \in PC(\mathbf{R}_+, \mathbf{R}_+)$, and $\beta_k \ge 0$, $p_k \ge 0$, γ_k, and a are constants.

Then for $t \in E[t_0,\infty)$,

$$
\begin{aligned}
u(t) \le & a \prod_{t_0<\bar{t}_k<t} (p_k + \beta_k) \exp\left(\int_{E[t_0,t)} b(s)\, ds\right) + \\
& + \int_{E[t_0,t)} c(s) \prod_{s<\bar{t}_k<t} (p_k + \beta_k) \exp\left(\int_{E[s,t)} b(\tau)\, d\tau\right) ds + \\
& + \sum_{t_0<\bar{t}_k<t} \gamma_k \prod_{\bar{t}_k<\bar{t}_j<t} (p_j + \beta_j) \exp\left(\int_{E[\bar{t}_k,t)} b(\tau)\, d\tau\right). \quad \square
\end{aligned}
\tag{16.30}
$$

COROLLARY 16.4. *If, in theorem 16.7, $b(t) \equiv b$, $c(t) \equiv 0$, $\gamma_k \equiv 0$, then (16.29) implies*

$$
u(t) \le a \prod_{k=1}^{n} (p_k + \beta_k) \exp\left\{ b \left[t - t_0 - \sum_{i=1}^{n} (\bar{t}_i - \underline{t}_i) \right] \right\}
\tag{16.31}
$$

for $t \ge t_0$, $t \in (\bar{t}_n, \underline{t}_{n+1}]$, $n \in \mathbf{N}_1$. $\square$

THEOREM 16.8. *Suppose that for $t \in E[t_0,\infty)$ the following inequality holds:*

$$
\begin{aligned}
u(t) \le & a \prod_{t_0<\bar{t}_k<t} p_k + \int_{E[t_0,t)} \left(\prod_{s<\bar{t}_k<t} p_k \right) k(s) g(u(s))\, ds + \\
& + \sum_{t_0<\bar{t}_k<t} \left(\prod_{\bar{t}_k<\bar{t}_j<t} p_j \right) \beta_k f(u(\underline{t}_k)),
\end{aligned}
\tag{16.32}
$$

where the functions u, k, g, f and constants a, p_k, β_k satisfy the conditions of theorem 16.6.

Then for $t \in E[t_0,\infty)$,

$$
u(t) \le G^{-1} \left[G(R(t)) + \int_{E[t_0,t)} \left(\prod_{s<\bar{t}_j<t} M_j \right) k(s)\, ds \right],
\tag{16.33}
$$

where $R(t) \equiv R_0 = a$ *for* $t \in [t_0, \underline{t}_1]$, $R(t) \equiv r_k(R_{k-1})$ *for* $t \in (\overline{t}_k, \underline{t}_{k+1}]$, $k \in \mathbf{N}_1$, *and*

$$T = \sup\left\{t \in E[t_0,\infty):\ G(R(t)) + \int_{E[t_0,t)} \left(\prod_{s<\overline{t}_j<t} M_j\right) k(s)\,ds \in \operatorname{Dom} G^{-1}\right\}. \qquad \square$$

17. Inequalities involving Stieltjes integrals

The classical inequalities of Gronwall–Bellman type can be generalized in several aspects. Here we consider the possibility of replacing the Riemann integral occurring in them by integrals of another type, e.g., Stieltjes integrals. The first results of this kind are the following two theorems of G.S. Jones. The integrals are ordinary Riemann–Stieltjes integrals, and the limit is taken in a refinement sense.

THEOREM 17.1. (Jones, 1964) *Let* u, a, b, q *be real-valued functions, defined in* $J = [\alpha,\beta]$, *which are either continuous or of bounded variation in* J, *with* b, q *nonnegative. Let* η *be a nondecreasing continuous function in* J. *If*

$$u(t) \le a(t) + q(t)\int_\alpha^t b(s)u(s)\,d\eta(s), \qquad t \in J, \tag{17.1}$$

then

$$u(t) \le a(t) + q(t)\int_\alpha^t a(s)b(s)\exp\left(\int_s^t q(\tau)b(\tau)\,d\eta(\tau)\right)d\eta(s), \qquad t \in J. \tag{17.2}$$

SKETCH OF THE PROOF. It should first be shown that the righthand side of (17.2), denoted by $v(t)$, satisfies the integral equation

$$v(t) = a(t) + q(t)\int_\alpha^t b(s)v(s)\,d\eta(s), \qquad t \in J. \tag{17.3}$$

To this end one integrates by parts the integral

$$\int_\alpha^t \left\{\int_\alpha^s a(\tau)b(\tau)\exp\left(-\int_\alpha^\tau q(r)b(r)\,d\eta(r)\right)d\eta(\tau)\right\} d\left\{\exp\left(\int_\alpha^s q(r)b(r)\,d\eta(r)\right)\right\}.$$

Setting $w(t) = v(t) - u(t)$, (17.1) and (17.3) imply

$$w(t) \ge q(t)\int_\alpha^t b(s)w(s)\,d\eta(s), \qquad t \in J. \tag{17.4}$$

For the righthand side of (17.4), denoted by $(Tw)(t)$, it follows that

$$w(t) \ge (Tw)(t) \ge (T^2w)(t) \ge \cdots \ge (T^nw)(t), \qquad t \in J, \qquad n \in \mathbf{N}_1. \tag{17.5}$$

If $M_1 = \max\{q(t)b(s)\colon t,s \in J\}$ and $M_2 = \max\{|w(t)|\colon t \in J\}$, then it follows by induction that

$$(T^nw)(t) \ge -M_1^nM_2\frac{[\eta(t)-\eta(\alpha)]^n}{n!}.$$

Hence, by (17.5), $w(t)$ is bounded below by a quantity that tends to zero as $n \to \infty$. Thus $w(t) \ge 0$, which proves (17.2). $\square$

We only quote the most general inequality proved in [101]. It generalizes both theorem 15.1 and theorem 17.1.

THEOREM 17.2. (Jones, 1964) *Let u, a, b, q be real-valued functions, defined in $J = [\alpha, \beta]$, which are either continuous or of bounded variation in J, with b, q nonnegative. Let μ be a nondecreasing left-continuous function in J. Suppose that u, a, b, q are right-continuous at the discontinuity points of μ. If*

$$u(t) \le a(t) + q(t) \int_\alpha^{t^-} b(s)u(s)\, d\mu(s), \qquad t \in J, \tag{17.6}$$

then

$$u(t) \le a(t) + q(t) \int_\alpha^{t^-} a(s)b(s) \exp\left(\int_s^{t^-} q(\tau)b(\tau)\, d\mu(\tau) \right) d\mu(s), \qquad t \in J. \tag{17.7}$$

Moreover, if $\tau_0 = \alpha$, and $\tau_1, \tau_2, \dots$ are the discontinuity points of μ in $(\alpha, \beta]$, and

$$\eta(t) \equiv \mu(t) - \sum_{\tau_i < t} \Delta\mu(\tau_i), \qquad t \in J, \qquad \Delta\mu(t) \equiv \mu(t^+) - \mu(t),$$

then

$$u(t) \le a(t) + q(t) \int_\alpha^{t^-} a(s)b(s) \exp\left(\int_s^t q(\tau)b(\tau)\, d\eta(\tau) \right) \times \\ \times \prod_{s < \tau_i < t} (1 + q(\tau_i)b(\tau_i)\Delta\mu(\tau_i))\, d\mu(s) \tag{17.8}$$

for all $t \in J$. Finally, if $|a(t)| \le A$ and $0 \le q(t) \le Q$ in J, then

$$u(t) \le A \exp\left(Q \int_\alpha^t b(s)\, d\eta(s) \right) \prod_{\alpha < \tau_i < t} (1 + Qb(\tau_i)\Delta\mu(\tau_i)), \qquad t \in J. \qquad \square \tag{17.9}$$

The next class of results deals with modified Stieltjes integrals $\int_\alpha^\beta u\, d\mu$, defined as follows. Let u, μ be real-valued functions in $J = [\alpha, \beta]$, and let w_1, w_2, w_3 be real numbers such that $w_1 + w_2 + w_3 = 1$. Let $P = \{t_0, t_1, \dots, t_n\}$ be a partition of J ($\alpha = t_0 < t_1 < \dots < t_n = \beta$), let $Q = (\tau_1, \dots, \tau_n)$, where $t_{i-1} < \tau_i < t_i$ for $1 \le i \le n$, and let

$$S(P, Q) = \sum_{i=1}^n \{w_1 u(t_{i-1}) + w_2 u(\tau_i) + w_3 u(t_i)\}[\mu(t_i) - \mu(t_{i-1})].$$

If $\lim S(P, Q) = I$ exists, where the limit is taken over all refinements (a 'limit in the refinement sense'), then we write $I = \int_\alpha^\beta u\, d\mu$. This integral is called a *weighted refinement integral* [237]. The case $w_2 = 0$, $w_1 = w_3 = 1/2$ is called a *mean sigma integral*, and the case $w_1 = w_3 = 0$ is called an *interior refinement integral*, or *Dushnik integral*.

THEOREM 17.3. (Wright, Klasi, and Kennenbeck, 1971) *Let μ be a nondecreasing function in $[\alpha, \beta]$, and b a bounded nonnegative function in $[\alpha, \beta]$ such that the weighted refinement integral $\int_\alpha^\beta b\, d\mu$ exists (for given weights w_1, w_2, w_3) and such that $A = \sup\{b(t)\colon \alpha \le t \le \beta\} > 0$. Suppose that*

$$\begin{cases} |1 - w_1| A[\mu(t^+) - \mu(t)] < 1 & \text{in } [\alpha, \beta), \\ |w_3| A[\mu(t) - \mu(t^-)] < 1 & \text{in } (\alpha, \beta]. \end{cases} \tag{17.10}$$

Let $a > 0$, and let $\mathcal{B}$ be the class of bounded nonnegative functions u in $[\alpha, \beta]$ for which $\int_\alpha^\beta bu\, d\mu$ exists and

$$u(\alpha) \leq a, \qquad u(t) \leq a + \int_\alpha^t b(s)u(s)\, d\mu(s), \qquad \alpha < t \leq \beta. \tag{17.11}$$

Then there is a $K \geq 0$ such that $u(t) \leq Ka$ for $\alpha \leq t \leq \beta$ and all $u \in \mathcal{B}$. □

Let $J = [\alpha, \beta]$ or $[\alpha, \infty)$. We denote by $BV(J, \mathbf{R})$ the space of real-valued functions of bounded variation in J that are right-continuous.

Let $\mu \in BV(J, \mathbf{R})$. A measurable function $f \colon J \to \mathbf{R}$ is said to be *locally $d\mu$-integrable in J* if for every $t \in J$ the Lebesgue–Stieltjes integral $\int_\alpha^t f(s)\, d\mu(s)$ is finite. Let $\{t_i\}$, $\alpha < t_1 < t_2 < \dots$, be the discontinuity points of μ. Suppose these are isolated. Then μ has a decomposition $\mu = \mu_1 + \mu_2$, where $\mu_1 \in BV(J, \mathbf{R})$ is an absolutely continuous function and μ_2 is the sum of jumps of μ. Hence μ' exists and equals μ_1' everywhere in J. Let $\lambda_k = \mu(t_k) - \mu(t_k^-)$ be the jump of μ at $t = t_k$.

THEOREM 17.4. (Pandit, 1978) *Suppose that*

$$u(t) \leq a(t) + q(t) \int_\alpha^t b(s)u(s)\, d\mu(s), \qquad t \in J, \tag{17.12}$$

where

i) *u, a, q, b are nonnegative, locally $d\mu$-integrable functions in $J = [\alpha, \infty)$, with a nondecreasing and $q \geq 1$;*

ii) *$\mu \in BV(J, \mathbf{R})$ is such that $\mu_1' \geq 0$ in J and*

$$\lambda_k q(t_k) b(t_k) < 1, \qquad k \in \mathbf{N}_1; \tag{17.13}$$

iii) *the series*

$$\sum_{k=1}^{\infty} \lambda_k q(t_k) b(t_k) \tag{17.14}$$

converges absolutely.

Then

$$u(t) \leq P^{-1} a(t) q(t) \exp\left(\int_\alpha^t q(s)b(s)\mu_1'(s)\, ds\right), \qquad t \in J, \tag{17.15}$$

where $P = \prod_{k=1}^{\infty} (1 - \lambda_k q(t_k) b(t_k))$.

PROOF. Since a is nondecreasing and $q \geq 1$ in J, (17.12) can be written as

$$\frac{u(t)}{a(t)} \leq q(t) \left[1 + \int_\alpha^t b(s) \frac{u(s)}{a(s)}\, d\mu(s)\right], \qquad t \in J. \tag{17.16}$$

Let $r(t)$ denote the expression at the righthand side of (17.16). First we assume that $\alpha \leq t < t_1$, Since μ is differentiable in $[\alpha, t_1)$, theorem 1.1 implies

$$r(t) \leq \exp\left(\int_\alpha^t q(s)b(s)\mu_1'(s)\, ds\right), \qquad t \in [\alpha, t_1). \tag{17.17}$$

For $t = t_1$ we have

$$r(t_1) = r(t_1 - \epsilon) + \int_{t_1-\epsilon}^{t_1} b(s)\frac{u(s)}{a(s)}\, d\mu(s),$$

where $\epsilon > 0$. Limit transition as $\epsilon \to 0_+$ and (17.17) give

$$r(t_1) \leq \exp\left(\int_\alpha^{t_1} q(s)b(s)\mu_1'(s)\, ds\right) + \lambda_1 g(t_1)b(t_1)r(t_1),$$

which, in view of (17.13), gives

$$r(t_1) \leq P_1^{-1}\exp\left(\int_\alpha^{t_1} q(s)b(s)\mu_1'(s)\, ds\right),$$

where $P_k = \prod_{i=1}^k(1-\lambda_i q(t_i)b(t_i))$, $k \in \mathbf{N}_1$. It follows by induction that

$$r(t_m) \leq P_m^{-1}\exp\left(\int_\alpha^{t_m} q(s)b(s)\mu_1'(s)\, ds\right), \qquad m \in \mathbf{N}_1. \tag{17.18}$$

Since $P_i \geq P_{i+1}$ for each $i \geq 1$ and $\lim_{i\to\infty} P_i = P$ (which exists by assumption iii)), we can write (17.18) as

$$r(t_m) \leq P^{-1}\exp\left(\int_\alpha^{t_m} q(s)b(s)\mu_1'(s)\, ds\right), \qquad m \in \mathbf{N}_1. \tag{17.19}$$

For $t \in [t_m, t_{m+1})$ we now have

$$\begin{aligned} r(t) &= r(t_m) + \int_{(t_m,t]} b(s)\frac{u(s)}{a(s)}\, d\mu(s) \leq \\ &\leq r(t_m)\exp\left(\int_{t_m}^{t} q(s)b(s)\mu_1'(s)\, ds\right), \end{aligned}$$

and (17.19) and (17.16) imply (17.15). □

THEOREM 17.5. (Das and Sharma, 1980) *Let $J = [\alpha,\beta]$ or $[\alpha,\infty)$, and $a, \mu \in BV(J,\mathbf{R})$, with μ nondecreasing. Let $f\colon J\times\mathbf{R}\to\mathbf{R}$ be such that*

$$x, y \in \mathbf{R},\ x \leq y \quad \text{implies} \quad 0 \leq f(t,y) - f(t,x) \leq L(t)(y-x), \quad t \in J$$

(i.e. $f(t,x)$ is nondecreasing and Lipschitz continuous with respect to x), and

$$L(t)[\mu(t) - \mu(t^-)] \leq 1, \qquad t \in J.$$

Further, suppose that $f(t,x(t))$ is $d\mu$-integrable for each $x \in BV(J,\mathbf{R})$. Let $\phi, \omega \in BV(J,\mathbf{R})$ be such that

$$\begin{aligned} \phi(\alpha) &< \omega(\alpha), \\ \phi(t) &\leq a(t) + \int_\alpha^t f(s,\phi(s))\, d\mu(s), \\ \omega(t) &> a(t) + \int_\alpha^t f(s,\omega(s))\, d\mu(s) \end{aligned}$$

for $t > \alpha$. Then $\phi(t) < \omega(t)$, $t \in J$. □

THEOREM 17.6. (Das and Sharma, 1980) *Let* $J = [\alpha, \beta]$ *or* $[\alpha, \infty)$, $a, \mu \in BV(J, \mathbf{R})$, $b \in BV(J, \mathbf{R}) \cap C(J, \mathbf{R})$, *with* μ *having isolated discontinuities.*

Then a solution $x \in BV(J, \mathbf{R})$ *of the equation*

$$x(t) = a(t) + \int_\alpha^t b(s)x(s)\,d\mu(s), \qquad t \in J,$$

is given by

$$x(t) = e^{B(t)}\left[a(\alpha) + \int_\alpha^t e^{-B(s)}\,da(s)\right], \qquad t \in J, \tag{17.20}$$

where $B(t) = \int_\alpha^t b(s)\,d\mu(s)$. □

Theorems 17.5 and 17.6 are proven in [62]. The following theorem is based on them.

THEOREM 17.7. (Das and Sharma, 1980) *Let* $J = [\alpha, \beta]$ *or* $[\alpha, \infty)$, $u, a, b, \mu \in BV(J, \mathbf{R})$, *with* μ *nondecreasing and having isolated discontinuities at* t_i*:* $\alpha < t_1 < t_2 < \ldots$, u *being* $d\mu$*-integrable in* J, b *being nonnegative and continuous in* J, *and*

$$b(t_k)[\mu(t_k) - \mu(t_k^-)] \le 1, \qquad k \in \mathbf{N}_1.$$

If

$$u(t) \le a(t) + \int_\alpha^t b(s)u(s)\,d\mu(s), \qquad t \in J,$$

then

$$u(t) \le e^{B(t)}\left[a(\alpha) + \int_\alpha^t e^{-B(s)}\,da(s)\right], \qquad t \in J, \tag{17.21}$$

where $B(t) = \int_\alpha^t b(s)\,d\mu(s)$.

PROOF. Consider the integral equation

$$\omega(t) = a(t) + \epsilon + \int_\alpha^t b(s)\omega(s)\,d\mu(s), \qquad t \in J, \tag{17.22}$$

where $\epsilon > 0$. Theorem 17.5 implies

$$u(t) \le \omega_\epsilon(t),$$

where $\omega_\epsilon(t)$ is a solution of (17.22). Theorem 17.6 then implies

$$u(t) < e^{B(t)}\left[a(\alpha) + \epsilon + \int_\alpha^t e^{-B(s)}\,da(s)\right], \qquad t \in J.$$

Letting $\epsilon \to 0_+$, (17.21) follows. □

18. Systems of integral inequalities

In this section we consider integral inequalities and comparison equations related to them. Before stating the results, we introduce some notation.

Let $x = (x_1, \dots, x_n) \in \mathbf{R}^n$ and $y = (y_1, \dots, y_n) \in \mathbf{R}^n$. Then $x \le y$ $(x < y)$ if and only if $x_i \le y_i$ $(x_i < y_i)$ for all $i = 1, \dots, n$, and we set, for $\alpha, \beta \in \mathbf{R}^n$ with $\alpha \le \beta$, $[\alpha; \beta] = \{x \in \mathbf{R}^n \colon \alpha \le x \le \beta\}$.

Let $|\cdot|$ be an arbitrary vector norm in $\mathbf{R}^n$, and let $\|M\| = \sup_{|x|=1} |Mx|$ be the norm of the $(m \times n)$-matrix M. Let ω_k be any combination of k of the integers $\{1, \dots, n\}$. We assume that in any combination $\omega_k = \{i_1, \dots, i_k\}$ the elements have been ordered (i.e. $i_1 < i_2 < \dots < i_k$). For a combination ω_k, we set $\omega_k' = \{1, \dots, n\} \setminus \omega_k$.

Let $\omega_k = \{i_1, \dots, i_k\}$. For $x \in \mathbf{R}^n$ we set $x_{\omega_k} = (x_{i_1}, \dots, x_{i_k})$. The multiple integral symbol $\int_{\alpha_{i_1}}^{x_{i_1}} \cdots \int_{\alpha_{i_k}}^{x_{i_k}}$ is also denoted by $\int_{\alpha_{\omega_k}}^{x_{\omega_k}}$, and the sequence of differentials $dr_{i_k}\, dr_{i_{k-1}} \dots dr_{i_1}$ is also denoted by dr_{ω_k}. For a function $g \colon \mathbf{R}^n \to \mathbf{R}^m$ we denote its partial derivative $g_{x_{i_1} \dots x_{i_k}}$ by $g_{x_{\omega_k}}$. For $x, y \in \mathbf{R}^n$ we set

$$w_i(x, y; \omega_k) = \begin{cases} x_i & \text{if } i \in \omega_k, \\ y_i & \text{if } i \notin \omega_k, \end{cases}$$

and $w(x, y; \omega_k) = (w_1(x, y; \omega_k), \dots, w_n(x, y; \omega_k))$.

Let $u, a \colon [\alpha; \beta] \to \mathbf{R}^m$, and, for each ω_k, $1 \le k \le n$, let $K_{\omega_k}(x, r_{\omega_k})$ be an $(m \times m)$-matrix-valued function, defined for $x \in [\alpha; \beta]$, $\alpha_{\omega_k} \le r_{\omega_k} \le x_{\omega_k} \le \beta_{\omega_k}$.

Now we can consider the following linear system of m Volterra integral equations in n independent variables:

$$u(x) = a(x) + \sum_{\substack{\omega_k \\ 1 \le k \le n}} \int_{\alpha_{\omega_k}}^{x_{\omega_k}} K_{\omega_k}(x, r_{\omega_k}) u(w(x, r; \omega_k))\, dr_{\omega_k}. \tag{18.1}$$

It can be shown that if $a(x)$ and the matrix-valued functions $K_{\omega_k}(x, r_{\omega_k})$ are continuous, then (18.1) has a unique continuous solution in $[\alpha; \beta]$. Also, we note that there are $2^n - 1$ integrals participating in (18.1). To simplify our notation even further, we write $\sum$ instead of $\sum_{\substack{\omega_k \\ 1 \le k \le n}}$.

Suppose the matrix function $A(x; \xi)$ satisfies the equation

$$A(x; \xi) = I + \sum \int_{\xi_{\omega_k}}^{x_{\omega_k}} K_{\omega_k}(x, r_{\omega_k}) A(w(x, r; \omega_k); \xi)\, dr_{\omega_k} \tag{18.2}$$

for $\alpha \le \xi \le x \le \beta$. Then $A(x; \xi)$ is said to be a *fundamental solution* of (18.1).

If the functions $K_{\omega_k}(x, r_{\omega_k})$ are continuous, it can be shown that for $\alpha \le \xi \le x \le \beta$, $A(x; \xi)$ exists, is unique, and is continuous. The following lemma, which can be verified directly, gives the solution of (18.1) in terms of the fundamental solution.

LEMMA 18.1. *Let $a(x)$ be a continuous function in $[\alpha; \beta]$, and let every matrix function $K_{\omega_k}(x, r_{\omega_k})$ be continuous for $x \in [\alpha; \beta]$, $\alpha_{\omega_k} \le r_{\omega_k} \le x_{\omega_k} \le \beta_{\omega_k}$.*

i) *If every* $A_{\xi_{\omega_k}}(x;\xi)$ *is continuous for* $\alpha \le \xi \le x \le \beta$, *then the unique continuous solution of* (18.1) *in* $[\alpha;\beta]$ *is*

$$u(x) = a(x) + \sum(-1)^k \int_{\alpha_{\omega_k}}^{x_{\omega_k}} A_{r_{\omega_k}}(x; w(x,r;\omega_k))a(w(x,r;\omega_k))\,dr_{\omega_k}. \tag{18.3}$$

ii) *If every* $a_{x_{\omega_k}}$ *is continuous in* $[\alpha;\beta]$, *then the unique continuous solution of* (18.1) *is*

$$u(x) = A(x;\alpha)a(\alpha) + \sum \int_{\alpha_{\omega_k}}^{x_{\omega_k}} A(x; w(\alpha,r;\omega_k))a_{r_{\omega_k}}(w(\alpha,r;\omega_k))\,dr_{\omega_k}. \quad \square \tag{18.4}$$

Let λ be a real number. The space $C_\lambda[\alpha;\beta]$ consists, by definition, of all continuous functions $g\colon [\alpha;\beta] \to \mathbf{R}^m$, normed by $\|g\|_\lambda = \sup_{x\in[\alpha;\beta]}\{|g(x)|\exp(-\lambda\sum_{i=1}^n x_i)\}$. It can be shown that $C_\lambda[\alpha;\beta]$ is a Banach space. Let $K \subset C_\lambda[\alpha;\beta]$ be the positive cone such that $a(x) = (a_1(x),\dots,a_m(x)) \in K$ if and only if $a_i(x) \ge 0$, $1 \le i \le m$. In $C_\lambda[\alpha;\beta]$ we define a partial order relation as follows: $g_1 \le g_2$ if and only if $g_2 - g_1 \in K$. Note that this relation has the property that if $\{g_n\}$ is an increasing sequence in $C_\lambda[\alpha;\beta]$ converging to g_0, then $g_n \le g_0$ for all n.

The following theorem holds.

Theorem 18.1. (De Franco, 1976) *Let* $a(x)$ *be a continuous function in* $[\alpha;\beta]$, *and let, for each* ω_k, $1 \le k \le n$, *the* $(m \times m)$*-matrix-function* $K_{\omega_k}(x, r_{\omega_k})$ *be continuous and have nonnegative entries. If* $u\colon [\alpha;\beta] \to \mathbf{R}^m$ *is a continuous function and*

$$u(x) \le a(x) + \sum \int_{\alpha_{\omega_k}}^{x_{\omega_k}} K_{\omega_k}(x, r_{\omega_k})u(w(x,r;\omega_k))\,dr_{\omega_k}$$

for $x \in [\alpha;\beta]$, *then* $u(x) \le v(x)$, *where* $v(x)$ *is the unique continuous solution of* (18.1) *in* $[\alpha;\beta]$.

In addition, if every $A_{\xi_{\omega_k}}(x;\xi)$ *is continuous on* $\alpha \le \xi \le x \le \beta$, *then*

$$u(x) \le a(x) + \sum(-1)^k \int_{\alpha_{\omega_k}}^{x_{\omega_k}} A_{r_{\omega_k}}(x; w(x,r;\omega_k))a(w(x,r;\omega_k))\,dr_{\omega_k}. \tag{18.5}$$

If, instead, we additionally assume that every $a_{x_{\omega_k}}(x)$ *is continuous in* $[\alpha;\beta]$, *then*

$$u(x) \le A(x;\alpha)a(\alpha) + \sum \int_{\alpha_{\omega_k}}^{x_{\omega_k}} A(x; w(\alpha,r;\omega_k))a_{r_{\omega_k}}(w(\alpha,r;\omega_k))\,dr_{\omega_k}. \tag{18.6}$$

Proof. The continuity of every $K_{\omega_k}(x, r_{\omega_k})$ on a compact domain implies the existence of a constant $M > 0$ such that $\|K_{\omega_k}(x, r_{\omega_k})\| \le M$ for every ω_k. We choose $\lambda_0 > 1$ so that $M(2^n - 1)/\lambda_0 < 1$. Then we define an operator T by

$$(Tg)(x) = a(x) + \sum \int_{\alpha_{\omega_k}}^{x_{\omega_k}} K_{\omega_k}(x, r_{\omega_k})g(w(x,r;\omega_k))\,dr_{\omega_k}$$

for $g \in C_{\lambda_0}[\alpha;\beta]$. The continuity of $a(x)$, $g(x)$, $K_{\omega_k}(x, r_{\omega_k})$ implies that $(Tg)(x)$ is continuous in $[\alpha;\beta]$.

Let $g_1, g_2 \in C_{\lambda_0}[\alpha;\beta]$. Then

$$\begin{aligned}
&|(Tg_1)(x) - (Tg_2)(x)| \leq \\
&\leq \sum \int_{\alpha_{\omega_k}}^{x_{\omega_k}} M|g_1(w(x,r;\omega_k)) - g_2(w(x,r;\omega_k))| \times \\
&\quad \times \exp\left[-\lambda_0 \sum_{i=1}^n w_i(x,r;\omega_k)\right] \exp\left[\lambda_0 \sum_{i=1}^n w_i(x,r;\omega_k)\right] dr_{\omega_k} \leq \\
&\leq M\|g_1 - g_2\|_{\lambda_0} \sum \int_{\alpha_{\omega_k}}^{x_{\omega_k}} \exp\left[\lambda_0 \sum_{i=1}^n w_i(x,r;\omega_k)\right] dr_{\omega_k} \leq \\
&\leq M\|g_1 - g_2\|_{\lambda_0} \sum \frac{1}{\lambda_0^k} \exp\left[\lambda_0 \sum_{i=1}^n x_i\right] \leq \\
&\leq \frac{M(2^n - 1)}{\lambda_0} \|g_1 - g_2\|_{\lambda_0} \exp\left[\lambda_0 \sum_{i=1}^n x_i\right].
\end{aligned}$$

Therefore $\|Tg_1 - Tg_2\|_{\lambda_0} \leq \frac{M(2^n-1)}{\lambda_0}\|g_1 - g_2\|$, and T is a contraction on $C_{\lambda_0}[\alpha;\beta]$.

Let now $g_1, g_2 \in C_{\lambda_o}[\alpha;\beta]$ with $g_1 \leq g_2$. Since the entries of every $K_{\omega_k}(x, r_{\omega_k})$ are nonnegative, we see that $Tg_1 \leq Tg_2$. The result now follows directly from lemma 18.1 and theorem 19.2. □

In §7 we have seen that 'solutions' of nonlinear integral inequalities can be estimated by extremal (maximal, minimal) solutions of corresponding comparison equations. For systems of nonlinear integral inequalities the situation is similar. In this respect, the following theorem is characteristic.

THEOREM 18.2. (Opial, 1957) *Let $J = [\alpha, \beta)$ and $f\colon J \times \mathbf{R}^n \to \mathbf{R}^n$ a continuous function satisfying $f(t,x) \leq f(t,y)$ for $t \in J$ and $x, y \in \mathbf{R}^n$ with $x \leq y$. If $u\colon J \to \mathbf{R}^n$ is a continuous function satisfying*

$$u(t) \leq c + \int_\alpha^t f(s, u(s))\, ds, \qquad t \in J,$$

for some $c \in \mathbf{R}^n$, and $\psi\colon [\alpha, \beta_1) \to \mathbf{R}^n$ (where $\beta_1 \leq \beta$) is the maximal solution of the system

$$\psi(t) = c + \int_\alpha^t f(s, \psi(s))\, ds, \qquad t \in [\alpha, \beta_1),$$

then $u(t) \leq \psi(t)$ for $t \in [\alpha, \beta_1)$. □

Before stating more general results, we give some definitions.

DEFINITION 18.1. A function $\psi\colon Q \to \mathbf{R}^n$ is said to be

i) *nondecreasing in Q* if $u, v \in Q$ and $u \leq v$ imply $\psi(u) \leq \psi(v)$;
ii) *monotonically indecreasing in Q* if $u, v \in Q$ and $u \leq v$ imply $\psi(u) \leq \psi(v)$, while $u < v$ implies $\psi(u) < \psi(v)$. □

DEFINITION 18.2. A function $F\colon \mathbf{R}_+ \times Q \to \mathbf{R}^n$ is said to be *quasimonotonically increasing in $\mathbf{R}_+ \times Q$* if for two arbitrary pairs of points $(t,u), (t,v) \in \mathbf{R}_+ \times Q$ and any $i = 1, \ldots, n$ we have that $u \leq v$ and $u_i = v_i$ imply $F_i(t,u) \leq F_i(t,v)$. □

Let $0 < c \le \infty$, $B_c = \{u \in \mathbf{R}^n\colon\ |u| < c\}$, and let $K(t,s,u)$ be a function defined for $\alpha \le s \le t < \beta \le \infty$, $u \in B_c$.

For $K(t,s,u)$ we introduce the following so-called *Carathéodory conditions*:

H18.1) $K(t,s,u)$ is continuous with respect to $u \in B_c$ for all t and almost all s, and it is measurable with respect to s for all t and u;

H18.2) $K(t,s,u)$ is nondecreasing in B_c for all t and almost all s;

H18.3) for any $d \in (0,c)$ there are functions $\mu_d(t,s)$ and $\nu_d(\tau,t,s)$, summable with respect to $s \in [\alpha,t]$ $(\alpha \le s \le t \le \tau < \beta)$ such that

$$\sup_{|u|\le d} |K(t,s,u)| \le \mu_d(t,s), \qquad \sup_{|u|\le d} |K(\tau,s,u) - K(t,s,u)| \le \nu_d(\tau,t,s);$$

H18.4) $\lim_{\tau-t\to 0_+}\left(\int_\tau^t \mu_d(\tau,s)\,ds\right) = 0$ for either t or τ fixed;

H18.5) $\lim_{\tau-t\to 0_+}\left(\int_\tau^t \nu_d(\tau,t,s)\,ds\right) = 0$ for either t or τ fixed;

H18.6) $\lim_{\tau-t\to 0_+}\left(\int_\alpha^t \nu_d(\tau,t,s)\,ds\right) = 0$ for either $t \ne \tau_k$ or $\tau \ne \tau_k$ fixed, where the sequence $\{\tau_k\}$ is such that $\alpha < \tau_1 < \tau_2 < \dots$, $\lim_{k\to\infty} \tau_k \ge \beta$.

The following theorem holds.

THEOREM 18.3. (Azbelev and Tsalyuk, 1962) *Suppose the following conditions are fulfilled:*

1) *conditions* H18.1–5 *hold;*
2) $a\colon [\alpha,\beta) \to \mathbf{R}^n$ *is a continuous function, and* $|a(\alpha)| < c$*;*
3) $v\colon [\alpha,\beta) \to B_c$ *is the maximal (minimal) maximally continued solution of the equation*

$$v(t) = a(t) + \int_\alpha^t K(t,s,v(s))\,ds, \qquad t \in [\alpha,\beta); \tag{18.7}$$

4) $u\colon [\alpha,\beta) \to B_c$ *is a continuous function satisfying the integral inequality*

$$\Delta(t) \equiv u(t) - a(t) - \int_\alpha^t K(t,s,u(s))\,ds \le 0 \qquad (0 \le \Delta(t)), \qquad t \in [\alpha,\beta). \tag{18.8}$$

Then for $t \in [\alpha,\beta)$,

$$u(t) \le v(t) \qquad (v(t) \le u(t)). \qquad \square \tag{18.9}$$

REMARK 18.1. Theorem 18.3 still holds if we modify its conditions to:

i) condition H18.5 is replaced by H18.6;
ii) the functions $a, u, v\colon [\alpha,\beta) \to \mathbf{R}^n$ (with $|a(\alpha)| < c$) are piecewise continuous, and their discontinuity points form a subsequence of the sequence $\{\tau_k\}$ of condition H18.6;
iii) for any $\gamma \in (\alpha,\beta)$,

$$\sup_{t\in[\alpha,\gamma]} |u(t)| < c, \qquad \sup_{t\in[\alpha,\gamma]} |v(t)| < c. \qquad \square$$

Using this remark we can readily prove the following comparison theorem.

THEOREM 18.4. (Simeonov and Bainov, 1987) *Suppose the following conditions are fulfilled:*

1) *conditions* H18.1–4 *and* H18.6 *hold;*
2) $\{t_k\}_1^\infty$ *is a sequence such that* $\alpha < t_1 < t_2 < \dots$, $\lim_{k\to\infty} t_k \geq \beta$;
3) $a\colon [\alpha,\beta) \to \mathbf{R}^n$ *is a piecewise continuous function, and* $|a(\alpha)| < c$;
4) $\psi_k\colon [\alpha,\beta]\times[\alpha,\beta]\times B_c \to \mathbf{R}^n$, $(t,s,u) \to \psi_k(t,s,u)$, $k \in \mathbf{N}_1$, *are functions which are nondecreasing in* B_c *for* t, s *fixed, and piecewise continuous with respect to* t *for* s, u *fixed;*
5) $v\colon [\alpha,\beta) \to \mathbf{R}^n$ *is the maximal (minimal) maximally continued solution of the equation*

$$v(t) = a(t) + \int_\alpha^t K(t,s,v(s))\,ds + \sum_{\alpha<t_k<t} \psi_k(t,t_k,v(t_k)), \qquad t \in [\alpha,\beta); \tag{18.10}$$

6) $u\colon [\alpha,\beta) \to \mathbf{R}^n$ *is a piecewise continuous function satisfying the inequality*

$$\Delta(t) \equiv u(t) - a(t) - \int_\alpha^t K(t,s,u(s))\,ds - \sum_{\alpha<t_k<t} \psi_k(t,t_k,u(t_k)) \leq 0 \tag{18.11}$$

$(0 \leq \Delta(t))$ *for* $t \in [\alpha,\beta)$;

7) *for any* $\gamma \in (\alpha,\beta)$,

$$\sup_{t\in[\alpha,\gamma]} |u(t)| < c, \qquad \sup_{t\in[\alpha,\gamma]} |v(t)| < c;$$

8) *the sequence* $\{t_k\}$ *and the sequence of discontinuity points of the functions* a, u, v *form a subsequence of the sequence* $\{\tau_k\}$.

Then, for $t \in [\alpha,\beta)$,

$$u(t) \leq v(t) \qquad (v(t) \leq u(t)). \qquad \square \tag{18.12}$$

REMARK 18.2. Theorem 18.4 still holds if in (18.10) and (18.11) the values $v(t_k)$ and $u(t_k)$ are replaced by $v(t_k^-)$ and $u(t_k^-)$, respectively. $\square$

In estimating solutions of equations of the form (18.7) and (18.10), the following theorem is sometimes useful.

THEOREM 18.5. (Simeonov and Bainov, 1988) *Suppose the following conditions hold:*

1) $F\colon \mathbf{R}_+ \times Q \to \mathbf{R}^n$ *is a quasimonotonically increasing function which is continuous inside the sets* $(\tau_{k-1},\tau_k] \times Q$, $k \in \mathbf{N}_1$, *and for any* $k \in \mathbf{N}_1$ *and* $v \in Q$ *the limit* $\lim_{\substack{(t,u)\to(\tau_k,v)\\ t>\tau_k}} F(t,u)$ *exists;*
2) $\psi_k\colon Q \to \mathbf{R}^n$, $k \in \mathbf{N}_1$, *are nondecreasing functions in* Q;
3) $v\colon (t_0,\omega) \to \mathbf{R}^n$ *is the maximal (minimal) solution of the equation*

$$\begin{aligned} v' &= F(t,v), \qquad t \neq \tau_k, \\ v(\tau_k^+) &= \psi_k(v(\tau_k)), \end{aligned} \tag{18.13}$$

such that $v(t_0^+) = v_0$, $(t_0, v_0) \in \mathbf{R}_+ \times Q$, *and* $v(\tau_k^+) \in Q$ *if* $\tau_k \in (t_0, \omega)$;

4) $u\colon (t_0, \tilde{\omega}) \to \mathbf{R}^n$ *is a continuous function for* $t \in (t_0, \tilde{\omega})$, $t \neq \tau_k$, *which is left-continuous at the points* τ_k *and is such that*
   i) $u(t) \in Q$ *for* $t \in (t_0, \tilde{\omega})$, *and* $u(\tau_k^+) \in Q$ *if* $\tau_k \in (t_0, \tilde{\omega})$;
   ii) $u(t_0^+) \le v_0$ $(v_0 \le u(t_0^+))$;
   iii) $Du(t) \le F(t, u(t))$ $(F(t, u(t)) \le Du(t))$ *for* $t \in (t_0, \tilde{\omega})$, $t \neq \tau_k$;
   iv) $u(\tau_k^+) \le \psi_k(u(\tau_k))$ $(\psi_k(u(\tau_k)) \le u(\tau_k^+))$ *if* $\tau_k \in (t_0, \tilde{\omega})$, *where* $Du(t)$ *is any Dini derivative of* u.

Then $u(t) \le v(t)$ $(v(t) \le u(t))$ *for* $t \in (t_0, \omega) \cap (t_0, \tilde{\omega})$. □

COROLLARY 18.1. *Let* $t_0 \ge 0$, $J = [t_0, \infty)$, $u \in PC(J, \mathbf{R}^n)$, *and*

$$\begin{aligned} Du(t) &\le A(t)u(t) + f(t), \qquad t > t_0, \qquad t \neq \tau_k, \\ u(\tau_k^+) &\le B_k u(\tau_k) + f_k, \qquad \tau_k > t_0, \end{aligned} \tag{18.14}$$

where $f \in PC(J, \mathbf{R}^n)$, $A = (a_{ij}) \in PC(J, \mathbf{R}^{n\times n})$, $B_k = (b_{ij}^{(k)}) \in \mathbf{R}^{n\times n}$, $f_k \in \mathbf{R}^n$, *and*

$$\begin{aligned} a_{ij}(t) &\ge 0, \qquad 1 \le i, j \le n, \quad i \neq j, \qquad t > t_0, \\ b_{ij}^{(k)} &\ge 0, \qquad 1 \le i, j \le n, \qquad \tau_k > t_0. \end{aligned} \tag{18.15}$$

If $u(t_0^+) \le v_0$, *then for* $t > t_0$,

$$u(t) \le W(t, t_0^+)v_0 + \int_{t_0}^{t} W(t,s) f(s)\, ds + \sum_{t_0 < \tau_k < t} W(t, \tau_k^+) f_k, \tag{18.16}$$

where $W(t,s)$ *is the Cauchy matrix of the impulse differential equation*

$$\begin{aligned} v'(t) &= A(t)v(t), \qquad t \neq \tau_k, \\ v(\tau_k^+) &= B_k v(\tau_k), \qquad \tau_k > t_0. \end{aligned} \tag{18.17}$$

□

REMARK 18.3. The righthand side of (18.16) coincides with the solution of the comparison differential equation corresponding to (18.14). More detailed information about solutions of linear inhomogeneous impulse differential equations and the Cauchy matrix $W(t,s)$ is given in §20. □

19. Inequalities in partially ordered spaces

In this section we consider inequalities in which the 'function to be estimated' is an element of a partially ordered space. Many inequalities studied in the previous sections can be investigated along the general scheme outlined below.

A binary relation R in a set X is said to be a *partial order relation* if R is reflexive, transitive, and antisymmetric, i.e. xRx for all $x \in X$, xRy and yRz imply xRz, and xRy and yRx imply $x = y$.

We denote a partial order relation by the symbol '$\leq$', and say that 'x is smaller than y' if $x \leq y$. The notation $x < y$ means that $x \leq y$ and $x \neq y$. A set X endowed with a partial order relation is called a *partially ordered set*.

REMARK 19.1. If X is a partially ordered set, then it is not always true that one of the relations $x \leq y$ or $y \leq x$ holds. □

If X is partially ordered by $\leq$, then an operator $T\colon X \to X$ is said to be *isotone* (*monotone*, or *order-preserving*) if $x \leq y$ implies $Tx \leq Ty$.

A linear space X with zero θ is said to be a *linear partially ordered space* if X is partially ordered by a relation $\leq$ and the following conditions hold:

$$\begin{gathered} \theta \leq x \quad \Rightarrow \quad \theta \leq \lambda x \qquad \text{for all } \lambda \geq 0, \\ x_1 \leq y_1, \quad x_2 \leq y_2 \quad \Rightarrow \quad x_1 + x_2 \leq y_1 + y_2, \end{gathered} \tag{19.1}$$

i.e. the partial order is compatible with the algebraic operations in X.

In a linear partially ordered space X the *interval* $[u; v] = \{x \in X\colon u \leq x \leq v\}$ is a convex set.

A closed subset K of a real Banach space X is said to be a *cone* if

$$x \in K, \quad x \neq \theta \quad \Rightarrow \quad \lambda x \in K \quad \text{for all } \lambda \geq 0 \qquad \text{and} \qquad -x \notin K.$$

EXAMPLE 19.1. In each of the spaces C, L_p, Orlicz spaces, the set of nonnegative functions is a cone. The set $\mathbf{R}^n_+ = \{x \in \mathbf{R}^n\colon x_i \geq 0,\ i = 1, \ldots, n\}$ is a cone in $\mathbf{R}^n$. □

A cone K in a space X induces a partial order in X by setting

$$x \leq y \quad \Leftrightarrow \quad y - x \in K$$

(also, $x < y \Leftrightarrow y - x \in \operatorname{Int} K$).

If X is a partially ordered Banach space with partial order $\leq$, then $X^+ = \{x \in X\colon \theta \leq x\}$ is called the *positive cone* of X. In this case an operator $S\colon X \to X$ is said to be *positive* if $S(X^+) \subset X^+$. If S is a linear operator, then S is positive if and only if S is isotone.

Let A be a nonempty subset of a linear partially ordered space X. An element $w \in X$ is the *least upper bound* of A if $x \leq w$ for all $x \in A$, and $x \leq y$ for all $x \in A$ implies $w \leq y$. The least upper bound w of A (if it exists) is also denoted by

$$w = \sup A.$$

Similarly we define the *greatest lower bound* $v = \inf A$.

A linear partially ordered space X is said to be a *lattice* (or *K-lineal*) if any two elements $x, y \in X$ have a least upper bound, denoted by

$$x \vee y = \sup\{x, y\}.$$

If X is a lattice, we can prove that any two elements $x, y \in X$ also have a greatest lower bound,

$$x \wedge y = \inf\{x, y\}.$$

In this case, any finite set $\{x_i\}_{i=1}^n$ of elements of X has a least upper bound $\vee_{i=1}^n x_i$ and a greatest lower bound $\wedge_{i=1}^n x_i$.

The relations $x \vee y$ and $x \wedge y$ define in the lattice X binary operations with the following properties:

$$\begin{array}{llll} 1) & x \vee x = x, & 1^*) & x \wedge x = x, \\ 2) & x \vee y = y \vee x, & 2^*) & x \wedge y = y \wedge x, \\ 3) & (x \vee y) \vee z = x \vee (y \vee z) & 3^*) & (x \wedge y) \wedge z = x \wedge (y \wedge z), \\ 4) & x \wedge (x \vee y) = x, & 4^*) & x \vee (x \wedge y) = x. \end{array} \tag{19.2}$$

Any element x of the lattice X has a representation $x = x^+ - x^-$, where $x^+ = \theta \vee x$, $x^- = \theta \vee (-x)$.

The element $|x| = x^+ + x^-$ is called the *modulus* of $x \in X$.

A linear normed space X is said to be a *KB-lineal* if it is a lattice and the following condition is met:

$$x, y \in X, \quad |x| \leq |y| \quad \Rightarrow \quad \|x\| \leq \|y\| \qquad \text{(monotonicity of the norm)}.$$

In a KB-lineal X we have $\|x\| = \||x|\|$ for all $x \in X$. An example of a KB-lineal is the space $C([a,b], \mathbf{R}^n)$ with norm $\|x\| = \sup_{t\in[a,b]} |x(t)|$ and positive cone $C^+ = C([a,b], \mathbf{R}^n_+)$.

A set X is said to be a *distributive lattice* if it is a lattice in which the distributive law holds:

$$\begin{aligned} &1) \quad (x \vee y) \wedge z = (x \wedge z) \vee (y \wedge z), \\ &2) \quad (x \wedge y) \vee z = (x \vee z) \wedge (y \vee z). \end{aligned} \tag{19.3}$$

Let X be a complete metric space with metric d. A mapping $T \colon X \to X$ is a *contraction* if there is a constant $q \in [0, 1)$ such that

$$d(Tx, Ty) \leq q d(x, y) \qquad \text{for all } x, y \in X.$$

THEOREM 19.1. (Banach fixed point theorem) *Let X be a complete metric space and $T \colon X \to X$. Suppose that for some $m \geq 1$ the iterate T^m is a contraction. Then T has a unique fixed point x_0, that is, there is a unique point $x_0 \in X$ such that $Tx_0 = x_0$. Moreover, for all $x \in X$, $x_0 = \lim_{n\to\infty} T^n x$.* □

THEOREM 19.2. *Let X be a complete metric space which is partially ordered in such a way that if $x_n \leq x_{n+1}$ for all $n \geq 1$ and $\lim x_n = x_0$ exists, then $x_n \leq x_0$ for all $n \geq 1$. Let $T \colon X \to X$ be an isotone operator for which T^m is a contraction, for some $m \geq 1$. If x_0 is the unique fixed point of T, then*

$$x \leq Tx \quad \Rightarrow \quad x \leq x_0. \tag{19.4}$$

Moreover, (19.4) *remains valid if $\leq$ is replaced by $<$, $\geq$, or $>$.*

PROOF. We prove the result for $<$. Since T is isotone, we have

$$x < Tx < T^2 x < \cdots < T^n x \to x_0.$$

By assumption, $T^n x \leq x_0$ for all $n \geq 1$, hence $x < Tx \leq x_0$, so that $x < x_0$. □

For proving our next theorem we need the following lemma (called the *Volterra fixed point theorem* by E. Hille [93], [94]).

LEMMA 19.1. *Let $S\colon X \to X$ be a bounded linear operator in a Banach space X such that*

$$\sum_{n=1}^{\infty} \|S^n\| < \infty. \tag{19.5}$$

We define, for each $a \in X$, the operator $T_a\colon X \to X$ by $T_a x = a + Sx$. Then each T_a has a unique fixed point, x_a, given by the Neumann series

$$x_a = \sum_{n=0}^{\infty} S^n a. \tag{19.6}$$

PROOF. For each $a \in X$, (19.5), the fact that

$$\left\| \sum_{k=m}^{n} S^k a \right\| \le \|a\| \sum_{k=m}^{n} \|S^k\|,$$

and the completeness of X imply that the series (19.6) converges. By direct substitution, $x = x_a$ is seen to be a solution of $x = T_a x$. On the other hand, if $x \in X$ satisfies $x = T_a x$, then repeated substitution leads to

$$x = a + S(a + Sx) = \sum_{k=0}^{1} S^k a + S^2 x = \cdots = \sum_{k=0}^{n-1} S^k a + S^n x.$$

Since $\|S^n x\| \le \|S^n\| \|x\| \to 0$ (by (19.5)), it follows that $x = x_a$. □

THEOREM 19.3. *Let X be a partially ordered Banach space whose positive cone X^+ is a closed set, and let $S\colon X \to X$ be a positive, bounded linear operator satisfying* (19.5). *Let $a \in X$, and let x_a be the unique solution of $x = a + Sx$. Then*

$$y \le a + Sy \quad \Rightarrow \quad y \le x_a. \tag{19.7}$$

Moreover (19.7) *still holds if $\le$ is replaced by $<$, $\ge$, or $>$. Finally, if $a \in X^+$, then $x_a \in X^+$.*

PROOF. Since S is isotone, as in the proof of lemma 19.1 it follows from $y \le a + Sy$ that

$$y \le \sum_{k=0}^{n-1} S^k a + S^n y \equiv y_n, \qquad n \ge 1. \tag{19.8}$$

hence $y_n - y \in X^+$, and since X^+ is closed,

$$x_a - y = \lim_{n\to\infty} (y_n - y) \in X^+,$$

that is, $\theta \le x_a - y$, so that (19.7) follows from (19.1). If $y < a + Sy$, then (19.8) becomes $y < a + Sy \le y_n$ for all $n \ge 1$, and so $y < a + Sy \le x_a$ follows.

If $a \in X^+$, then $S^k a \in X^+$ for all $k \ge 0$, and then (19.1) implies that

$$\sum_{k=0}^{n} S^k a \in X^+, \qquad n \ge 0.$$

Since X^+ is closed, (19.6) implies that $x_a \in X^+$. □

Theorem 19.3 is an abstract version of the general linear Gronwall inequality of theorem 1.5, and, in essence, includes this theorem as a special case. Indeed, let $J = [\alpha, \beta]$ and $X = C(J)$ (the space of continuous functions in J) with norm $\|x\| = \max\{|x(t)|\colon t \in J\}$. For $x, y \in C(J)$ we set

$$x \le y \quad \Leftrightarrow \quad x(t) \le y(t) \qquad \text{for } t \in J.$$

Then $X^+ = C^+(J) = \{x \in C(J)\colon x(t) \ge 0,\ t \in J\}$ is clearly a closed set. For $b, q \in C^+(J)$ we define $S\colon C(J) \to C(J)$ by

$$(Sx)(t) = q(t)\int_\alpha^t b(s)x(s)\,ds, \qquad t \in J \qquad (x \in C(J)).$$

It is readily verified that

$$|(S^n x)(t)| \le \frac{\|x\|(\|b\|\,\|q\|(t-\alpha))^n}{n!}, \qquad t \in J, \qquad n \ge 1.$$

Hence, if $M = \|b\|\,\|q\|(\beta - \alpha)$, then $\|S^n x\| \le \|x\| M^n/n!$ for all $x \in C(J)$, so that

$$\|S^n\| \le \frac{M^n}{n!}, \qquad n \ge 1.$$

Thus, (19.5) holds, and hence theorem 19.3 implies that for $x, a \in C(J)$ with

$$x(t) \le a(t) + q(t)\int_\alpha^t b(s)x(s)\,ds, \qquad t \in J,$$

we have $x(t) \le x_a(t)$, $t \in J$, where $y = x_a$ is the unique solution of the equation

$$y(t) = a(t) + q(t)\int_\alpha^t b(s)y(s)\,ds, \qquad t \in J.$$

Setting $U(t) \equiv \int_\alpha^t b(s)y(s)\,ds$, this equation is equivalent to the initial value problem

$$U' - qbU = ab, \qquad U(\alpha) = 0.$$

The latter has the solution $U(t) = \int_\alpha^t a(s)b(s)\exp(\int_s^t q(\tau)b(\tau)\,d\tau)\,ds$. Hence

$$x(t) \le x_a(t) = a(t) + q(t)\int_\alpha^t a(s)b(s)\exp\left(\int_s^t q(\tau)b(\tau)\,d\tau\right)ds, \qquad t \in J.$$

Note that theorem 19.3 also includes as special cases theorems 15.1, 17.1, 2.1, 2.2, 11.9, 12.19, 15.3, and 18.1.

To obtain another consequence of theorem 19.3 we take $X = M_n C(J)$, the space of $(n \times n)$-matrices x with entries $x_{ij} \in C(J)$, with norm

$$\|x\| = \sup_{t \in J} \max_{1 \le i \le n} \sum_{j=1}^n |x_{ij}(t)|$$

and partial order relation

$$x \le y \quad \Leftrightarrow \quad x_{ij}(t) \le y_{ij}(t), \qquad 1 \le i, j \le n, \qquad t \in J.$$

The positive cone $X^+ = M_nC^+(J)$ consists of those matrices whose entries are all nonnegative in J; it is clearly a closed set. Let $\kappa = (\kappa_{ij}) \in X^+$, and define $S\colon M_nC(J) \to M_nC(J)$ by

$$(Sx)(t) = \kappa \int_\alpha^t x\,ds = \left(\sum_{r=1}^n \kappa_{ir}(t) \int_\alpha^t x_{rj}(s)\,ds\right), \qquad t \in J,$$

for each $x \in M_nC(J)$. Clearly, S is a positive linear operator, and

$$\begin{aligned}|(Sx)_{ij}(t)| &\le \sum_{r=1}^n |\kappa_{ir}(t)|(t-\alpha)\sup_{s\in J}|x_{rj}(s)| \le \\ &\le \sum_{r=1}^n |\kappa_{ir}(t)|(t-\alpha)\|x\| \le \|\kappa\|\,\|x\|(t-\alpha)\end{aligned}$$

for $1 \le i,j \le n$, $t \in J$. By induction,

$$|(S^m x)_{ij}(t)| \le \frac{\|\kappa\|^m \|x\|(t-\alpha)^m}{m!}, \qquad 1 \le i,j \le n, \qquad t \in J,$$

so that

$$\|S^m x\| \le \frac{n\|\kappa\|^m\|x\|(\beta-\alpha)^m}{m!}, \qquad m \ge 1, \qquad x \in X,$$

and hence

$$\|S^m\| \le \frac{n\|\kappa\|^m(\beta-\alpha)^m}{m!}, \qquad m \ge 1.$$

Therefore S satisfies (19.5), and we have proved

COROLLARY 19.1. *Let $a = (a_{ij})$, $\kappa = (\kappa_{ij})$ be constant matrices with all $\kappa_{ij} \ge 0$. If $x = (x_{ij}) \in M_nC(J)$ and*

$$x(t) \le a + \kappa \int_\alpha^t x(s)\,ds, \qquad t \in J,$$

then $x(t) \le e^{\kappa(t-\alpha)}\ \alpha, t \in J$. □

The following two theorems are generalizations of theorem 19.3 to the case of non-linear operators.

THEOREM 19.4. (Losonczi, 1973) *Let X be a Banach space with partial order relation $\le$ such that the positive cone X^+ is closed. Let A and B be operators from X into X such that*

i) *$x, y \in X$, $x \le y$ implies $Ax \le By$;*
ii) *the equations $x = g + Ax$, $y = h + By$ have unique solutions x_g, y_h, for arbitrary $g, h \in X$, and these solutions can be obtained as limits of the corresponding successive approximations beginning with $x_1 = g$, $y_1 = h$.*

Then

$$u - Au \le v - Bv \quad \Rightarrow \quad u \le v. \tag{19.9}$$

Moreover, (19.9) *also holds if $\le$ is replaced by $\ge$.*

PROOF. We set $g = u - Au$, $h = v - Bv$, and define sequences of successive approximations to u, v by

$$u_1 = g, \qquad u_{n+1} = g + Au_n, \qquad v_1 = h, \qquad v_{n+1} = h + Bv_n, \qquad n \geq 1.$$

By hypothesis, $u_1 \leq v_1$. Moreover, if $u_n \leq v_n$, then (by i))

$$u_{n+1} = g + Au_n \leq g + Bv_n \leq h + Bv_n = v_{n+1},$$

so that $u_n \leq v_n$, or $(v_n - u_n) \in X^+$, for all $n \geq 1$. Since X^+ is closed,

$$v - u = \lim_{n\to\infty} (v_n - u_n) \in X^+,$$

so that $u \leq v$. □

REMARK 19.2. Taking $Bx \equiv Ax \equiv a + Sx$, where S is a positive, bounded linear operator on X satisfying (19.5), we see that conditions i) and ii) are satisfied. hence (19.9) gives

$$u \leq a + Su \quad \Rightarrow \quad u \leq x_a \qquad (x_a = a + Sx_a),$$

and theorem 19.4 includes theorem 19.3. □

REMARK 19.3. Condition i) holds if $Ax \leq Bx$ for all $x \in X$ and either A or B is isotone. □

REMARK 19.4. Condition ii) holds if both A and B are contractions (not necessarily linear), or if both A and B are linear operators such that A^n, B^m are contractions, for certain integers $n, m \geq 1$. □

REMARK 19.5. More generally, by a theorem of Boyd and Wong [38], condition ii) is satisfied (for A, say) if there is a continuous function $\omega\colon \mathbf{R}_+ \to \mathbf{R}_+$ such that $\omega(r) < r$ for $r > 0$, and

$$\|Ax - Ay\| \leq \omega(\|x - y\|) \qquad \text{for all } x, y \in X. \tag{19.10}$$

For other conditions ensuring ii), see Geraghty [77] and Kannan [102]. □

REMARK 19.6. If A is isotone and satisfies (19.10) (with $\omega(r) < r$ for $r > 0$), then

$$u \leq q + Au \quad \Rightarrow \quad u \leq x_q \qquad (x_q = q + Ax_q). \tag{19.11}$$

This includes part of theorem 2 of Chandra and Fleishman [48]. □

The following theorem weakens the condition that $\omega(r) < r$ for all $r > 0$.

THEOREM 19.5. (Chandra and Fleishman, 1970) *Let X be a partially ordered Banach space with closed positive cone X^+, and let $A\colon X \to X$ be an isotone, compact (=completely continuous) operator satisfying condition* (19.10), *where $\omega\colon \mathbf{R}_+ \to \mathbf{R}_+$ is a nonnegative, nondecreasing function such that*

$$\omega(r) + \|g\| + \|A\theta\| \leq r \qquad \textit{for } r \geq r_0 > 0. \tag{19.12}$$

Suppose that

$$u \leq q + Au. \tag{19.13}$$

Then the equation $v = q + Av$ has a solution, given by $v = \lim_{n\to\infty} v_n$, where

$$v_1 = u, \qquad v_{n+1} = q + Av_n, \qquad n \geq 1, \tag{19.14}$$

and, moreover, $u \leq v$.

In particular, if $v = q + Av$ has a maximal solution ϕ, then $u \leq \phi$.

PROOF. Using (19.13) and the fact that A is isotone, it is readily proved by induction that $v_n \leq v_{n+1}$ for all $n \geq 1$. Let $R = \max(\|u\|, r_0)$. Then $\|v_1\| \leq R$, and if $\|v_n\| \leq R$, then

$$\begin{aligned}\|v_{n+1}\| \leq &\|q\| + \|Av_n\| \leq \|q\| + \|A\theta\| + \|Av_n - A\theta\| \leq \\ \leq &\|q\| + \|A\theta\| + \omega(\|v_n\|) \leq \|q\| + \|A\theta\| + \omega(R) \leq R,\end{aligned}$$

using (19.12) and the nondecreasing nature of ω. Thus, $\|v_n\| \leq R$ for all $n \geq 1$. Since A is compact, $\{Av_n\}$ has a convergent subsequence $\{Av_{n_k}\}$. Then

$$v = \lim_{k\to\infty} v_{n_k+1}$$

exists, by (19.14). Suppose $\lim v_{n_k} \neq v$. Then there is a subsequence, say $\{v_{n'_k}\}$, of $\{v_{n_k}\}$ with the property that no subsequence of it converges to v. However, as above, $\{Av_{n'_k-1}\}$ has a convergent subsequence, and so has $\{v_{n'_k}\}$:

$$v' = \lim v_{n'_{k_j}}.$$

Now

$$v_{n_k+1} - v_{n'_j} \in X^+ \qquad \text{or} \qquad v_{n'_{k_j}} - v_{n_k+1} \in X^+$$

for sufficiently large k or j, respectively, since $\{v_n\}$ is monotone increasing. Letting $k \to \infty$ or $j \to \infty$, we find

$$v - v_{n'_{k_j}} \in X^+ \qquad \text{or} \qquad v' - v_{n_k+1} \in X^+$$

for arbitrary j, k. Hence also $v - v' \in X^+$, $v' - v \in X^+$, or $v' \leq v$, $v \leq v'$, i.e. $v' = v$. This contradiction proves that $\lim v_n = v$ exists. Since A is continuous, it follows from (19.14) that $v = q + Av$. Moreover, $u = v_1 \leq v_n \leq v$. □

REMARK 19.7. Note that the conditions involving ω can be replaced (with almost no change in the proof above) by:

> There are numbers r_0, R such that $0 < r_0 \leq R$, and a nonnegative nondecreasing function $\omega\colon [0, R] \to \mathbf{R}_+$ such that

$$\begin{aligned}&\|Ax\| \leq \omega(\|x\|) \qquad \text{for } \|x\| \leq R, \\ &\omega(r) + \|q\| \leq r \qquad \text{for } r_0 \leq r \leq R.\end{aligned} \tag{19.15}$$

The conclusion of the theorem is then valid for all $u \in X$ with $\|u\| \leq R$. □

Let $X = S_nC(J)$ be the set of all real symmetric $(n \times n)$-matrices $x = (x_{ij})$ for which every entry $x_{ij} \in C(J)$, and $J = [\alpha, \beta]$ a compact interval. Then X is a Banach space with norm

$$\|x\| = \sup_{t \in J} \max_{1 \le i \le n} \sum_{j=1}^{n} |x_{ij}(t)| \equiv \sup_{t \in J} |x(t)|$$

(as before), and it is partially ordered by the relation $\le$:

$$x, y \in X, \quad x \le y \quad \Leftrightarrow \quad y(t) - x(t) \in S_n^+$$

for all $t \in J$, where S_n^+ is the set of all nonnegative-definite symmetric matrices. The cone $X^+ = S_n^+C(J)$ is the set of all matrices in $S_nC(J)$ that are nonnegative-definite for all $t \in J$. It is clearly a closed set. Let S_n be the set of constant matrices in $S_nC(J)$. Note that if $x \in S_n$, then $\|x\| = |x| = \sup_i \sum_{j=1}^{n} |x_{ij}|$.

THEOREM 19.6. (Beesack, 1975) *Let $a = (a_{ij}) \in S_nC(J)$, and let $G\colon S_n \to S_n$ be an isotone operator satisfying a Lipschitz condition:*

$$|G(x) - G(y)| \le K|x - y| \qquad \text{for } x, y \in S_n. \tag{19.16}$$

If $x \in S_nC(J)$ and

$$x(t) \le a(t) + \int_\alpha^t G(x(s))\, ds, \qquad t \in J, \tag{19.17}$$

then $x(t) \le y(t)$, $t \in J$, where $y(t)$ is the unique solution in J of the corresponding equation.

PROOF. For arbitrary $u, v \in S_nC(J)$,

$$\left|\int_\alpha^t G(u(s))\, ds - \int_\alpha^t G(v(s))\, ds\right| \le K \int_\alpha^t |u(s) - v(s)|\, ds, \qquad t \in J. \tag{19.18}$$

Hence, if u and v are both solutions of the equation

$$y(t) = a(t) + \int_\alpha^t G(y(s))\, ds, \qquad t \in J, \tag{19.19}$$

then $|u(t) - v(t)| \le K\|u - v\|(t - \alpha)$, and repeated substitution in (19.18) leads to

$$|u(t) - v(t)| \le \frac{\|u - v\| K^n (t - \alpha)^n}{n!}, \qquad t \in J \qquad n \ge 1.$$

It follows that (19.19) has at most one solution.

Now we define the successive approximations

$$v_0(t) = x(t), \qquad v_{n+1} = a(t) + \int_\alpha^t G(v_n(s))\, ds, \qquad n \ge 0, \qquad t \in J.$$

Using (19.18), by induction it follows that the sequence $\{v_n(t)\}$ converges uniformly on J, and that $y = \lim v_n$ is a solution of (19.19). Since G is isotone,

$$v_0 = x \le a + \int_\alpha^t G(x)\, ds = v_1 \quad \Rightarrow \quad G(v_0) \le G(v_1),$$

and hence $v_1 = a + \int_\alpha^t G(v_0)\,ds \le a + \int_\alpha^t G(v_1)\,ds = v_2$. By induction,

$$x \le v_0 \le \cdots \le v_n \le v_{n+1}$$

for all n. Since $S_n^+ C(J)$ is closed, $x \le \lim v_n = y$. □

THEOREM 19.7. *Let X be a KB-lineal, and $F \subset X$ a closed and bounded set. Suppose that $P\colon F \to F$ is a compact isotone operator such that there is a $u \in F$ for which*

$$u \le P(u) \qquad (P(u) \le u). \tag{19.20}$$

Then the equation

$$v = P(v) \tag{19.21}$$

has a solution v^* such that

$$u \le v^* \qquad (v^* \le u). \tag{19.22}$$

Moreover, v^* can be obtained as the limit of the successive approximations

$$v_0 = u, \qquad v_{n+1} = P(v_n), \qquad n \ge 0. \tag{19.23}$$

PROOF. Since $v_0 \le v_1$ ($v_1 \le v_0$) and P is isotone, it follows that the sequence v_n is nondecreasing (nonincreasing). Since P is compact, $PF \subset F$, and F is bounded, the sequence $\{v_n\}$ is compact and contained in F. We prove that there is a v^* such that

$$\lim_{n\to\infty} \|v_n - v^*\| = 0,$$

and

$$u \le v^* \qquad (v^* \le u).$$

For the sake of being specific we assume that the sequence v_n is nondecreasing. The compactness of v_n implies the existence of a subsequence v_{n_k} converging to some element $v^* \in F$. In this case, for $z_n = v^* - v_n$ we have $z_m \le z_n$, $m \ge n$. We show that $0 \le z_{n_k}$. Assuming this to be not true, there is an l such that $0 < z_{n_l}^-$. For $m \ge l$ the inequality

$$z_{n_l}^- \le z_{n_m}^-$$

implies that $\lim_{m\to\infty} \|z_{n_m}^-\| > 0$, contradicting

$$\lim_{m\to\infty} \|z_{n_m}^-\| \le \lim_{m\to\infty} \||z_{n_m}|\| = \lim_{m\to\infty} \|z_{n_m}\| = 0.$$

Hence $0 \le z_{n_k}$, which implies that $0 \le z_n$, $n \ge 0$, and

$$\lim_{n\to\infty} \|z_n\| \le \lim_{k\to\infty} \|z_{n_k}\| = 0.$$

The relations $u \le v^*$, $v^* = P(v^*)$ are obvious now. If v_n is nonincreasing, the arguments are similar. □

THEOREM 19.8. *Let X be a KB-lineal, and $F \subset X$ a closed set. Suppose that $P\colon F \to F$ is an isotone contraction, and that there is a $u \in F$ such that $u \le P(u)$ ($P(u) \le u$). Then $u \le v$ ($v \le u$), where v is the unique solution of the equation $v = P(v)$ in F.*

PROOF. Set $v_0 = u$, $v_{n+1} = P(v_n)$. By the fixed point theorem, $\lim_{n\to\infty} v_n = v \in F$, and $v = P(v)$. Moreover, $v_n \le v_{n+1}$ ($v_{n+1} \le v_n$), since P is isotone. The remaining arguments are as in the proof of theorem 19.7, and we omit them. □

20. Applications

20.1. Asymptotic behaviour of solutions. In this part we apply some of the results of §10 to the study of the asymptotic behaviour of solutions of the equation

$$y^{(n)} = f(t, y, y', \dots, y^{(n-1)}). \tag{20.1}$$

THEOREM 20.1. (Pinto, 1990) *Let $f = f(t, y_1, \dots, y_n)$ be a continuous function in $[1, \infty) \times \mathbf{R}^n$ such that*

$$|f(t, y_1, \dots, y_n)| \le \sum_{i=1}^{n} \kappa_i(t) g_i \left(\frac{|y_i|}{t^{n-i}} \right), \tag{20.2}$$

where

i) *g_i, $i = 1, \dots, n$, are nondecreasing continuous functions in $[0, \infty)$ which are positive in $(0, \infty)$, and such that $g_1 \propto g_2 \propto g_2 \propto \cdots \propto g_n$ (see §10);*

ii) *κ_i, $i = 1, \dots, n$, are nonnegative, integrable, continuous functions in $[1, \infty)$.*

Suppose that there is a constant $c > 0$ such that

$$\int_1^\infty \kappa_i(s)\, ds \le \int_{M_{i-1}}^\infty \frac{dz}{g_i(z)}, \qquad i = 1, \dots, n, \tag{20.3}$$

where $M_0 = c$, $M_i = G_i^{-1}[G_i(M_{i-1}) + \int_1^\infty \kappa_i(s)\, ds]$, $i = 1, \dots, n$, and $G_i(u) = \int_{u_0}^u dz/g_i(z)$, $u > 0$ ($u_0 > 0$).

Then all solutions of (20.1) such that $\sum_{i=0}^{n-1} |y^{(i)}(1)| < c$ are defined in $[1, \infty)$ and satisfy

$$y = \sum_{i=0}^{n-1} \delta_i t^i + o(t^{n-1})$$

as $t \to \infty$, where δ_i, $i = 0, \dots, n-1$, are constants. Moreover, the limit

$$\lim_{t\to\infty} \int_1^t f(s, y(s), y'(s), \dots, y^{(n-1)}(s))\, ds = \delta$$

always exists, and $(n-1)!\delta_{n-1} = y^{(n-1)}(1) + \delta$.

PROOF. Let $y(t)$ be a solution of (20.1) satisfying $\sum_{i=0}^{n-1} |y^{(i)}(1)| = a < c$ and defined for $t \in [1, T) \subset [1, \infty)$. Successively integrating (20.1) n times, we obtain

$$y^{(k)}(t) = \sum_{\nu=k}^{n-1} \frac{y^{(\nu)}(1)}{(\nu - k)!} (t-1)^{\nu-k} + \int_1^t \frac{(t-s)^{n-k-1}}{(n-k-1)!} g(s)\, ds,$$

$$y(t) = \sum_{\nu=0}^{n-1} \frac{y^{(\nu)}(1)}{\nu!} (t-1)^{\nu} + \int_1^t \frac{(t-s)^{n-1}}{(n-1)!} g(s)\, ds,$$

where $k = 1, \ldots, n-1$, $t \in [1, T)$, and $g(s) \equiv f(s, y(s), y'(s), \ldots, y^{(n-1)}(s))$. Then

$$|y(t)| \le t^{n-1} \left(\sum_{\nu=0}^{n-1} |y^{(\nu)}(1)| + \int_1^t |g(s)|\, ds \right),$$

and, by (20.2),

$$\frac{|y(t)|}{t^{n-1}} \le a + \sum_{i=1}^{n} \int_1^t \kappa_i(s) g_i \left(\frac{|y^{(i-1)}(s)|}{s^{n-i}} \right).$$

Similarly, for $k = 0, \ldots, n-1$,

$$\frac{|y^{(k)}(t)|}{t^{n-k-1}} \le a + \sum_{i=1}^{n} \int_1^t \kappa_i(s) g_i \left(\frac{|y^{(i-1)}(s)|}{s^{n-i}} \right). \tag{20.4}$$

Hence, if z denotes the righthand side of (20.4), we have

$$z'(t) \le \sum_{i=1}^{n} \kappa_i(t) g_i(z(t)), \qquad z(1) = a,$$

or, after integration over $[1, t]$,

$$z(t) \le a + \sum_{i=1}^{n} \int_1^t \kappa_i(s) g_i(z(s))\, ds.$$

Theorem 10.3 then implies

$$z(t) \le c_n(t),$$

where $c_0(t) \equiv a$, $c_j(t) = G_j^{-1}[G_j(c_{j-1}(t)) + \int_1^t \kappa_j(s)\, ds]$, $j = 1, \ldots, n$. Since κ_i are integrable functions in $[1, \infty)$, (20.3) and remark 5.7 imply that the $c_j(t)$, $j = 1, \ldots, n$, are functions defined in $[1, \infty)$ such that $c_j(t) < M_j$ for $t \ge 1$. Consequently,

$$z(t) < M_n, \qquad t \in [1, T). \tag{20.5}$$

Inequalities (20.4) and (20.5) imply that $y(t)$ can be continued up to ∞, i.e. $T = \infty$ in (20.5). Moreover, $g(t)$ is integrable in $[1, \infty)$. This implies that the limit

$$\lim_{t \to \infty} y^{(n-1)}(t) = y^{(n-1)}(1) + \delta = a_{n-1}$$

exists, and by l'Hospital's rule, $\lim_{t\to\infty} y^{(n-k-1)}(t)/t^k = a_{n-1}/k!$. Now,

$$y^{(n-2)}(t) = a_{n-2} + a_{n-1}t + o(t), \qquad y^{(n-3)}(t) = a_{n-3} + a_{n-2}t + a_{n-1}\frac{t^2}{2!} + o(t^2), \ldots,$$

and if a_i, $i = 0, \ldots, n-1$, are constants,

$$y(t) = a_0 + a_1 t + a_2 \frac{t^2}{2!} + \cdots + a_{n-1}\frac{t^{n-1}}{(n-1)!} + o(t^{n-1}) \qquad \text{as } t \to \infty \qquad \square.$$

20.2. Integro-differential equations. Consider the following initial value problem for an integro-differential equation:

$$x'(t) = A(t)x(t) + \int_{t_0}^{t} B(t,s)x(s)\,ds + P(t,x(t)), \qquad x(t_0) = x_0, \tag{20.6}$$

where $t, t_0 \in \mathbf{R}_+$, $x_0 \in \mathbf{R}^n$, and $A\colon \mathbf{R}_+ \to \mathbf{R}^{n\times n}$, $B\colon \mathbf{R}_+^2 \to \mathbf{R}^{n\times n}$, $P\colon \mathbf{R}_+ \times \mathbf{R}^n \to \mathbf{R}^n$ are continuous functions in their domains.

The solution $x(t)$ of (20.6) satisfies the Volterra-type integral equation

$$x(t) = \Phi(t,t_0)x_0 + \int_{t_0}^{t} \Phi(t,s)P(s,x(s))\,ds, \tag{20.7}$$

where $\Phi(t,s)$ is the unique solution of the initial value problem

$$\frac{\partial \Phi}{\partial t}(t,s) = A(t)\Phi(t,s) + \int_{s}^{t} B(t,\tau)\Phi(\tau,s)\,d\tau, \qquad \Phi(s,s) = I. \tag{20.8}$$

In particular, if $A(t) \equiv A$, $B(t,s) = b(t-s)$, then

$$\Phi(t,s) = R(t-s), \tag{20.9}$$

where $R(t)$ satisfies the initial value problem

$$R'(t) = AR(t) + \int_{0}^{t} b(t-\tau)R(\tau)\,d\tau, \qquad R(0) = I. \tag{20.10}$$

(20.7) enables us to use integral inequalities in the study of various properties of solutions of an initial value problem (20.6).

Consider the nonlinear integro-differential equation

$$x'(t) = f(t,x(t)) + \int_{t_0}^{t} g(t,s,x(s))\,ds, \tag{20.11}$$

and the corresponding perturbed equation

$$y'(t) = f(t,y(t)) + \int_{t_0}^{t} g(t,s,y(s))\,ds + h\left(t, y(t), \int_{t_0}^{t} k(t,s,y(s))\,ds\right), \tag{20.12}$$

where $t, t_0 \in \mathbf{R}_+$, $x, y, f, g, h \in \mathbf{R}^n$.

Suppose that the following condition holds:

H20.1) $f\colon \mathbf{R}_+ \times \mathbf{R}^n \to \mathbf{R}^n$, $f_x\colon \mathbf{R}_+ \times \mathbf{R}^n \to \mathbf{R}^{n\times n}$, $g, k\colon \mathbf{R}_+ \times \mathbf{R}_+ \times \mathbf{R}^n \to \mathbf{R}^n$, $g_x\colon \mathbf{R}_+ \times \mathbf{R}_+ \times \mathbf{R}^n \to \mathbf{R}^{n\times n}$, and $h\colon \mathbf{R}_+ \times \mathbf{R}^n \times \mathbf{R}^n \to \mathbf{R}^n$ are continuous functions in their domains.

Let $x(t) = x(t;t_0,x_0)$ and $y(t) = y(t;t_0,x_0)$ be the solutions of (20.11) and (20.12), respectively, for which $x(t_0) = y(t_0) = x_0$. It is well known [40] that

$$\frac{\partial x}{\partial x_0}(t;t_0,x_0) = \Phi(t,t_0,x_0)$$

exists and satisfies the variational equation

$$z'(t) = f_x(t,x(t;t_0,x_0))z(t) + \int_{t_0}^{t} g_x(t,s,x(s;t_0,x_0))z(s)\,ds, \qquad z(t_0) = I, \tag{20.13}$$

and

$$\frac{\partial x}{\partial t_0}(t; t_0, x_0) + \Phi(t, t_0, x_0) f(t_0, x_0) + \int_{t_0}^{t} \Phi(t, s, x_0) g(s, t_0, x_0)\, ds = 0. \quad (20.14)$$

Now the solutions $x(t)$ and $y(t)$ are related by

$$y(t) = x(t) + \int_{t_0}^{t} \Phi(t, s, y(s)) h\left(s, y(s), \int_{t_0}^{s} k(s, \tau, y(\tau))\, d\tau\right) ds. \quad (20.15)$$

THEOREM 20.2. (Pachpatte, 1977) *Suppose that the condition* H20.1 *holds, and that the inequalities*

$$|\Phi(t, s, y)| \leq M e^{-\alpha(t-s)}, \quad (20.16)$$

$$|\Phi(t, s, y) h(s, y, z)| \leq p(s)(|y| + |z|), \quad (20.17)$$

$$|k(t, s, y)| \leq q(s)|y|, \quad (20.18)$$

hold for $t \geq s \geq 0$, $y, z \in \mathbf{R}^n$, *where* $M \geq 1$, $\alpha > 0$ *are constants,* $p, q \in C(\mathbf{R}_+, \mathbf{R}_+)$, *and*

$$\int^{\infty} p(s)\, ds < \infty, \qquad \int^{\infty} q(s)\, ds < \infty. \quad (20.19)$$

Then for any bounded solution $x(t; t_0, x_0)$ *of* (20.11) *in* $\mathbf{R}_+$, *the corresponding solution* $y(t; t_0, x_0)$ *of* (20.12) *is bounded in* $\mathbf{R}_+$.

PROOF. In view of (20.16)–(20.18), (20.15) implies that

$$|y(t)| \leq M|x_0| + \int_{t_0}^{t} p(s)|y(s)|\, ds + \int_{t_0}^{t} p(s) \left(\int_{t_0}^{s} q(\tau)|y(\tau)|\, d\tau\right) ds.$$

Consequently, by corollary 11.2,

$$|y(t)| \leq M|x_0| \left[1 + \int_{t_0}^{t} p(s) \exp\left(\int_{t_0}^{s} (p(\tau) + q(\tau))\, d\tau\right) ds\right],$$

which implies that $y(t)$ is bounded, because of (20.19). □

20.3. Partial differential equations. We consider some simple applications of the results in §12 to the study of hyperbolic partial differential equations.

EXAMPLE 20.1. Consider the nonlinear hyperbolic partial differential equation

$$u_{xy} = f(x, y, u), \quad (20.20)$$

where f is a continuous function in $\mathbf{R}^3$ satisfying a Lipschitz condition with respect to u, i.e. there is a constant $K > 0$ such that $|f(x, y, u) - f(x, y, v)| \leq K|u - v|$ for all $x, y, u, v \in \mathbf{R}$.

Suppose we have imposed boundary conditions such that (20.20) is equivalent to the Volterra integral equation

$$u(x, y) = g(x, y) + \int_{x_0}^{x} \int_{y_0}^{y} f(s, t, u(s, t))\, ds\, dt,$$

where g is a continuous function in $\mathbf{R}^2$. If u, v are two solutions of this boundary value problem, then

$$u - v = \int_{x_0}^{x} \int_{y_0}^{y} [f(s,t,u) - f(s,t,v)]\, ds\, dt.$$

Consequently, if $(x - x_0)(y - y_0) \geq 0$, then

$$|u - v| \leq \int_{x_0}^{x} \int_{y_0}^{y} K|u - v|\, ds\, dt,$$

and by corollary 12.1, $|u - v| \leq 0$. Thus, $u \equiv v$, i.e. there is at most one solution of this problem. □

EXAMPLE 20.2. Consider the characteristic initial value problem for a linear hyperbolic partial differential equation

$$u_{xy} - b(x,y)u = f(x,y), \qquad b \geq 0,$$

u being prescribed on $x = x_0$ and on $y = y_0$. This problem is equivalent to a Volterra integral equation

$$u = F(x,y) + \int_{x_0}^{x} \int_{y_0}^{y} bu\, ds\, dt,$$

where $F(x,y)$ can be computed from f and the boundary conditions.

Suppose that v, w satisfy

$$v \leq F + \int_{x_0}^{x} \int_{y_0}^{y} bv\, ds\, dt,$$
$$w \geq F + \int_{x_0}^{x} \int_{y_0}^{y} bw\, ds\, dt.$$

Then corollary 12.1 guarantees that $v \leq u \leq w$. Hence we obtain a comparison theorem for the solution u of the characteristic initial value problem. □

EXAMPLE 20.3. Consider the two boundary value problems

$$\begin{gathered} u_{xy} = f(x,y,u), \\ u(x_0,y) = g(y), \qquad u(x,y_0) = h(x), \qquad g(y_0) = h(x_0), \end{gathered} \tag{20.21}$$

and

$$\begin{gathered} U_{xy} = F(x,y,U), \\ U(x_0,y) = G(y), \qquad U(x,y_0) = H(x), \qquad G(y_0) = H(x_0), \end{gathered}$$

where all functions are continuous and f satisfies a Lipschitz condition with constant K. Example 20.1 shows that there is at most one solution $u(x,y)$ of (20.21). The equivalent integral equations are:

$$u(x,y) = g(y) + h(x) - g(y_0) + \int_{x_0}^{x} \int_{y_0}^{y} f(s,t,u(s,t))\, ds\, dt$$

and

$$U(x,y) = G(y) + H(x) - G(y_0) + \int_{x_0}^{x}\int_{y_0}^{y} F(s,t,U(s,t))\,ds\,dt.$$

Then

$$\begin{aligned} u - U =& (g - G) + (h - H) + [G(y_0) - g(y_0)] + \\ &+ \int_{x_0}^{x}\int_{y_0}^{y} [f(s,t,u) - F(s,t,U)]\,ds\,dt. \end{aligned}$$

Adding and subtracting $f(s,t,U)$ in the integrand leads to

$$\begin{aligned} |u - U| \leq& \\ &\leq |g - G| + |h - H| + |G(y_0) - g(y_0)| + \\ &+ \int_{x_0}^{x}\int_{y_0}^{y} |f(s,t,u) - f(s,t,U)|\,ds\,dt + \int_{x_0}^{x}\int_{y_0}^{y} |f(s,t,U) - F(s,t,U)|\,ds\,dt, \end{aligned}$$

if $(x - x_0)(y - y_0) \geq 0$.

If $|g - G| \leq \epsilon$, $|h - H| \leq \epsilon$, $|f(s,t,U) - F(s,t,U)| \leq \epsilon$, then

$$|u - U| \leq 3\epsilon + \epsilon(x - x_0)(y - y_0) + \int_{x_0}^{x}\int_{y_0}^{y} K|u - U|\,ds\,dt.$$

By theorem 12.12,

$$\begin{aligned} |u - U| \leq \epsilon\Bigg\{& 3 + (x - x_0)(y - y_0) + \\ &+ \int_{x_0}^{x}\int_{y_0}^{y} [3 + (s - x_0)(t - y_0)]Kv(s,t;x,y)\,ds\,dt\Bigg\}, \end{aligned}$$

if only $(x - x_0)(y - y_0) \geq 0$. On a compact set, the quantity between brackets is bounded by a constant M. Hence $|u - U| \leq M\epsilon$ on this set. Consequently, the solution of such a boundary value problem (20.21) depends continuously on f and the boundary data. If $\epsilon \to 0$, then $|u - U| \to 0$ on the compact set. □

20.4. Functional differential equations. Let $r_0 > 0$ be a given real number, and let, for $t \geq t_0 \geq 0$, $x \in C([t_0 - r_0, \infty), \mathbf{R}^n)$, the function $x_t \in C = C([-r_0, 0], \mathbf{R}^n)$ be defined by $x_t(s) = x(t + s)$, $-r_0 \leq s \leq 0$. For a fixed $r \in [0, r_0]$, $t \geq 0$, and $x \in C$ we set $\|x_t\|_r = \max_{-r \leq s \leq 0} |x(t + s)|$.

Consider the functional differential equation

$$\dot{x}(t) = F(t, x_t), \tag{20.22}$$

and suppose that the following conditions hold:

H20.2) $F\colon \mathbf{R}_+ \times C \to \mathbf{R}^n$ is a continuous function,

$$F(t,0) = 0, \qquad t \geq 0,$$

and for any initial point $t = t_0 \geq 0$ and initial function $\phi \in C$, any solution $x(t_0, \phi)$ of (20.22) corresponding to the initial condition $x_{t_0}(t_0, \phi) = \phi$ can be continued up to $+\infty$;

H20.3) $\gamma\colon \mathbf{R}_+ \to \mathbf{R}$ is a strictly increasing continuous function such that

$$0 \le r(t) = t - \gamma(t) \le r_0, \qquad t \ge 0,$$

and there are continuous functions $k\colon \mathbf{R}_+ \times [-r_0, 0] \to \mathbf{R}_+$, $a\colon \mathbf{R}_+ \to [1,\infty)$, and $p\colon \mathbf{R}_+ \to \mathbf{R}_+$ such that

$$\int_s^{\gamma^{-1}(s)} k(t, s-t)a(t)\,dt \le a(s) - 1, \qquad s \ge 0, \tag{20.23}$$

and

$$|F(t,\phi)| \le \int_{-r(t)}^0 k(t,s)|\dot\phi(s)|\,ds + p(t)\|\phi\|_{r(t)}, \qquad t \ge 0,$$

whenever $\phi \in C$ is absolutely continuous in $[\gamma(t), t]$.

THEOREM 20.3. (Arino and Györi, 1984) *Assume that the conditions* H20.2–3 *hold. Then*

i) *for any solution $x(t_0,\phi)$ of* (20.22) *passing through $(t_0,\phi) \in \mathbf{R}_+ \times C$ we have*

$$|x(t_0,\phi)(t)| \le M(t_0, x(t_0,\phi))e^{\int_{\gamma^{-1}(t_0)}^t a(\tau)p(\tau)\,d\tau}, \tag{20.24}$$

where $t \ge \gamma^{-1}(t_0)$ and

$$M(t_0, x(t_0,\phi)) = \max_{t_0 \le t \le \gamma^{-1}(t_0)} |x(t_0,\phi)(t)| + \int_{t_0}^{\gamma^{-1}(t_0)} [a(s)-1]|F(s, x_s(t_0,\phi))|\,ds;$$

ii) *if*

$$\int_0^\infty a(t)p(t)\,dt < \infty, \tag{20.25}$$

then

$$\int_{t_0}^\infty |\dot x(t_0,\phi)(t)|\,dt < \infty, \tag{20.26}$$

i.e. the limit $x(t_0,\phi)(\infty) = \lim_{t\to\infty} x(t_0,\phi)(t) \in \mathbf{R}^n$ exists.

PROOF. We set $\tau_0 = \gamma^{-1}(t_0)$ and $x(t) = x(t_0,\phi)(t)$, $\tau_0 - r_0 \le t < \infty$. Then $x(t)$ is continuously differentiable in $[\gamma(\tau_0), \infty)$, since $\gamma(\tau_0) = t_0$. Thus,

$$|F(t,x_t)| \le \int_{-r(t)}^0 k(t,s)|\dot x_t(s)|\,ds + p(t)\|x_t\|_{r(t)} =$$

$$= \int_{\gamma(t)}^t k(t,s-t)|\dot x(s)|\,ds + p(t)\|x_t\|_{r(t)}, \qquad t \ge \tau_0.$$

Using (20.22) we obtain

$$|\dot x(t)| \le \int_{\gamma(t)}^t k(t,s-t)|\dot x(s)|\,ds + p(t)\|x_t\|_{r(t)}, \qquad t \ge \tau_0.$$

By condition (20.23), theorem 14.10 implies

$$|x(t)| \le \left[\|x_{\tau_0}\|_{r(\tau_0)} + \int_{\gamma(\tau_0)}^{\tau_0} [a(s)-1]|\dot x(s)|\,ds\right] e^{\int_{\tau_0}^t a(\tau)p(\tau)\,d\tau}, \qquad t \ge \tau_0,$$

and hence, if (20.24) holds, then $\int_{\tau_0}^{\infty} |\dot{x}(t)|\, dt < \infty$.

From the definition of $x = x(t, \phi)$ we see that the inequalities (20.24) and (20.26) hold, since

$$\|x_{\tau_0}\|_{r(\tau_0)} + \int_{\gamma(\tau_0)}^{\tau_0} [a(s) - 1]|\dot{x}(s)|\, ds \le$$

$$\le \max_{t_0 \le t \le \gamma^{-1}(t_0)} |x(t_0, \phi)(t)| + \int_{t_0}^{\gamma^{-1}(t_0)} [a(s) - 1]|F(s, x_s(t_0, \phi))|\, ds. \quad \square$$

DEFINITION 20.1. The solution $x \equiv 0$ of (20.22) is said to be

i) *stable* if for every $\epsilon > 0$ and $t_0 \ge 0$ there is a positive number $\delta = \delta(t_0, \epsilon)$ such that $|x(t_0, \phi)(t)| < \epsilon$ for all $\phi \in C$, $\|\phi\| < \delta(t_0, \epsilon)$ and $t \ge t_0$;

ii) *uniformly stable* in $[0, \infty)$ if the number δ in i) can be chosen independently of $t_0 \in [0, \infty)$.

From theorem 20.3 we can derive the following stability result.

THEOREM 20.4. (Arino and Győri, 1984) *Suppose that the conditions* H20.2 *and* H20.3 *hold, as well as* (20.25). *Suppose, moreover, that there is a locally Lebesgue integrable function* $a\colon \mathbf{R}_+ \to \mathbf{R}_+$ *such that*

$$|F(t, \phi)| \le a(t)\|\phi\|_{r(t)}$$

for all $(t, \phi) \in \mathbf{R}_+ \times C$. *Then*

i) *the solution* $x \equiv 0$ *of* (20.22) *is stable in* $[0, \infty)$;

ii) *if*

$$A = \sup_{t_0 \ge 0} \int_{t_0}^{\gamma^{-1}(t_0)} a(s)\, ds < \infty, \qquad B = \sup_{t_0 \ge 0} \int_{t_0}^{\gamma^{-1}(t_0)} [a(s) - 1]a(s)\, ds < \infty,$$

then the solution $x \equiv 0$ *of* (20.22) *is uniformly stable in* $[0, \infty)$. $\square$

COROLLARY 20.1. (Arino and Győri, 1984) *Let* $\mu > 0$ *and* $r_0 > 0$ *be constants and* $p\colon \mathbf{R}_+ \to \mathbf{R}_+$ *a continuous function. Then the solution* $x \equiv 0$ *of the scalar equation*

$$\dot{x}(t) = \mu[x(t) - x(t - r_0)] + p(t)x(t) \tag{20.27}$$

is stable if and only if

$$\mu r_0 < 1 \qquad \text{and} \qquad \int_0^{\infty} p(t)\, dt < \infty. \quad \square \tag{20.28}$$

20.5. Difference equations. Consider the linear difference equation

$$x(t + 1) = A(t)x(t), \tag{20.29}$$

and the corresponding perturbed equation

$$y(t + 1) = A(t)y(t) + g(t, y(t)), \tag{20.30}$$

where $t \in \mathbf{N}_0$, $x \in \mathbf{R}^n$, $A(t) \in \mathbf{R}^{n \times n}$, $g\colon \mathbf{N}_0 \times \mathbf{R}^n \to \mathbf{R}^n$.

Direct verification shows that the solution $x(t)$ of (20.29) has the form

$$x(t) = \Phi(t, t_0)x(t_0), \qquad t, t_0 \in \mathbf{N}_0, \qquad t \ge t_0,$$

where $\Phi(t_0, t_0) = I$ and

$$\Phi(t, t_0) = A(t-1)A(t-2)\dots A(t_0) \qquad \text{for } t > t_0.$$

$\Phi(t, t_0)$ is called the *Cauchy matrix* of (20.29).

Since $\Phi(t+1, t_0) = A(t)\Phi(t, t_0)$ for $t \geq t_0$, we find that the solution $y(t)$ of (20.30) satisfies the equation

$$y(t) = \Phi(t, t_0)y(t_0) + \sum_{s=t_0}^{t-1} \Phi(t, s+1)g(s, y(s)), \qquad t \geq t_0. \tag{20.31}$$

Consider now the nonlinear difference equation

$$x(t+1) = f(t, x(t)), \tag{20.32}$$

and corresponding perturbed equation

$$y(t+1) = f(t, y(t)) + g(t, y(t)), \tag{20.33}$$

where $f, g\colon \mathbf{N}_0 \times \mathbf{R}^n \to \mathbf{R}^n$.

Suppose that f has a continuous partial derivative $f_x\colon \mathbf{N}_0 \times \mathbf{R}^n \times \mathbf{R}^{n\times n}$, while the solution $x(t) = x(t; t_0, x_0)$ with $x(t_0) = x_0$ is defined for $t \geq t_0$. Then the derivative

$$\Phi(t, t_0, x_0) = \frac{\partial x}{\partial x_0}(t; t_0, x_0)$$

exists and satisfies for $t \geq t_0$ the variational equation

$$z(t+1) = f_x(t, x(t; t_0, x_0))z(t), \qquad z(t_0) = I. \tag{20.34}$$

Moreover,

$$x(t; t_0, y_0) = x(t; t_0, x_0) + \int_0^1 \Phi(t, t_0, x_0 + \tau(y_0 - x_0))\, d\tau \cdot (y_0 - x_0). \tag{20.35}$$

If $x(t)$ and $y(t)$ are the solutions of (20.32) and (20.33), respectively, and $x(t_0) = y(t_0)$, then the following analogue of Alekseev's formula holds:

$$y(t) = x(t) + \sum_{s=t_0}^{t-1} \int_0^1 \Phi(t, s+1, f(s, y(s)) + \tau g(s, y(s)))\, d\tau \cdot g(s, y(s)). \tag{20.36}$$

Indeed, let $p(s) = x(t; s, y(s))$. Using (20.35),

$$\begin{aligned}\Delta p(s) &= p(s+1) - p(s) = x(t; s+1, y(s+1)) - x(t; s, y(s)) = \\ &= x(t; s+1, f(s, y(s)) + g(s, y(s))) - x(t; s+1, f(s, y(s))) = \\ &= \int_0^1 \Phi(t, s+1, f(s, y(s)) + \tau g(s, y(s)))\, d\tau \cdot g(s, y(s)).\end{aligned}$$

Summation from t_0 to $t-1$ and taking into account that $p(t) = x(t; t, y(t)) = y(t)$ and $p(t_0) = x(t; t_0, y(t_0)) = x(t)$, we arrive at (20.36).

Consider now equation (20.30) under the following conditions:

$$\begin{aligned} &|\Phi(t,s)| \le Kq^{t-s}, \qquad t,s \in \mathbf{N}_0, \qquad t \ge s, \\ &|f(t,x)| \le \beta(t)|x|^m, \qquad t \in \mathbf{N}_0, \qquad x \in \mathbf{R}^n, \end{aligned} \tag{20.37}$$

where $K \ge 1$, $m \ge 1$, and $q > 0$ are constants and $\beta\colon \mathbf{N}_0 \to \mathbf{R}_+$.

Now (20.31) implies

$$|y(t)| \le Kq^{t-t_0}|y(t_0)| + \sum_{s=t_0}^{t-1} Kq^{t-s-1}\beta(s)|y(s)|^m. \tag{20.38}$$

We set $u(t) = |y(t)|q^{-t}$. Then (20.38) implies

$$u(t) \le Kq^{-t_0}|y(t_0)| + \sum_{s=t_0}^{t-1} \frac{K}{q} q^{ms-s}\beta(s)u^m(s).$$

By applying theorem 15.6 we arrive at the following estimates:

i) for $m = 1$,

$$|y(t)| \le K|y(t_0)| \prod_{s=t_0}^{t-1} (q + K\beta(s)), \qquad t \ge t_0; \tag{20.39}$$

ii) for $m > 1$,

$$|y(t)| \le K|y_0|q^{t-t_0} \left\{ 1 - (m-1)(K|y_0|q^{-t_0})^{m-1} \sum_{s=t_0}^{t-1} Kq^{(m-1)s-1}\beta(s) \right\}^{1/(1-m)} \tag{20.40}$$

for $t \ge t_0$, if only $(m-1)(K|y_0|q^{-t_0})^{m-1} \sum_{s=t_0}^{t-1} Kq^{(m-1)s-1}\beta(s) < 1$.

In particular, suppose that (20.30) has the form

$$y(t+1) = A(t)y(t) + B(t)y(t), \tag{20.41}$$

where $B(t) \in \mathbf{R}^{n\times n}$ is such that there is a function $\beta\colon \mathbf{N}_0 \to \mathbf{R}_+$ such that

$$|B(t)| \le \beta(t), \qquad t \in \mathbf{N}_0. \tag{20.42}$$

Let $\Psi(t,t_0)$ be the Cauchy matrix of (20.41). Since every solution $y(t)$ of (20.41) satisfies (20.39), we have

$$|\Psi(t,t_0)| \le K \prod_{s=t_0}^{t-1} (q + K\beta(s)), \qquad t \ge t_0. \tag{20.43}$$

This implies the following theorem.

THEOREM 20.5. *Suppose* (20.37) *and* (20.42) *hold. Then*

i) *the condition* $|\beta(s)| \le \delta$, $s \in \mathbf{N}_0$, *implies*

$$|\Psi(t,t_0)| \le K(q+K\delta)^{t-t_0}, \qquad t,t_0 \in \mathbf{N}_0, \qquad t \ge t_0; \tag{20.44}$$

ii) *the condition* $\sum_{s=0}^{\infty} \beta(s) \le M$ *implies*

$$|\Psi(t,t_0)| \le Ke^{KMq^{-1}} \cdot q^{t-t_0}, \qquad t, t_0 \in \mathbf{N}_0, \qquad t \ge t_0. \tag{20.45}$$

PROOF. (20.44) follows directly from (20.43) and (20.45), in view of the inequality

$$\prod_{s=0}^{t-1} \left(1 + \frac{K}{q}\beta(s)\right) \le \exp\left(\sum_{s=t_0}^{t-1} \frac{K}{q}\beta(s)\right). \qquad \square$$

COROLLARY 20.2. i) *If* (20.29) *is asymptotically stable, i.e.* $q < 1$ *in* (20.37), *then the condition* $|B(s)| \le \delta < (1-q)/K$ *implies that* (20.41) *is asymptotically stable.*

ii) *If* (20.29) *is uniformly stable, i.e.* $q = 1$ *in* (20.37), *then the condition* $\sum_{s=0}^{\infty} |B(s)| < \infty$ *implies that* (20.41) *is uniformly stable.* $\square$

EXAMPLE 20.4. Consider the discrete Volterra-type equation

$$x_1(t) = c_1 + \sum_{s=0}^{t-1} [F_1(t,s,x_1(s),x_2(s)) + H(t,s,x_1(s))], \tag{20.46}$$

$$x_2(t) = c_2 + \sum_{s=0}^{t-1} F_2(t,s,x_1(s)), \tag{20.47}$$

where F_1, F_2, and H are functions satisfying

$$|F_1(t,s,x_1,x_2)| \le a(s)|x_1| + b(s)|x_2|, \tag{20.48}$$

$$|F_2(t,s,x_1)| \le c(s)|x_1|, \tag{20.49}$$

$$|H(t,s,x_1)| \le h(s)W(|x_1|), \tag{20.50}$$

for $t \in \mathbf{N}_0$ and $c_1 \ge 0$, $c_2 \ge 0$, and $W \in \mathcal{F}$.

(20.47) and (20.49) imply

$$|x_2(t)| \le c_2 + \sum_{s=0}^{t-1} c(s)|x_1(s)|, \tag{20.51}$$

and (20.46), (20.48), (20.50), and (20.51) imply

$$|x_1(t)| \le c_1 + \sum_{s=0}^{t-1} a(s)|x_1(s)| + \sum_{s=0}^{t-1} b(s)\left[c_2 + \sum_{\tau=0}^{s-1} c(\tau)|x_1(\tau)|\right] + \sum_{s=0}^{t-1} h(s)W(|x_1(s)|).$$

This inequality is of the same type as that in theorem 15.15. Thus, it is possible to find estimates for $|x_1(t)|$ and $|x_2(t)|$ in terms of the given functions. $\square$

EXAMPLE 20.5. Consider the difference equation

$$\Delta x(n) = H\left(n, x(n), \sum_{s=0}^{n-1} K(n,s,\Delta x(s))\right), \qquad x(0) = x_0, \tag{20.52}$$

where $K\colon \mathbf{N}_0 \times \mathbf{N}_0 \times \mathbf{R} \to \mathbf{R}$, $H\colon \mathbf{N}_0 \times \mathbf{R} \times \mathbf{R} \to \mathbf{R}$ are functions satisfying the estimates

$$|H(n,x,y)| \le p(n)(|x|+|y|), \tag{20.53}$$
$$|K(n,s,y)| \le q(s)|y|, \tag{20.54}$$

for $n \in \mathbf{N}_0$, $s \in \mathbf{N}_0$, $s \le n$, $x \in \mathbf{R}$, $y \in \mathbf{R}$, and $p(n) \ge 0$, $q(n) \ge 0$ for $n \in \mathbf{N}_0$.

Then it follows from (20.52)–(20.54) that

$$|\Delta x(n)| \le p(n)\left[|x(n)| + \sum_{s=0}^{n-1} q(s)|\Delta x(s)|\right], \qquad n \in \mathbf{N}_0. \tag{20.55}$$

Let $y(n) = |x(0)| + \sum_{s=0}^{n-1} |\Delta x(s)|$. Then

$$\Delta y(n) = |\Delta x(n)|, \qquad |x(n)| \le y_n, \tag{20.56}$$

and (20.55) implies

$$\Delta y(n) \le p(n)\left[y(n) + \sum_{s=0}^{n-1} q(s)\Delta y(s)\right]. \tag{20.57}$$

We apply theorem 15.16 to this inequality, and obtain, in view of (20.56),

$$|x(n)| \le |x_0|\left[1 + \sum_{s=0}^{n-1} p(s) \prod_{\tau=0}^{s-1} (1 + p(\tau) + p(\tau)q(\tau))\right], \qquad n \in \mathbf{N}_0. \qquad \square$$

EXAMPLE 20.6. Consider the $(k+1)$-st order difference equation

$$\Delta^{k+1} y(t) = f(t, y(t), \Delta y(t), \ldots, \Delta^k y(t)), \tag{20.58}$$

where $f\colon \mathbf{N}_0 \times \mathbf{R}^{k+1} \to \mathbf{R}$ is a function satisfying

$$|f(t, u_0, u_1, \ldots, u_k)| \le \sum_{j=0}^{k} h_j(t)|u_j|. \tag{20.59}$$

Every solution of (20.58) also satisfies

$$\Delta^k y(t) = \Delta^k y(0) + \sum_{s=0}^{t-1} f(s, y(s), \Delta y(s), \ldots, \Delta^k y(s)).$$

From (20.59) we obtain

$$|\Delta^k y(t)| \le |\Delta^k y(0)| + \sum_{j=0}^{k} \sum_{s=0}^{t-1} h_j(s)|\Delta^j y(s)|.$$

Hence theorem 15.18 implies

$$|\Delta^k y(t)| \le |\Delta^k y(0)| + \sum_{s=0}^{t-1} \phi_1^*(s) \prod_{\tau=s+1}^{t-1} [1 + \phi_2^*(\tau)],$$

where $\phi_1^*(t)$ and $\phi_2^*(t)$ are the same as $\phi_1(t)$ and $\phi_2(t)$ in theorem 15.18 with $p(t) = |\Delta^k y(0)|$ and $q(t) = 1$. $\square$

20.6. Impulse equations. Here we consider the problem of stability of the zero solution of the impulse differential equation

$$\begin{aligned} x' &= A(t)x + g(t,x), \qquad t \neq \tau_k, \\ \Delta x &= A_k x + g_k(x), \qquad t = \tau_k, \end{aligned} \tag{20.60}$$

where $g\colon \mathbf{R}_+ \times \Omega \to \mathbf{R}^n$, $g_k\colon \Omega \to \mathbf{R}^n$, $B_k, A(t) \in \mathbf{R}^{n\times n}$, $\Omega = \{x \in \mathbf{R}^n\colon |x| < H \le \infty\}$, $0 = \tau_0 < \tau_1 < \tau_2 < \dots$, $\lim_{k\to\infty} \tau_k = \infty$.

Next to (20.60) we consider the equations

$$\begin{aligned} x' &= A(t)x, \qquad t \neq \tau_k, \\ \Delta x &= A_k x, \qquad t = \tau_k, \end{aligned} \tag{20.61}$$

and

$$x' = A(t)x, \qquad \tau_{k-1} < t \le \tau_k, \qquad k = 1, 2, \dots. \tag{20.62}$$

Let $U_k(t,s)$ be the Cauchy matrix of equation (20.62).

Suppose the following conditions hold:

H20.4) the matrix-valued function $A(t) \in PC(\mathbf{R}_+, \mathbf{R}^{n\times n})$;

H20.5) $|U_k(t,s)| \le \phi(t)\psi(s)$ for $\tau_{k-1} < s \le t \le \tau_k$, $k = 1, 2, \dots$, where $\phi, \psi \in PC(\mathbf{R}_+, \mathbf{R}_+)$ and $\phi(t) > 0$, $\psi(t) > 0$, $\phi(\tau_k^+) > 0$, $\psi(\tau_k^+) > 0$ for $t \in \mathbf{R}_+$, $\tau_k \in \mathbf{R}_+$;

H20.6) g is a continuous function which is locally Lipschitz continuous with respect to x in the sets $(\tau_{k-1}, \tau_k] \times \Omega$, $k = 1, 2, \dots$, and for any $k = 1, 2, \dots$ and $x \in \Omega$ the limit of $g(t,y)$ exists as $(t,y) \to (\tau_k, x)$, $t > \tau_k$;

H20.7) for $t \in \mathbf{R}_+$ and $x \in \Omega$,

$$|g(t,x)| \le a(t)|x|^m,$$

where $m \ge 1$ is a constant and $a \in PC(\mathbf{R}_+, \mathbf{R}_+)$;

H20.8) $|I + A_k| \le \gamma_k$, $k = 1, 2, \dots$, where $\gamma_k \ge 0$ are constants;

H20.9) $g_k(x)$ are continuous functions in Ω such that

$$|g_k(x)| \le \rho_k |x| \qquad \text{for } x \in \Omega,\ k = 1, 2, \dots,$$

where $\rho_k \ge 0$ are constants.

Let $W(t,s)$ be the Cauchy matrix of (20.61), i.e.

$$W(t,s) = \begin{cases} U_k(t,s), \\ \qquad t, s \in (\tau_{k-1}, \tau_k], \\ U_{k+1}(t, \tau_k^+)(I + A_k)U_k(\tau_k, s), \\ \qquad \tau_{k-1} < s \le \tau_k < t \le \tau_{k+1}, \\ U_{k+1}(t, \tau_k^+) \prod_{j=k}^{i+1} (I + A_j) U_j(\tau_j, \tau_{j-1}^+)(I + A_i)U_i(\tau_i, s), \\ \qquad \tau_{i-1} < s \le \tau_i < \tau_k < t \le \tau_{k+1}. \end{cases} \tag{20.63}$$

Conditions H20.4–5, H20.8, and (20.63) imply that

$$|W(t,s)| \leq \phi(t)\psi(s) \prod_{s\leq\tau_k<t} \gamma_k\phi(\tau_k)\psi(\tau_k^+). \tag{20.64}$$

Let $t_0 \in \mathbf{R}_+$ and $x_0 \in \Omega$, and let $x(t) = x(t;t_0,x_0)$ be the solution of (20.60) for which $x(t_0^+) = x_0$. For this solution we have

$$\begin{aligned} x(t) =& W(t,t_0^+)x_0 + \int_{t_0}^t W(t,s)g(s,x(s))\,ds + \\ &+ \sum_{t_0<\tau_k<t} W(t,\tau_k^+)g_k(x(\tau_k)). \end{aligned} \tag{20.65}$$

Conditions H20.7–9 and (20.64), (20.65) imply

$$\begin{aligned} |x(t)| \leq& |x_0|\phi(t)\psi(t_0^+) \prod_{t_0<\tau_k<t} \gamma_k\phi(\tau_k)\psi(\tau_k^+) + \\ &+ \int_{t_0}^t \phi(t)\psi(s) \prod_{s\leq\tau_k<t} \gamma_k\phi(\tau_k)\psi(\tau_k^+)a(s)|x(s)|^m\,ds + \\ &+ \sum_{t_0<\tau_k<t} \phi(t)\psi(\tau_k^+) \prod_{\tau_k<\tau_i<t} \gamma_i\phi(\tau_i)\psi(\tau_i^+)\rho_k|x(\tau_k)|, \end{aligned}$$

which, in turn, implies

$$\begin{aligned} \frac{|x(t)|}{\phi(t)} \leq& \psi(t_0^+)|x_0| \prod_{t_0<\tau_k<t} \gamma_k\phi(\tau_k)\psi(\tau_k^+) + \\ &+ \int_{t_0}^t \prod_{s\leq\tau_k<t} \gamma_k\phi(\tau_k)\psi(\tau_k^+)\phi^m(s)\psi(s)a(s)\left(\frac{|x(s)|}{\phi(s)}\right)^m ds + \\ &+ \sum_{t_0<\tau_k<t} \prod_{\tau_k<\tau_i<t} \gamma_i\phi(\tau_i)\psi(\tau_i^+)\rho_k\phi(\tau_k)\psi(\tau_k^+)\frac{|x(\tau_k)|}{\phi(\tau_k)}. \end{aligned} \tag{20.66}$$

To this inequality we apply corollary 16.3 (if $m > 1$) or theorem 16.5, and obtain

I) if $m > 1$, then

$$|x(t)| \leq \phi(t)\psi(\tau_k^+)|x_0| \prod_{t_0<\tau_k<t} r_k[1-(m-1)\psi^{m-1}(t_0^+)|x_0|^{m-1}D(t_0,t)]^{1/(1-m)} \tag{20.67}$$

for all $t > t_0$ such that

$$(m-1)\psi^{m-1}(t_0^+)|x_0|^{m-1}D(t_0,t) < 1,$$

where $r_k = (\gamma_k + \rho_k)\phi(\tau_k)\psi(\tau_k^+)$,

$$D(t_0,t) = \int_{t_0}^t \left(\prod_{t_0<\tau_k\leq s} r_k\right)^{m-1} \phi^m(s)\psi(s)a(s)\,ds; \tag{20.68}$$

II) if $m = 1$, then for $t > t_0$,

$$|x(t)| \leq \phi(t)\psi(t_0^+)|x_0| \prod_{t_0<\tau_k<t} r_k \exp\left(\int_{t_0}^{t} \phi(s)\psi(s)a(s)\,ds\right). \quad (20.69)$$

Let $m > 1$, i.e. $|g(t,x)| \leq a(t)|x|^m$ for $(t,x) \in \mathbf{R}_+ \times \Omega$.
We introduce the following conditions:

H20.10) for any $t_0 \geq 0$ there is an $N > 0$ such that

$$\phi(t) \prod_{t_0<\tau_k<t} r_k < N \qquad \text{for } t > t_0;$$

H20.11) $D(0,\infty) = \int_0^\infty (\prod_{0<\tau_k\leq s} r_k)^{m-1}\phi^m(s)\psi(s)a(s)\,ds < \infty$;

H20.12) there is an $N > 0$ such that

$$\phi(t)\psi(t_0^+) \prod_{t_0<\tau_k<t} r_k < N \qquad \text{for } 0 \leq t_0 < t < \infty,$$

$$\psi^{m-1}(t_0^+)D(t_0,\infty) < N \qquad \text{for } t_0 \geq 0;$$

H20.13) for any $\epsilon > 0$ and $t_0 \geq 0$ there is a $\sigma > 0$ such that

$$\phi(t) \prod_{t_0<\tau_k t} r_k < \epsilon \qquad \text{for } t \geq t_0 + \sigma;$$

H20.14) for any $\epsilon > 0$ there is a $\sigma > 0$ such that for any $t_0 \geq 0$ and $t \geq t_0 + \sigma$,

$$\phi(t)\psi(t_0^+) \prod_{t_0<\tau_k<t} r_k < \epsilon.$$

The following theorem holds.

THEOREM 20.6. *Suppose the conditions* H20.4–9 *hold with* $m > 1$. *Then the solution* $x \equiv 0$ *of* (20.6) *is:*

i) *stable if conditions* H20.10–11 *hold;*
ii) *uniformly stable if condition* H20.12 *holds;*
iii) *asymptotically stable if conditions* H20.11 *and* H20.13 *hold;*
iv) *uniformly asymptotically stable if conditions* H20.12 *and* H20.14 *hold.*

PROOF. For $\epsilon > 0$ and $t_0 \geq 0$ we set

$$\delta(\epsilon,t_0) = \min\{[2(m-1)\psi^{(m-1)}(t_0^+)D(t_0,\infty)]^{1/(1-m)}, \epsilon N^{-1}2^{1/(1-m)}\} > 0. \quad (20.70)$$

i) If $|x_0| < \delta$, then (20.67), (20.68), and (20.70) imply that $|x(t;t_0,x_0)| < \epsilon$ for $t > t_0$, i.e. the solution $x \equiv 0$ of (20.60) is stable.
ii) Condition H20.12 implies that we can choose $\delta(\epsilon) > 0$ independently of t_0. Hence the solution $x \equiv 0$ of (20.60) is uniformly stable.
iii) Condition H20.13 implies H20.10, and theorem 20.6i) implies that the solution $x \equiv 0$ of (20.60) is stable. Now (20.67), H20.10, and H20.13 imply that if $|x_0| < \delta(\epsilon,t_0)$, then $x(t;t_0,x_0) \to 0$ as $t \to \infty$, i.e. the solution $x \equiv 0$ is attractive. Hence the solution $x \equiv 0$ of (20.60) is asymptotically stable.

iv) By theorem 20.6ii), the solution $x \equiv 0$ of (20.60) is uniformly stable, and (20.67), H20.12, and H20.14 imply that it is uniformly attractive. Hence the solution $x \equiv 0$ of (20.60) is uniformly asymptotically stable. □

Let $m = 1$, i.e. $|g(t,x)| \le a(t)|x|$ for $(t,x) \in \mathbf{R}_+ \times \Omega$.
We introduce the function

$$M(t_0,t) = \phi(t)\psi(t_0^+) \prod_{t_0<\tau_k<t} (\gamma_k + \rho_k)\phi(\tau_k)\psi(\tau_k^+) \exp\left(\int_{t_0}^t \phi(s)\psi(s)a(s)\,ds\right),$$

and the following conditions:

H20.15) for any $t_0 \ge 0$ there is an $N > 0$ such that

$$M(t_0,t) < N \qquad \text{for } t > t_0;$$

H20.16) there is an $N > 0$ such that for all $t_0 \ge 0$,

$$M(t_0,t) < N \qquad \text{for } t > t_0;$$

H20.17) for any $\epsilon > 0$ and $t_0 \ge 0$ there is a $\sigma > 0$ such that

$$M(t_0,t) < \epsilon \qquad \text{for } t \ge t_0 + \sigma;$$

H20.18) for any $\epsilon > 0$ there is a $\sigma > 0$ such that for all $t_0 \ge 0$ and any $t \ge t_0 + \sigma$,

$$M(t_0,t) < \epsilon.$$

THEOREM 20.7. *Suppose the conditions* H20.4–9 *hold with* $m = 1$. *Then the solution* $x \equiv 0$ *of* (20.6) *is:*

i) *stable if condition* H20.15 *holds;*
ii) *uniformly stable if condition* H20.16 *holds;*
iii) *asymptotically stable if condition* H20.17 *holds;*
iv) *uniformly asymptotically stable if conditions* H20.16 *and* H20.18 *hold.*

PROOF. The proof of this theorem is similar to that of theorem 20.6, and is based on the estimate

$$|x(t;t_0,x_0)| \le |x_0| M(t_0,t),$$

which follows from (20.69). □

Finally we consider the nonlinear impulse equations

$$\begin{aligned} x' &= f(t,x), \qquad t \ne \tau_k, \\ \Delta x &= I_k(x), \qquad t = \tau_k, \qquad x(t_0^+) = x_0, \end{aligned} \tag{20.71}$$

and

$$\begin{aligned} y' &= f(t,y) + g(t,y), \qquad t \ne \tau_k, \\ \Delta y &= I_k(y) + h_k(y), \qquad t = \tau_k, \qquad y(t_0^+) = x_0, \end{aligned} \tag{20.72}$$

where $t \in \mathbf{R}_+$, $x, y \in \mathbf{R}^n$, $f, g \in C(\mathbf{R}_+ \times \mathbf{R}^n, \mathbf{R}^n)$, $\partial f/\partial x \in C(\mathbf{R}_+ \times \mathbf{R}^n, \mathbf{R}^{n\times n})$, $I_k, h_k \in C(\mathbf{R}^n, \mathbf{R}^n)$, $\partial I_k/\partial x \in C(\mathbf{R}^n, \mathbf{R}^{n\times n})$.

For the solutions $x(t) = x(t; t_0, x_0)$ and $y(t) = y(t; t_0, x_0)$ of (20.71) and (20.72), respectively, the following analogue of Alekseev's formula [208] holds:

$$\begin{aligned} y(t) =& x(t) + \int_{t_0}^{t} \Phi(t, s, y(s)) g(s, y(s))\, ds + \\ &+ \sum_{t_0<\tau_k<t} \int_0^1 \Phi(t, \tau_k, y(\tau_k) + I_k(y(\tau_k)) + \tau h_k(y(\tau_k)))\, d\tau \cdot h_k(y(\tau_k)), \end{aligned} \tag{20.73}$$

where $t > t_0$ and $\Phi(t, t_0, x_0) = \partial x/\partial x_0(t; t_0, x_0)$ is the solution of the variational equation

$$\begin{aligned} z' &= \frac{\partial f}{\partial x}(t, x(t; t_0, x_0))z, \qquad t \neq \tau_k, \\ \Delta z &= \frac{\partial I_k}{\partial x}(x(\tau_k; t_0, x_0)z, \qquad t = \tau_k, \qquad z(t_0^+) = I. \end{aligned} \tag{20.74}$$

Moreover, $\partial x/\partial t_0(t; t_0, x_0) = -\Phi(t, t_0, x_0) f(t_0, x_0)$.

Formula (20.73) enables us to apply the method of integral inequalities to the study of equation (20.72).

Chapter III: notes and comments

Lemma 10.1 and theorems 10.2–4 were proved by P.S. Simeonov. For similar results see P.R. Beesack [27], M. Pinto [168], F.M. Dannan [61], U.D. Dhongade and S.G. Deo [70].

Theorems 11.1–3 were taken from [74]. Theorem 11.5 is close to [177], Thm. 2. Theorem 11.8 is new, and includes E.C. Young [247], Thm. 2. Theorem 11.9 is adapted from [203], Lemma 2. Theorem 11.11 is new, and improves En Hao Yang [238], Thm. 1. For other results see [145], [150], [151], [152], [238].

We note that the method of single differentiation, used to prove theorems 12.2-11, was proposed by D.D. Bainov [28], p. 1484. Theorem 12.11 has been taken from [245]. Lemmas 12.1–2 have been taken from [218]. Theorem 12.14 (with $n = 2$) was proved by D.Y. Kasture and S.G. Deo [103], and is given in E.C. Young [246], Lemma 1. Theorem 12.19 generalizes results of S.G. Chu and F.T. Metcalf [52] (for $n = 1$) and Lt. D.L. Rasmussen [181] (for $n = 2$). In [181], a nonlinear integral inequality is considered, and a corresponding comparison theorem is proved (in terms of the maximal solution). For other comparison results see V.B. Headley [88]. The estimate (12.85) (for $Q(x; y) = [x; y]$) was obtained by P.R. Beesack [25]. For results similar to corollary 12.7 see P.R. Beesack [26] and A.M. Fink [75]. Lemma 12.4 is new, and was proved in [199] under more restrictive assumptions. Theorem 12.20 and corollaries 12.8–9 were proved by P.S. Simeonov. Comparison results generalizing theorem 12.9 have been obtained by D.D. Bainov, A.I. Zahariev, and A.D. Myshkis [19] (for functions defined in compact metric spaces), and by A.B. Ronkov and D.D.

Bainov [186], [187] (for functions defined in partially ordered spaces). For other results see [35], [36], [37], [96], [97], [157], [158], [159], [160], [162], [220]. The results connected with theorems 13.1–3 are based on [99], [113], [114], and [197]. For other results see [3], [153], [171].

Theorem 14.1 has been taken from [230]. Theorems 14.7–9 and corollaries 14.2–4 are new, and were proved by P.S. Simeonov. These extend results of M.M. Konstantinov [106]. For other results see [10] and [135].

The papers of B.G. Pachpatte [146] and S. Sugiyama [216] include results which are close to lemma 15.2. Lemma 15.6 is a discrete analogue of theorem 6.4, and was proved by P.S, Simeonov. Theorems 15.2–3 are new. For results close to theorem 15.3 see J. Dixon and McKee [72]. Under more restrictive assumptions, theorem 15.7 has been proved in [113] and [175], in another manner. Theorems 15.8–9 and corollary 15.9 improve results of J. Popenda and Werbowski [175]. Theorems 15.10–11 and lemma 15.7 were proved by P.S. Simeonov. Corollary 15.11 includes [149], Thm. 2. Theorems 15.14–15 are close to [7], Thm. 1. Theorem 15.17 was proved by P.S. Simeonov as a generalization of [169], Thm. 1. Theorem 15.20 is stated in [223]. For other discrete inequalities see [5], [6], [7], [9], [155], [163], [170], [171], [173], [174], [183], [219], [242].

For more information about impulse differential equations see [112], [192]. Lemma 16.1 and corollary 16.1 have been taken from [207]. Theorems 16.2–3 and corollary 16.2 have been taken from [112]. Theorems 16.4–6 are new, and were proved by P.S. Simeonov. Theorem 16.6 improves [207], Thm. 1. See also [206], [209].

For Gronwall-type inequalities involving a Stieltjes integral see [20], [89], [90], [92], [133], [195].

Lemma 18.1 has been taken from De Franco [63]. Corollary 18.1 is new. A special case of theorem 18.5 is given in [188], Comparison lemma IX, 2.6 ($\psi_k \equiv 0$). For other results see [28], [46], [54], [66], [78], [139], [204], [212].

Theorem 19.1 can be found in [94]. For theorem 19.2 and lemma 19.1 see [93], and for theorem 19.3 see [94]. Generalizations of theorem 19.3 involving nonlinear operators are given by F. Chandra and B.A. Fleishman [47], and by L. Losonczi [120]. Theorems 19.7–8 have been taken from [122], and were obtained by N.B. Azbelev and Z.B. Tsalyuk. For other results see [38], [48], [65], [87], [102], [107], [108], [167], [196], [224], [225].

The contents of §20 has been adapted from [168], [154], [211], [13], [15], [113], [7], [155], [6], [201]. Formula (20.36) was proved by P.S. Simeonov. For more information about functional-differential equations see R. Bellman and K.L. Cooke [31] and J.K. Hale [86]. For other applications see also [32], [41], [82], [91], [98], [111], [132], [144], [153], [154], [156], [178], [200], [205], [209], [210], [228], [231].

Bibliography

[1] Abramovich, J., On Gronwall and Wendroff type inequalities, *Proc. Amer. Math. Soc.* **87** (1983), pp. 481–486.

[2] Akhmedov, K.T., Yakubov, M.A., Veisov, I.A., Some integral inequalities, *Izv. Akad. Nauk UzSSR* **1** (1972), pp. 18–24. (In Russian.)

[3] Agarwal, R.P., On integrodifferential inequalities in two independent variables, *JMAA* **89:2** (1982), pp. 581-597.

[4] -, On an integral inequality in n independent variables, *JMAA* **85** (1982), pp. 192–196.

[5] -, Systems of multidimensional discrete inequalities, *JMAA* **140** (1989), pp. 241–250.

[6] -, Thandapani, E., On some new discrete inequalities, *Appl. Math. Comp.* **7** (1980), pp. 205–224.

[7] -, -, On nonlinear discrete inequalities of Gronwall type, *Ann. Ştiint. Univ. 'Al.I. Cuza' din Iaşi Ser. Ia* **XXVII: 1** (1981), pp. 139–144.

[8] -, -, Some inequalities of Gronwall type, *Ann. Ştiint. Univ. 'Al.I. Cuza' din Iaşi Ser. Ia* **XXVIII: 2** (1982), pp. 71–75.

[9] -, Wilson, S.J., On discrte inequalities involving higher order partial differences, *Ann. Ştiint. Univ. 'Al.I. Cuza' din Iaşi Ser. Ia* **XXX: 1** (1984), pp. 41–50.

[10] Akinyele, O., On Gronwall–Bellman–Bihari type integral inequalities in several variables with retardation, *JMAA* **104** (1984), pp. 1–24.

[11] Alekseev, V.M., An estimate for the perturbations of solutions of ordinary differential equations, *Vestnik Moskov. Gos. Univ. Ser. I Math. Mekh.* **2** (1961), pp. 8–36. (In Russian.)

[12] -, A theorem on an integral inequality and some of its applications, *Mat. Sb.* **68 (110)** (1965), pp. 251–273. Translation: *Transl. Amer. Math. Soc.* **89** (1970), pp. 61–88.

[13] Arino, O., Györi, I, Stability results based on Gronwall type inequalities for some functional-differential systems, *Coll. Math. Soc. J. Bolyai* **47** Differential equations. Qualitative theory. Szeged (Hungary)(1984), pp. 37-59.

[14] Atkinson, F.V., Haddock, J.R., Criteria for asymptotic constancy of solutions of functional differential equations, *JMAA* **91** (1983), pp. 410–423.

[15] Aulbach, B., On linearly perturbed linear systems, *JMAA* **112** (1985), pp. 317–327.

[16] Azbelev, N.B., Tsalyuk, Z.B., On integral inequalities I, *Mat. Sb.* **56 (98): 3** (1962), pp. 325–342. (In Russian.)

[17] Babkin, B.N., On a generalization of a theorem of Academician S.A. Chaplygin on a differential inequaltity, *Molotov. Gos. Univ.* **8** (1953), pp. 3–6. (In Russian.)

[18] Baiada, E., Confronto e dipendenza dai parametri degli integrali delle equazioni differenziali I, *Rend. Acad. (8)* **3** (1947), pp. 258–263.

[19] Bainov, D.D., Zahariev, A.I., Myshkis, A.D., Asymptotic properties of a class of operator-differential inequalities, *Publ. RIMS Kyoto Univ.* **20: 5** (1984), pp. 903–911.

[20] Barbashin, E.L., On stability with respect to impulse effects, *Diff. Eq.* **11: 7** (1966), pp. 863–871. (In Russian.)

[21] Beckenbach, E.F., Bellman, R., *Inequalities*, Springer, 1965. Second revised edition.

[22] Beesack, P.R., Comparison theorems and integral inequalities for Volterra integral equations, *Proc. Amer. Math. Soc.* **20: 1** (1969), pp. 61-66.

[23] -, Gronwall inequalities, *Cavleton Math. Lecture Notes* **11** (1975).

[24] -, On Lakshmikantham's comparison method for Gronwall inequalities, *Ann. Polon. Mat.* **35** (1977), pp. 187–222.

[25] -, Riemann's function has an exponential bound, *Proc. Amer. Math. Soc.* **88: 2** (1983), pp. 313–316.

[26] -, On some Gronwall-type integral inequalities in n independent variables, *JMAA* **100** (1984), pp. 393–408.

[27] -, Asymptotic behaviour of solutions of some general nonlinear differential equations and integral inequalities, *Proc. R. Soc. Edinburgh* **98A** (1984), pp. 49–67.

[28] -, Systems of multidimensional Volterra integral equations and inequalities, *Nonlinear Anal. Theory, Methods & Appl.* **9: 12** (1985), pp. 1451–1486.

[29] Bellman, R., The stability of solutions of linear differential equations, *Duke Math. J.* **10** (1943), pp. 643–647.

[30] -, *Stability theory of differential equations*, McGraw-Hill, 1953.

[31] -, Cooke, K.L., *Differential-difference equations*, McGraw-Hill, 1963.pp. 238–257.

[32] Bernfeld, S.R., Lord, M.E., A nonlinear variation of constants method for integro-differential and integral equations, *T. R. Dept. Math. Univ. Texas at Arlington* **38** (1976).

[33] Bihari, I., A generalization of a lemma of Bellman and its application to uniqueness problems of differential equations, *Acta Math. Acad. Sci. Hungr.* **7: 1** (1956), pp. 71–94.

[34] Bobrowski, D., Popenda, J., Werbowski, J., On the system of integral inequalities with delay of Gronwall-Bellman type, *Fasc. Math.* **10** (1978), pp. 97–104.

[35] Bondge, B.K., Pachpatte, B.G., On Wendroff type integral inequalities in n independent variables, *Chinese J. Math.* **7** (1979), pp. 37–46.

[36] -, -, On nonlinear integral inequalities of Wendroff type, *JMAA* **70** (1979), pp. 161–169.

[37] -, -, Walter, W., On generalized Wendroff type inequalities and their applications, *Nonlinear Anal.* **4** (1980), pp. 491–495.
[38] Boyd, D.W., Wong, J.S.W., On nonlinear contractions, *Proc. Amer. Math. Soc.* **20** (1969), pp. 458–464.
[39] Brauer, F., Bounds for solutions of ordinary differential equations, *Proc. Amer. Math. Soc.* **14** (1963), pp. 36-43.
[40] -, A nonlinear variation of constants formula for Volterra equations, *Mat. Systems Th.* **6** (1972), pp. 226–234.
[41] Burton, T.A., *Volterra integral and differential equations*, Acad. Press, 1983.
[42] Butler, G., Rogers, T.A., A generalization of a lemma of Bihari and applications to pointwise estimates for integral equations, *JMAA* **33: 1** (1971), pp. 77–81.
[43] Bykov, Ya.V., Salpagarov, Kh.M., On the theory of integro-differential equations, In: Investigations in Integro-Differential Equations in Khirghizia **2** Izd. Akad. Nauk KirghiziaSSR, 1962. (In Russian.)
[44] Cafiero, F., Su due teoremi di confronto relativi ad un'equazioni differenziale ordinaria del primo ordine, *Boll. Union. Mat. Italiana (3)* **3** (1948), pp. 124–128.
[45] Chandirov, G.I., On a generalization of Gronwall's inequality and its applications, *Uchen. Zap. Azerb. Univ. Ser. Fiz.-Mat. Nauk* **6** (1956), pp. 3–11. (In Russian.)
[46] Chandra, J., Davis, P.W., Linear generalizations of Gronwall's inequality, *Proc. Amer. Math. Soc.* **60** (1976), pp. 157–160.
[47] -, Fleishman, B.A., Bounds and maximal solutions of nonlinear functional equations, *Bull. Amer. Math. Soc.* **74** (1968), pp. 512–516.
[48] -, -, On a generalization of the Gronwall–Bellman lemma in partially ordered Banach spaces, *JMAA* **31** (1970), pp. 668–681.
[49] -, Lakshmikantham, V., Some pointwise estimates for solutions of a class of nonlinear functional-integral inequalities, *Ann. Mat. Pura Appl. (4)* **94** (1972), pp. 63–74.
[50] Chaplygin, S.A., A new method of approximate integration of differential equations, *Trudy TsAGI* **130** (1932), pp. 5–17. (In Russian.)
[51] -, A new method of approximate integration of differential equations, In: Selected Works, Moscow, 1976. pp. 307–360. (In Russian.)
[52] Chu, S.C., Metcalf, F.T., On Gronwall's inequality, *Proc. Amer. Math. Soc.* **18: 3** (1967), pp. 439–440.
[53] Conlan, J., Daiz, J.B., Existence of solutions of an n-th order hyperbolic partial differential equation, *Contr. Diff. Eq.* **2** (1963), pp. 277–289.
[54] -, Wang, C.-L., Gronwall–Bellman-type inequalities for systems, *JMAA* **120** (1986), pp. 622–630.
[55] Cooke, K.L., A non-local existence theorem for systems of ordinary differential equations, *Rend. Circ. Mat. Palermo* **4** (1955), pp. 301–308.
[56] Coppel, W.A., *Stability and asymptotic behaviour of differential equations*, Heath, 1965.
[57] -, *Dichotomies in stability theory*, Springer, 1978.

[58] Corduneanu, A., A note on the Gronwall inequality in two independent variables, *J. Integral Eq.* **4** (1982), pp. 261–276.

[59] -, Integral inequalities in two independent variables, *Rev. Roumaines Math. Pures Appl.* **32: 4** (1987), pp. 331–341.

[60] Daletskii, Yu.L., Krein, M.G., *Stability of solutions of differential equations in Banach space*, Amer. Math. Soc., 1974. (Translated from the Russian.)

[61] Dannan, F.M., Integral inequalities of Gronwall–Bellman–Bihari type and asymptotic behaviour of certain second order nonlinear differential equations, *JMAA* **108** (1985), pp. 151–164.

[62] Das, P.C., Sharma, R.R., Some Stieltjes integral inequalities, *JMAA* **73** (1980), pp. 423–433.

[63] De Franco, Gronwall's inequality for systems of multiple Volterra integral equations, *Funkcialaj Ekvacioj* **19** (1976), pp. 1–9.

[64] Demidovich, B.P., *Lectures on mathematical stability theory*, Nauka, Moscow, 1967. (In Russian.)

[65] Deo, S.G., Functional inequalities and nonlinear control systems, *Boll. Union Mat. Italia (4)* **6** (1972), pp. 193–200.

[66] -, Murdeshwar, M.G., On a system of integral inequalities, *Proc. Amer. Math. Soc.* **26** (1970), pp. 141–144.

[67] -, -, A note on Gronwall's inequality, *Bull. London Math. Soc.* **3: 1** (1971), pp. 34–36.

[68] Dhongade, U.D., Deo, S.G., Some generalizations of Bellman–Bihari integral inequalities, *JMAA* **44: 1** (1973), pp. 216–218.

[69] Dhongade, U.D., Deo, S.G., Pointwise estimates of solutions of some Volterra integral equations, *JMAA* **45: 3** (1974), pp. 615–628.

[70] -, -, A nonlinear generalization of Bihari's inequality, *Proc. Amer. Math. Soc.* **54** (1976), pp. 211–216.

[71] Dixon, J., McKee, S., Repeated integral inequalities, *IMA J. Numer. Anal.* **4** (1984), pp. 99–107.

[72] -, -, Weakly singular discrete Gronwall inequalities, *ZAMM* **66: 11** (1986), pp. 535–544.

[73] Filatov, A.N., *Averaging methods for differential and integro-differential equations*, Fan, Tashkent, 1971. (In Russian.)

[74] -, Sharova, L.V., *Integral inequalities and the theory of nonlinear oscillations*, Nauka, Moscow, 1976. (In Russian.)

[75] Fink, A.M., Wendroff's inequalities, *Nonlinear Anal. Theory, Methods & Appl.* **5: 8** (1981), pp. 873–874.

[76] Gamidov, Sh.T., Some integral inequalities for boundary value problems for differential equations, *Diff. Eq.* **5: 3** (1969), pp. 463–472. (In Russian.)

[77] Geraghty, M.A., On contractive mappings, *Proc. Amer. Math. Soc.* **40** (1973), pp. 604–608.

[78] Ghoshal, S.K., Ghoshal, A., Masood, M.A., Generalized n-dimensional Gronwall's inequality and its applications to non-self-adjoint and non-linear hyperbolic equations, *Ann. Bolonici Mat.* **XXXIII** (1977), pp. 223–233.

[79] Giuliano, L., Generalizzazione di un lemma di Gronwall e di una diseguaglianza di Peano, *Rend. Acad. Lincei* **1: 8** (1946), pp. 1264–1271.
[80] Gollwitzer, H.E., A note on a functional inequality, *Proc. Amer. Math. Soc.* **23: 3** (1969), pp. 642–647.
[81] Gronwall, T.H., Note on the derivatives with respect to a parameter of the solutions of a system of differential equations, *Ann. of Math.* **20: 2** (1919), pp. 292–296.
[82] Grossman, S.I., Miller, R.K., Perturbation theory for Volterra integrodifferential systems, *J. Diff. Eq.* **8** (1970), pp. 457–474.
[83] Gutowski, R., Etude d'une inégalité intégrale non linéaire en deux variables, *Ann. Polon. Mat.* **25: 3** (1978), pp. 247–252.
[84] -, Radziszewski, B., Asymptotic behaviour and properties of solutions of a system of nonlinear second order ordinary differential equations describing motion of mechanical systems, *Arch. Mech.* **22: 6** (1970), pp. 576–594.
[85] -, -, Asymptotic behaviour and properties of solutions of nonlinear differential equations of first order describing the motion of mechanical systems, *Arch. Mech.* **23: 1** (1971), pp. 17–25.
[86] Hale, J.K., *Theory of functional differential equations*, Springer, 1977.
[87] Hanson, D.L., Waltman, P., A note on a functional equation, *JMAA* **10** (1965), pp. 330–333.
[88] Headley, V.B., A multidimensional nonlinear Gronwall inequality, *JMAA* **47** (1974), pp. 250–255.
[89] Helton, B.W., A product integral representation for a Gronwall inequality, *Proc. Amer. Math. Soc.* **23** (1969), pp. 493–500.
[90] -, The solution of a nonlinear Gronwall inequality, *Proc. Amer. Math. Soc.* **38** (1973), pp. 337–342.
[91] Henrici, P., *Discrete variables methods in ordinary differential equations*, Wiley, 1962.
[92] Herod, J.V., A Gronwall inequality for linear Stieltjes integrals, *Proc. Amer. Math. Soc.* **23** (1969), pp. 34–36.
[93] Hille, E., *Lectures on ordinary differential equations*, Addison-Wesley, 1969.
[94] -, *Methods in classical and functional analysis*, Addison-Wesley, 1972.
[95] Hristova, S.G., Bainov, D.D., On some integral inequalities, *Glasnik Mat.* **13 (33)** (1978), pp. 249–253. (In Russian.)
[96] -, -, On some integral inequalities for a scalar function of a vector argument, *Univ. Ann. Appl. Math.* **17** (1981), pp. 67–70.
[97] -, -, Some generalizations of integral inequalities of Bellman–Bihari type for functions of several variables, *Rend. Circ. Mat. Palermo Ser. II* **XXX** (1981), pp. 378–384.
[98] Hu, S., Lakshmikantham, V., Rama Mohan Rao, Nonlinear variation of parameters formula for integro-differential equations of Volterra type, *JMAA* **129** (1988), pp. 223-230.
[99] -, Zhuang, W., Khavanin, M., On the existence and uniqueness of nonlinear integro-differential equations, *J. Math. Phys. Sci.* **21** (1987), pp. 93–103.

[100] Hull, T.E., Luxemburg, W.A.J., Numerical methods and existence theorems for ordinary differential equations, *Numer. Math.* **2** (1960), pp. 34–41.

[101] Jones, G.S., Fundamental inequalities for discrete and discontinuous functional equations, *SIAM J.* **12** (1964), pp. 43–57.

[102] Kannan, R., Construction of fixed points of aclass of nonlinear mappings, *JMAA* **41** (1973), pp. 430–438.

[103] Kasture, D.Y., Deo, S.G., Inequalities of Gronwall type in two independent variables, *JMAA* **58** (1977), pp. 361–372.

[104] Kharlamov, P.V., On the estimate of the solutions of a system of differential equations, *Ukr. Mat. Zh.* **7: 4** (1955), pp. 471–473. (In Russian.)

[105] Kong Qingkai, Zhang Binggen, Some generalizations of Gronwall–Bihari integral inequalities and their applications, *China Ann. Math.* **10B (3)** (1989), pp. 371–385.

[106] Konstantinov, M.M., On a functional inequality, *Ukr. Mat. Zh.* **30: 2** (1978), pp. 226–228. (In Russian.)

[107] Krasnosel'skii, M.A., Vainikko, G.M., Zabreiko, P.P., et al., *Approximate solution of operator equations*, Noordhoff, 1972. (Translated from the Russian.)

[108] Kurpel', N.S., Shuvar, B.A., *Two-sided operator inequalities and their application*, Nauk. Dumka, Kiev, 1980. (In Russian.)

[109] Lakshmikantham, V., A variation of constants formula and Bellman–Gronwall–Reid inequalities, *JMAA* **41: 1** (1973), pp. 199-204.

[110] -, Leela, S., *Differential and integral inequalities*, Vols. 1–2, Acad. Press, 1969.

[111] -, Trigiante, D., *Theory of difference equations and numerical analysis*, Acad. Press, 1987.

[112] -, Bainov, D.D., Simeonov, P.S., *Theory of impulsive differential equations*, World Sci., 1989.

[113] -, Leela, S., Martynyuk, A.A., *Stability analysis of nonlinear systems*, M. Dekker, 1989.

[114] -, -, Rama Mohan Rao, Integral and integro-differential inequalities, *Appl. Anal.* **24** (1987), pp. 157–164.

[115] Langenhop, C.E., Bounds on the norm of a solution of a general differential equation, *Proc. Amer. Math. Soc.* **11: 1** (1960), pp. 795–799.

[116] Lees, M., Approximate solutions of parabolic equations, *SIAM J.* **7** (1959), pp. 167–183.

[117] Levinson, N., The asymptotic behavior of a system of linear differential equations, *Amer. J. Math.* **68** (1946), pp. 1–6.

[118] Li Yue-Sheng, The bound, stability and error estimates for the solution of nonlinear differential equations, *Acta Math. Sinica* **12** (1962), pp. 32–39.

[119] Lord, M.E., Mitchell, A.R., A new approach to the method of nonlinear variation of parameters, *J. Appl. Math. Comp.* **4** (1978), pp. 95–105.

[120] Losonczi, L., A generalization of the Gronwall–Bellman lemma and its applications, *JMAA* **44** (1973), pp. 701–709.

[121] Lusin, N.N., On the method of approximate integration of Academician S.A. Chaplygin, *Uspekhi Mat. Nauk* **6: 6** (1951), pp. 3–27. (In Russian.)

[122] Mamedov, Ya.D., Ashirov, S., Atdaev, S, *Theorems on inequalities*, Ilim, Ashkabad, 1980. (In Russian.)

[123] Marlin, J.A., Struble, R.A., Asymptotic equivalence of nonlinear systems, *J. Diff. Eqs.* **6** (1969), pp. 578–596.

[124] Maroni, P., Une généralisation non linéaire de l'inégualité de Gronwall, *Bull. Acad. Polon. Sci. Ser. Sci. Math. Astron. et Phys.* **16** (1968), pp. 703–709.

[125] Martynyuk, A.A., Gutowski, R., *Integral inequalities and stability of motion*, Nauk. Dumka, Kiev, 1979. (In Russian.)

[126] -, Kosolapov, V.I., Comparison principle and averaging method in the problem of stability of nonasymptotically stable motions under persistent disturbances, *Preprint Akad. Nauk UkrSSR Inst. Mat.* **78. 33** (1978), (In Russian.)

[127] -, Lakshmikantham, V., Leela, S., *Stability of motion: the method of integral inequalities*, Nauk. Dumka, Kiev, 1989. (In Russian.)

[128] Marušiak, P., The differential equation with retarded argument asymptotically equivalent to the equation $y^{(n)} = 0$, *Mat. Časopis* **23: 1** (1973), pp. 45–54.

[129] Matrosov, V.M., Anapol'skii, Yu., Vasil'ev, S.N., *Comparison method in mathematical system theory*, Nauka, Novosibirsk, 1980. (In Russian.)

[130] McKee, S., Generalised discrete Gronwall lemmas, *ZAMM* **62** (1982), pp. 429–434.

[131] Megahed, F., Hamad, G., Salem, Sh., On some integrodifferential inequalities in one variable defined on an infinite interval, *Arch. Math. (Brno)* **23: 1** (1987), pp. 23–28.

[132] Miller, P.K., *Nonlinear Volterra equations*, Benjamin, 1971.

[133] Mingarelli, A.B., On a Stieltjes version of Gronwall's inequality, *Proc. Amer. Math. Soc.* **82: 2** (1981), pp. 249–252.

[134] Movlyankulov, Kh., Filatov, A.N., On an approximate method for constructing solutions of integral equations, *Trudy Inst. Kibernet. Akad. Nauk UzbSSR* **2** (1972), pp. 11–18. (In Russian.)

[135] Morchalo, J., A functional differential inequality for linear integro-differential equations with retardation, *Ann. Soc. Math. Polon. Ser. I: Comment. Math.* **XXI** (1979), pp. 177–186.

[136] Muldowney, J.S., Wong, J.S.W., Bounds for solutions of nonlinear integrodifferential equations, *JMAA* **23** (1968), pp. 487–499.

[137] Nguyen The Hoan, On the asymptotic behaviour of solutions of nonlinear systems of differential equations, *Diff. Eq.* **XVII: 4** (1981), pp. 624–628. (In Russian.)

[138] Norbury, J., Stuart, A.M., Volterra integral equations and a new Gronwall inequality. Part I: the linear case, *Proc. Royal Soc. Edinburgh* **106A** (1987), pp. 361–373.

[139] Olech, C., On a system of integral inequalities, *Colloq. Math.* **16** (1967), pp. 137–139.

[140] Opial, Z., Sur une système d'inégualités intégrales, *Ann. Polon. Math.* **7** (1960), pp. 247–254.

[141] Osztaszewski, K., Sochacki, J., Gronwall's inequality and the Henstock integral, *JMAA* **127** (1987), pp. 370–374.

[142] Oziraner, A.S., On the stability of motion by linear approximation, *Prikl. Mat. Mekh.* **41: 3** (1977), pp. 413–421. (In Russian.)

[143] Pachpatte, B.G., Finite-diference inequalities and an extension of Lyapunov's method, *Michigan Math. J.* **18** (1971), pp. 385–391.

[144] -, Perturbations of nonlinear integrodifferential equations, *The Math. Student* **XL: 3** (1972), pp. 292–296.

[145] -, A note on Gronwall–Bellman inequality, *JMAA* **44** (1973), pp. 758–762.

[146] -, On the discrete generalizations of Gronwall's inequality, *J. Indian Math. Soc.* **37** (1973), pp. 147–156.

[147] -, Integral inequalities of Gronwall–Bellman type and their applications, *J. Math. Phys. Sci.* **8** (1974), pp. 309–318.

[148] -, Stability and asymptotic behaviour of perturbed nonlinear systems, *J. Diff. Eqs.* **16** (1974), pp. 14–25.

[149] -, Discrete generalization of Gronwall's inequality, *Math. Education* **VIII: 4** (1974), pp. 73–75.

[150] -, On some generalization of Bellman's lemma, *JMAA* **51** (1975), pp. 141–150.

[151] -, A note on integral inequalities of the Bellman–Bihari type, *JMAA* **49** (1975), pp. 295–301.

[152] -, On some integral inequalities similar to Bellman–Bihari inequalities, *JMAA* **49** (1975), pp. 794–802.

[153] -, On the stability and asymptotic behavior of solutions of integrodifferential equations in Banach spaces, *JMAA* **53** (1976), pp. 604–617.

[154] -, On perturbations of Volterra integrodifferential equations, *Rev. Roumaine Math. Pures Appl.* **XXII: 6** (1977), pp. 831–839.

[155] -, On some fundamenntal finite difference inequalities, *Univ. Beograd Publ. Elektrotekhn. Fak. Ser. Mat. Fiz.* **577–598** (1977), pp. 65–73.

[156] -, On the behavior of solutions of a certain class of nonlinear integrodifferential equations, *Ann. Ştiint. Univ. 'Al.I. Cuza' din Iaşi Ser. Ia* **XXIV: 1** (1978), pp. 77–86.

[157] -, On some new integral and integrodifferential inequalities in two independent variables and their applications, *J. Diff. Eqs.* **33** (1979), pp. 249–272.

[158] -, On some new integral inequalities for non-self-adjoint hyperbolic partial integrodifferential equations, *JMAA* **76** (1980), pp. 58–71.

[159] -, On some fundamental partial integral inequalities, *JMAA* **73** (1980), pp. 238–251.

[160] -, On some new integrodifferential inequalities of the Wendroff type, *JMAA* **73** (1980), pp. 491–500.

[161] -, On some partial integral inequalities in n independent variables, *JMAA* **79** (1981), pp. 256–272.

[162] -, On certain integral inequalities for partial differential and integral equations, *Ann. Ştiint. Univ. 'Al.I. Cuza' din Iaşi Ser. Ia* **XXVII: 2** (1981), pp. 365–374.

[163] -, Singare, S.M., Discrete generalized Gronwall inequlities in three independent variables, *Pacific J. Math.* **82: 1** (1979), pp. 197–210.

[164] Palmer, K., On the stability of the center manifold, *J. Appl. Math. Phys. (ZAMP)* **38** (1987), pp. 273–278.

[165] Pandit, S.G., On Stieltjes–Volterra integral equations, *Bull. Austalian Math. Soc.* **18** (1978), pp. 321–334.

[166] Peano, G., Sull'integrabilità delle equazioni differenziali del primo ordine, *Atti R. Accad. Sci. Torino* **21** (1985/6), pp. 677–685.

[167] Pelczar, A., On invariant points of monotone transformations in partially ordered spaces, *Ann. Polon. Math.* **17** (1965), pp. 49–53.

[168] Pinto, M., Integral inequalities of Bihari-type and applications, *Funkcial. Ekv.* **33** (1990), pp. 387–404.

[169] Popenda, J., Finite difference inequalities, *Fasc. Math.* **13 (154)** (1981), pp. 70–87.

[170] -, On some discrete Gronwall type inequalities, *Fasc. Math.* **14** (1985), pp. 109–114.

[171] -, A note on Gronwall–Bellman inequality, *Fasc. Math.* **16 (191)** (1986), pp. 29–42.

[172] -, Finite difference inequalities, *Ann. Soc. Math. Polon. Ser. I* **3** (1986), pp. 89–96.

[173] -, On the discrete analogy of Gronwall inequality, *Fasc. Math.* **17 (205)** (1987), pp. 65–78.

[174] -, Finite difference inequalities, *J. Math. Phys. Sci.* **23: 2** (1989), pp. 107–116.

[175] -, Werbowski, J., On the discrete analogy of Gronwall lemma, *Fasc. Math.* **11 (119)** (1979), pp. 143–154.

[176] Quade, W., Ein neues Verfahren der schrittweisen Näherungen zur Lösung von $\dot{x} = f(t,x)$, *Math. Z.* **48: 3** (1942), pp. 324–368.

[177] Ráb, M., Linear integral inequalities, *Arch. Math. 1. Scripta Fac. Sci. Nat. Ujep Brunensis* **XV** (1979), pp. 37–46.

[178] Rabczuk, R., *Elementy nierównosci różniczkowych*, PWN, Warszawa, 1976.

[179] Rakhmatulina, L.F., On an application of the solvability conditions of Chaplygin's problem to questions of boundedness and stability of solutions of differential equations, *Izv. Vuzov. Mat.* **2** (1959), pp. 198–201. (In Russian.)

[180] Rama Mohan Rao, M., A note on an integral inequality, *Indian Math. Soc.* **27** (1963), pp. 65–71.

[181] Rasmussen Lt., D.L., Gronwall's inequality for functions of two independent variables, *JMAA* **55** (1976), pp. 407–417.

[182] Redheffer, R.M., Differential and integral inequalities, *Proc. Amer. Math. Soc.* **15** (1964), pp. 715–716.

[183] -, Walter, W., A comparison theorem for difference inequalities, *J. Diff. Eqs.* **44** (1982), pp. 111–117.

[184] Reid, W.T., Properties of solutions of an infinite system of ordinary linear differential equations of thefirst order with auxiliary boundary conditions, *Trans. Amer. Math. Soc.* **32** (1930), pp. 284–318.

[185] -, *Ordinary differential equations*, Springer, 1971.
[186] Ronkov, A.B., Bainov, D.D., Integral equations and inequalities of Volterra type for functions defined in partially ordered spaces, *JMAA* **125: 2** (1987), pp. 483–507.
[187] -, -, Nonlinear integral inequalities for functions defined in partially ordered topological spaces, *Nonlinear Anal. Theory, Methods & Appl.* **11: 3** (1987), pp. 297–304.
[188] Rouche, N., Habets, P., Laloy, M., *Stability theory by Lyapunov's direct method*, Springer, 1987.
[189] Rudakov, V.P., On a two-sided estimate of the norms of solutions of a class of nonlinear systems, *Diff. Eq.* **VI: 2** (1970), pp. 371–372. (In Russian.)
[190] -, On an effective two-sided estimate of the norms of solutions of nonlinear perturbed systems, *Izv. Vuzov. Mat.* **103: 12** (1970), pp. 78–83. (In Russian.)
[191] Samoilenko, A.M., Perestyuk, N.A., Stability of solutions of differential equations with impulse effect, *Diff. Eq.* **13: 11** (1977), pp. 1981–1992. (In Russian.)
[192] -, -, *Differential equations with impulse effect*, Vysha Shkola, Kiev, 1987. (In Russian.)
[193] Sardarly, S.M., Investigation of the solutions of nonlinear integral equations on the halfline with respect to a parameter, *Uchen. Zap. Azerb. Univ.* **2** (1965), pp. 24–30. (In Russian.)
[194] Sato, T., Sur l'équation intégrale non linéaire de Volterra, *Comp. Math.* **11** (1953), pp. 271–289.
[195] Schmaedeke, W.W., Sell, G.R., The Gronwall inequality for modified Stieltjes integrals, *Proc. Amer. Math. Soc.* **19** (1968), pp. 1217–1222.
[196] Schröder, J., *Operator inequalities*, Acad. Press, 1980.
[197] Shendge, G.R., A new approach to the stability theory of functional differential systems, *JMAA* **95** (1983), pp. 319–334.
[198] Shih, M.-H., Yeh, C.-C., Some integral inequalities in n independent variables, *JMAA* **84** (1981), pp. 569–593.
[199] Simeonov, P.S., Bainov, D.D., On some generalizations of the Gronwall–Bellman integral inequality for scalar functions of many variables, *Rend. Circ. Mat. Palermo Ser. II* **XXXII** (1983), pp. 27–38.
[200] -, -, Asymptotic equivalence of two systems of differential equations with impulse effect, *Systems & Control Lett.* **3** (1983), pp. 297–301.
[201] -, -, Stability in linear approximation of systems with impulse effect, *Rend. Sem. Mat. Univ. Politecn. Torino* **43: 2** (1985), pp. 303–322.
[202] -, -, On the stability under persistent disturbances for systems with impulse effect, *Riv. Mat. Univ. Parma (4)* **11** (1985), pp. 31–41.
[203] -, -, Stability of the solutions of a system of integro-differential equations with impulse effect, *Math. Report Toyama Univ.* **9** (1986), pp. 1–24.
[204] -, -, Stability with respect to part of the variables in systems with impulse effect, *JMAA* **117: 1** (1986), pp. 247–263.
[205] -, -, Asymptotic equivalence of a linear and nonlinear system with impulse effect, *Riv. Mat. Univ. Parma (4)* **12** (1986), pp. 1–8.

[206] -, -, On an integral inequality for piecewise continuous functions, *J. Math. Phys. Sci.* **21: 4** (1987), pp. 315–323.

[207] -, -, Integral and differential inequalities for a class of piecewise continuous functions, *Ann. Polon. Math.* **XLVIII** (1988), pp. 207–216. (In Russian.)

[208] -, -, Differentiability of solutions of systems with impulse effect with respect to initial data and parameter, *Proc. Edinburgh Math. Soc.* **31** (1988), pp. 353–368.

[209] -, -, Perturbation theorems for systems with impulse effect, *Int. J. Systems Sci.* **19: 7** (1988), pp. 1213–1223.

[210] -, -, Asymptotic equivalence of nonlinear systems with impulse effect, *Bull. Inst. Math. Acad. Sinica* **117: 1** (1989), pp. 27–40.

[211] Snow, D.R., A two independent variable Gronwall type inequality, In: Inequalities **III** Acad. Press, 1972. pp. 333–340.

[212] -, Gronwall's inequality for systems of partial differential equations in two independent variables, *Proc. Amer. Math. Soc.* **33** (1972), pp. 46–54.

[213] Stachurska, B., On a nonlinear integral inequality, *Zeszyty Nauk. Uniw. Jagiellońsk.* **252** *Prace Mat.* **15** (1971), pp. 151–157.

[214] Sternberg, H.M., The solution of the characteristic and the Cauchy boundary value problem for the Bianchi partial differential equation in n independent variables by a generalization of Riemann's method, *Ph.D. Thesis Univ. Maryland* (1960).

[215] Strauss, A., Yorke, J., Perturbation theorems for ordinary differentia equations, *J. Diff. Eq.* **3** (1967), pp. 15–30.

[216] Sugiyama, S., On the stability problems of difference equations, *Bull. Sci. Engin. Research Lab. Waseda Univ.* **45** (1969), pp. 140–144.

[217] Szarski, J., *Differential inequalities*, PWN, 1965.

[218] Thandapani, E., Agarwal, R.P., On some new integrodifferential inequalities: theory and applications, *Tamkang J. Math.* **11: 2** (1980), pp. 169–184.

[219] -, -, Some new discrete inequalities in two independent variables, *Ann. Ştiint. Univ. 'Al.I. Cuza' din Iaşi Ser. Ia* **XXVII: 2** (1981), pp. 269–278.

[220] -, -, On some new inequalities in n independent variables, *JMAA* **86** (1982), pp. 542–561.

[221] Trench, W.F., On the asymptotic behavior of solutions of second order linear differential equations, *Proc. Amer. Math. Soc.* **14: 1** (1963), pp. 12–14.

[222] -, On t_∞ quasi-similarity of linear systems, *Ann. Mat. Pura Applic. (IV)* **CXLII** (1985), pp. 293–302.

[223] Tsalyuk, Z.B., On some methods for obtaining estimates of solutions of inequalities, *Diff. Eqs.* **24: 2** (1988), pp. 250–258. (In Russian.)

[224] Turinici, M., Abstract Gronwall–Bellman inequalities in ordered metrizable uniform spaces, *J. Integral Eq.* **6** (1984), pp. 105–117.

[225] -, Abstract comparison principle and multivariate Gronwall–Bellman inequalities, *JMAA* **117** (1986), pp. 100-127.

[226] Ved', Yu.A., On perturbations of linear homogeneous differential equatios with variable coefficients, In: Issled. Integro-Differents. Uravn. Kirghizia **3** Ilim, Frunze, 1965. (In Russian.)

[227] Viswanatham, B., A generalization of Bellman's lemma, *Proc. Amer. Math. Soc.* **14** (1963), pp. 15–18.

[228] Walter, W., *Differential and integral inequalities*, Springer, 1970.

[229] Ważewski, T., Remarque sur un système d'inégualités intégrales, *Ann. Polon. Math.* **3** (1957), 210–212.

[230] Werbowski, J., On some asymptotic behaviour of solutions of a Volterra integral equation with delay, *Demonst. Math.* **XIII: 3** (1980), pp. 579–584.

[231] Westphal, H., Zur Abschätzung der Lösungen nichtlinearer parabolischer Differentialgleichungen, *Math. Z.* **51** (1949), pp. 690–695.

[232] Willett, D., Nonlinear vector integral equations as contraction mappings, *Arch. Rat. Mech. Anal.* **15** (1964), pp. 79–86.

[233] -, A linear generalization of Gronwall's inequality, *Proc. Amer. Math. Soc.* **16** (1965), pp. 774–778.

[234] -, Wong, J.S.W., On the discrete analogues of some generalizations of Gronwall's inequality, *Monatsh. Math.* **69: 4** (1965), pp. 362–367.

[235] Wintner, A., An abelian lemma concerning asymptotic equilibria, *Amer. J. Math.* **78** (1946), pp. 451–454.

[236] Wong, J.S.W., On an integral inequality of Gronwall, *Rev. Roumaine Math. Pures Appl.* **12** (1967), pp. 1519–1522.

[237] Wright, F.M., Klasi, M.L., Kennenbeck, D.R., The Gronwall inequality for weighted integrals, *Proc. Amer. Math. Soc.* **30** (1971), pp. 504–510.

[238] Yang, E.-H., On the most general form of Bellman-type linear inequalities involving multiple-fold integral functionals, *JMAA* **103** (1984), pp. 184–197.

[239] Yang, E.-H., On some new integral inequalities in n independent variables, *JMAA* **109** (1985), pp. 171–181.

[240] Yeh, C.-C., Wendroff's inequality in n independent variables, *JMAA* **78** (1980), pp. 78–87.

[241] -, Bellman–Bihari integral inequalities in several independent variables, *JMAA* **87** (1982), pp. 311–321.

[242] -, Discrete inequalities of the Gronwall–Bellman type in n independent variables, *JMAA* **105** (1985), pp. 322–332.

[243] -, Shih, M.-H., The Gronwall –Bellman inequality in several variables, *JMAA* **86** (1982), pp. 157–167.

[244] Young, E.C., Gronwall's inequality in n indepndent variables, *Proc. Amer. Math. Soc.* **41: 1** (1973), pp. 241–244.

[245] -, On Bellman–Bihari integral inequalities, *Internat. J. Math. Math. Sci.* **5: 1** (1982), pp. 97–103.

[246] -, Functional integral inequalities in N variables, *Chinese J. Math.* **10: 1** (1982), pp. 1–7.

[247] -, On integral inequalities of Gronwall-Bellman type, *Proc. Amer. Math. Soc.* **94: 4** (1985), pp. 636–640.

[248] Zadiraka, K.V., Investigation of irregularly perturbed differential equations, In: Questions of the theory and history of differential equations, Nauk. Dumka, Kiev, 1968. pp. 81–108 (In Russian.)

[249] Zahariev, A.I., Bainov, D.D., A note on Bellman–Gronwall's inequality, *JMAA* **80: 1** (1981), pp. 147–149.
[250] Ziebur, A.D., On the Gronwall–Bellman lemma, *JMAA* **22** (1968), pp. 92–95.

Author Index

Subject Index

MIX
Papier aus verantwortungsvollen Quellen
Paper from responsible sources
FSC® C105338

If you have any concerns about our products,
you can contact us on
ProductSafety@springernature.com

In case Publisher is established outside the EU,
the EU authorized representative is:
Springer Nature Customer Service Center GmbH
Europaplatz 3, 69115 Heidelberg, Germany

Printed by Libri Plureos GmbH
in Hamburg, Germany